Theory of Stabilization for Linear Boundary Control Systems

Theory of Stabilization for Linear Boundary Control Systems

Takao Nambu
Professor Emeritus
Department of Applied Mathematics
Kobe University
Kobe, Japan

CRC Press
Taylor & Francis Group
Boca Raton London New York

CRC Press is an imprint of the
Taylor & Francis Group, an **informa** business
A SCIENCE PUBLISHERS BOOK

CRC Press
Taylor & Francis Group
6000 Broken Sound Parkway NW, Suite 300
Boca Raton, FL 33487-2742

Printed on acid-free paper
Version Date: 20160708

International Standard Book Number-13: 978-1-4987-5847-5 (Hardback)

Visit the Taylor & Francis Web site at
http://www.taylorandfrancis.com

and the CRC Press Web site at
http://www.crcpress.com

To my family,
Mariko, Ryutaro, and Hiromu.

Preface

This monograph studies the stabilization theory for linear systems governed by partial differential equations of parabolic type in a unified manner. As long as controlled plants are relatively small, such as electric circuits and mechanical oscillations/rotations of rigid bodies, ordinary differential equations, abbreviated as *ode(s)*, are suitable mathematical models to describe them. When the controlled plants are, e.g., chemical reactors, wings of aircrafts, or other flexible systems such as robotics arms, plates, bridges, and cranes, however, effects of space variables are essential and non-neglegeble terms. For the set up of mathematical models describing these plants, partial differential equations, abbreviated as *pde(s)*, are a more suitable language. It is generally expected that control laws based on more accurate pde models would work effectively in actual applications.

The origin of control theory is said to be the paper, "On governors" by J.C. Maxwell (1868). For many years, control theory has been studied mainly for systems governed by odes in which controlled plants are relatively small. Control theory for pdes began in 60's of the 20th century, and the study of stabilization in mid 70's to cope with much larger systems. Fundamental concepts of control such as controllability, observability, optimality, and stabilizability are the same as in those of odes, and translated by the language of pdes. The essence of pdes consists in their infinite-dimensional properties, so that control problems of pdes face serious difficulties in respective aspects, which have never been experienced in the world of odes: However, these difficulties provide us rich and challenging fields of study both from mathematical and engineering viewpoints.

Among other control problems of pdes such as optimal control problems, etc., we concentrate ourselves on the topic of stabilization problems. Stabilization problems of pdes have a new aspect of pdes in the framework of *synthesis* (or design) of a desirable spectrum by involving the concept of observation/control, and are connected not only with functional analysis but also non-harmonic analysis and classical Fourier analysis, etc. The monograph consists of eight chapters which strongly reflects the author's works over thirty years except for Chapter 2: Some were taught in graduate courses at Kobe University. The organization of the monograph is stated as follows: It begins with the linear tabilization problem of finite dimension in Chapter 1. Finite-dimensional models constitute *pseudo*-internal structures of pdes. Although the problem is entirely solved by W. M. Wonham in 1967 [70], we develop a much easier new approach, which has never appeared even among the community of finite-dimensional control theory: It is based on Sylvester's equation. Infinite-dimensional versions of the equation appear in later chapters as an essential tool for stabilization problems throughout the monograph. Chapter 2 is a brief introduction of basic results on standard elliptic differential operators L and related Sobolev spaces necessary for our control problems: These results are well known among the pdes community, but proofs of some results are stated for the readers' convenience. As for results requiring much preparation we only provide some references instead of proofs. In Chapters 3 through 7, the main topics discussed are, where stabilization problems of linear parabolic systems are successfully solved in the boundary observation/boundary feedback scheme. The elliptic operator L is derived from a pair of standard (but general enough) differential operators $(\mathscr{L}, \tau)$, and forms the coefficient of our control systems, where $\mathscr{L}$ denotes a uniformly elliptic differential operator and τ a boundary operator. The operator L is sectorial, and thus $-L$ turns out to be an infinitesimal generator of an analytic semigroup. One of important issues is certainly the existence or non-existence of Riesz bases associated with L: When an associated Riesz basis exists, a sequence of finite-dimensional approximation models of the original pde is quantitatively justified, so that the control laws based on the approximated finite-dimensional models effectively works. There is an attempt to draw out a class of elliptic operators with Riesz bases (see the footnote in the beginning of Chapter 4). However, is the class of pdes admitting associated Riesz bases general enough or much narrower than expected? We do not have a satisfactory solution to the question yet. Based on these observations, our feedback laws are constructed so that they are applied to a general class of pdes, without assuming Riesz bases.

There are two kinds of feedback schemes: One is a *static feedback* scheme, and the other a *dynamic feedback* scheme. In Chapter 3, the stabilization problem and related problems are discussed in the static feedback scheme, in which the outputs of the system are directly fed back into the system through the actuators. While the scheme has difficulties in engineering implementations, it works as an auxiliary means in the dynamic feedback schemes. In Chapter 4, we establish stabilization in the scheme of boundary observation/boundary feedback. The feedback scheme is the dynamic feedback scheme, in which the outputs on the boundary are fed back into the system through another differential equation described in another abstract space. This differential equation is called a *dynamic compensator*, the concept of which originates from D. G. Luenberger's paper [33] in 1966 for linear odes. In his paper, two kinds of compensators are proposed: One is an *identity* compensator, and the other a compensator of *general type*. We formulate the latter compensator in the feedback loop to cope with the stabilization problem, and finally reduce the compensator to a finite-dimensional one. All arguments are algebraic, and do not depend on the kind of boundary operators τ. In Chapter 5, the problem is discussed from another viewpoint when the system admits a Riesz basis. Since a finite-dimensional approximation to the pde is available as a strongly effective means, an identity compensator is installed in the feedback loop. Most stabilization results in the literature are based on identity compensators, but have difficulty in terms of mathematical generality. In Chapters 4 and 5, observability and controllability conditions on sensors and actuators, respectively, are assumed on the *pseudo*-internal substructure of finite dimension. We then ask in Chapter 6 the following: What can we claim when the observability and controllability conditions are lost? *Output stabilization* is one of the answers: Assuming an associated Riesz basis, we propose sufficient conditions on output stabilization. A related problem is also discussed, which leads to a new problem, that is, the problem of *pole allocation with constraints*. To show mathematical generality of our stabilization scheme, we generalize in Chapter 7 the class of operators L, in which $-L$ is a generator of *eventually differentiable* semigroups: A class of delay-differential equations generates such operators L.

In our general stabilization scheme, we solve an inverse problem associated with the infinite-dimensional Sylvester's equation. The problem forms a so called *ill-posed* problem lacking of continuity property. Finally in Chapter 8, we propose a numerical approximation algorhism to the inverse problem, the solution of which is mathematically ensured. The algorhism consists of a simple idea, but needs tedious calculations. Although the algorhism has some restrictions at present, it is expected that it would work in more general settings of the parameters. Numerical approximation itself is a problem independent of our stabilization problem. However, the latter certainly leads to a development of new problems in numerical analysis. The author hopes that willing readers could open a new area in effective numerical algorhisms.

The author in his graduate school days had an opportunity to read papers by Y. Sakawa, by H. O. Fattorini, and by S. Agmon and L. Nirenberg ([2, 17, 18, 57]) among others, and learned about the close relationships lying in differential equations, functional analysis, and the theory of functions. Inspired by these results, he had a hope to contribute to deep results of such nature, since then. He is not certain now, but would be happy, if the monograph cooould reflect his hope even a little.

Takao Nambu

December, 2015
Kobe

Contents

Chapter 1

Preliminary results - Stabilization of linear systems of finite dimension

1.1 Introduction

We develop in this chapter the basic problem arising from stabilization problems of finite-dimension. Since the celebrated pole assignment theory [70] (see also [56, 68]) for linear control systems of finite dimension appeared, the theory has been applied to various stabilization problems both of finite dimension and infinite dimension such as the one with boundary output/boundary input scheme (see, e.g., [12, 13, 28, 37 – 40, 42 – 45, 47 – 50, 53, 58, 59] and the references therein). The symbol H_n, $n = 1, 2, \dots$, hereafter will denote a finite-dimensional Hilbert space with $\dim H_n = n$, equipped with inner product $\langle \cdot, \cdot \rangle_n$ and norm $\|\cdot\|$. The symbol $\|\cdot\|$ is also used for the $\mathscr{L}(H_n)$-norm. Let L, G, and W be operators in $\mathscr{L}(H_n)$, $\mathscr{L}(\mathbb{C}^N; H_n)$, and $\mathscr{L}(H_n; \mathbb{C}^N)$, respectively. Here and hereafter, the symbol $\mathscr{L}(R; S)$, R and S being linear spaces of finite or infinite dimension, means the set of all linear bounded operators mapping R into S. The set $\mathscr{L}(R; S)$ forms a linear space. When $R = S$, $\mathscr{L}(R; R)$ is abbreviated simply as $\mathscr{L}(R)$. Given L, W, and any set of n complex numbers, $Z = \{\zeta_i\}_{1 \leqslant i \leqslant n}$, the problem is to seek a suitable G such that $\sigma(L - GW) = Z$, where $\sigma(L - GW)$ means the spectrum of the operator

$L-GW$. Or alternatively, given L and G, its algebraic counterpart is to seek a W such that $\sigma(L-GW)=Z$. Stimulated by the result of [70], various approaches and algorhisms for computation of G or W have been proposed since then (see, e.g., [7, 10, 14]). However, each approach needs much preparation and a deep background in linear algebra to achieve stabilization and determine the necessary parameters. Explicit realizations of G or W sometimes seem complicated. One for this is no doubt the complexity of the process in determining G or W *exactly* satisfying the relation, $\sigma(L-GW)=Z$.

Let us describe our control system: Our system, consisting of a state $u(\cdot)\in H_n$, output $y=Wu\in\mathbb{C}^N$, and input $f\in\mathbb{C}^N$, is described by a linear differential equation in H_n,

$$\frac{du}{dt}+Lu=Gf,\quad y=Wu,\quad u(0)=u_0\in H_n. \tag{1.1}$$

Here,

$$\begin{aligned} Gf&=\sum_{k=1}^{N} f_k g_k \quad \text{for } f=(f_1\ \dots\ f_N)^{\mathrm{T}}\in\mathbb{C}^N,\\ Wu&=\left(\langle u,w_1\rangle_n\ \ \dots\ \ \langle u,w_N\rangle_n\right)^{\mathrm{T}} \quad \text{for } u\in H_n, \end{aligned} \tag{1.2}$$

$(\dots)^{\mathrm{T}}$ denoting the transpose of vectors or matrices throughout the monogtaph. The vectors $w_k\in H_n$ denote given weights of the observation (output); and $g_k\in H_n$ are actuators to be constructed. By setting $f=y$ in (1.1), the control system yields a feedback system,

$$\frac{du}{dt}+(L-GW)u=0,\quad u(0)=u_0\in H_n. \tag{1.3}$$

According to the choice of a basis for H_n, the operators L, G, and W are identified with matrices of respective size. We hereafter employ the above symbols somewhat different from those familiar in the control theory community of finite dimension, in which state of the system, for example, would be often represented as $x(\cdot)$; output Cx; input u; and equation

$$\frac{dx}{dt}=Ax+Bu=(A+BC)x,\quad u=Cx.$$

The reason for employing present symbols is that they are consistent with those in systems of infinite dimension discussed in later chapters.

Let us assume that $\sigma(L)\cap\mathbb{C}_-\neq\varnothing$, so that the system (1.1) with $f=0$ is unstable. Given a $\mu>0$, the *stabilization problem* for the finite dimensional control system (1.3) is to seek a G or W such that

$$\left\|e^{-t(L-GW)}\right\|\leqslant \text{const}\, e^{-\mu t},\quad t\geqslant 0. \tag{1.4}$$

The pole assignment theory [70] plays a fundamental role in the above problem, and has been applied so far to various linear systems. The theory is concretely stated as follows: *Let* $Z=\{\zeta_i\}_{1\leqslant i\leqslant n}$ *be any set of n complex numbers, where some* ζ_i *may coincide. Then, there exists an operator G such that* $\sigma(L-GW)=Z$, *if and only if the pair* (W,L) *is observable.* Thus, if the set Z is chosen such that $\min_{\zeta\in Z}\mathrm{Re}\,\zeta$, say μ $(=\mathrm{Re}\,\zeta_1)$ is positive, and if there is *no* generalized eigenspace of $L-GW$ corresponding to ζ_1, we obtain the decay estimate (1.4).

Now we ask: Do we need *all* information on $\sigma(L-GW)$ for stabilization? In fact, to obtain the decay estimate (1.4), it is not necessary to designate *all* elements of the set Z: What is really necessary is the number, $\mu=\min_{\zeta_i\in Z}\mathrm{Re}\,\zeta_i$, say $=\mathrm{Re}\,\zeta_1$, and the spectral property that ζ_1 does not allow any generalized eigenspace; the latter is the requirement that *no* factor of algebraic growth in time is added to the right-hand side of (1.4). In fact, when an algebraic growth is added, the decay property becomes a little worse, and the gain constant $(\geqslant 1)$ in (1.4) increases. The above operator $L-GW$ also appears, as a *pseudo*-substructure, in the stabilization problems of infinite dimensional linear systems such as parabolic systems and/or retarded systems (see, e.g., [16]): These systems are decomposed into two, and understood as composite systems consisting of two states; one belonging to a finite dimensional subspace, and the other to an infinite dimensional one. It is impossible, however, to manage the infinite dimensional substructures. Thus, no matter how *precisely* the finite dimensional spectrum $\sigma(L-GW)$ could be assigned, it does not exactly dominate the whole structure of infinite dimension. In other words, the assigned spectrum of finite dimension is not necessarily a subset of the spectrum of the infinite-dimensional feedback control system.

In view of the above observations, our aim in this chapter is to develop a new approach much simpler than those in existing literature, which allows us to construct a desired operator G or a set of actuators g_k ensuring the decay (1.4) in a simpler and more explicit manner (see (2.10) just below Lemma 2.2). The result is, however, not as sharp as in [70] in the sense that it does not generally provide the precise location of the assigned eigenvalues. From the above viewpoint of infinite-dimensional control theory, however, the result would be meaningful enough, and satisfactory for stabilization. We note that our result exactly coincides with the standard pole assignment theory in the case where we can choose $N=1$ (see Proposition 2.3 in Section 2). The results of this chapter are based on those discussed in [48, 51, 52].

Our approach is based on Sylvester's equation of finite dimension. Sylvester's equation in infinite-dimensional spaces has also been studied extensively (see, e.g., [6] for equations involving only bounded operators), and even the unboundedness of the given operators are allowed [37, 39, 40, 42 – 45, 47, 49, 50, 53]. Sylvester's equation in this chapter is of finite dimension, so that there arises no difficulty caused by the complexity of infinite dimension. Its infinite-dimensional version and the properties are discussed later in Chapters 4,

6, and 7. Given a positive integer s and vectors $\xi_k \in H_s$, $1 \leqslant k \leqslant N$, let us consider the following Sylvester's equation in H_n:

$$
\begin{aligned}
&XL - MX = -\Xi W, \quad \Xi \in \mathscr{L}(\mathbb{C}^N; H_s), \quad \text{where} \\
&\Xi z = \sum_{k=1}^{N} z_k \xi_k \quad \text{for } z = (z_1 \ \dots \ z_N)^{\mathrm{T}} \in \mathbb{C}^N.
\end{aligned}
\tag{1.5}
$$

Here, M denotes a given operator in $\mathscr{L}(H_s)$, and ξ_k vectors to be designed in H_s. A possible solution X would belong to $\mathscr{L}(H_n; H_s)$. An approach via Sylvester's equations is found, e.g., in [7, 10], in which, by setting $n = s$, a condition for the existence of the bounded inverse $X^{-1} \in \mathscr{L}(H_n)$ is sought. Choosing an M such that $\sigma(M) \subset \mathbb{C}_+$, it is then proved that

$$
L + (X^{-1}\Xi)W = X^{-1}MX, \quad \sigma(X^{-1}MX) = \sigma(M) \subset \mathbb{C}_+,
$$

the left-hand side of which means a desired perturbed operator. The procedure of its derivation is, however, rather complicated, and the choice of the ξ_k is unclear. In fact, X^{-1} might not exist sometimes for some ξ_k.

The approach in this chapter is new and rather different. Let us characterize the operator L in (1.5). There is a set of generalized eigenpairs $\{\lambda_i, \ \varphi_{ij}\}$ with the following properties:

(i) $\sigma(L) = \{\lambda_i; \ 1 \leqslant i \leqslant \nu \, (\leqslant n)\}, \quad \lambda_i \neq \lambda_j$ for $i \neq j$; and

(ii) $L\varphi_{ij} = \lambda_i \varphi_{ij} + \sum_{k<j} \alpha^i_{jk} \varphi_{ik}, \quad 1 \leqslant i \leqslant \nu, \ 1 \leqslant j \leqslant m_i$.

Let P_{λ_i} be the projector in H_n corresponding to the eigenvalue λ_i. Then, we see that $P_{\lambda_i} u = \sum_{j=1}^{m_i} u_{ij} \varphi_{ij}$ for $u \in H_n$. The restriction of L onto the invariant subspace $P_{\lambda_i} H_n$ is, in the basis $\{\varphi_{i1}, \dots, \varphi_{im_i}\}$, is represented by the $m_i \times m_i$ upper triangular matrix Λ_i, where

$$
\Lambda_i \big|_{(j,k)} = \begin{cases} \alpha^i_{kj}, & j < k, \\ \lambda_i, & j = k, \\ 0, & j > k. \end{cases} \tag{1.6}
$$

If we set $\Lambda_i = \lambda_i + N_i$, the matrix N_i is nilpotent, that is, $N_i^{m_i} = 0$. The minimum integer n such that $\ker N_i^n = \ker N_i^{n+1}$, denoted as l_i, is called the *ascent* of $\lambda_i - L$. It is well known that the ascent l_i coincides with the order of the pole λ_i of the resolvent $(\lambda - L)^{-1}$. Laurent's expansion of $(\lambda - L)^{-1}$ in a neighborhood of the pole $\lambda_i \in \sigma(L)$ is expressed as

$$
\begin{aligned}
&(\lambda - L)^{-1} = \sum_{j=1}^{l_i} \frac{K_{-j}}{(\lambda - \lambda_i)^j} + \sum_{j=0}^{\infty} (\lambda - \lambda_i)^j K_j, \quad \text{where} \\
&l_i \leqslant m_i, \quad K_j = \frac{1}{2\pi i} \int_{|\zeta - \lambda_i| = \delta} \frac{(\zeta - L)^{-1}}{(\zeta - \lambda_i)^{j+1}} \, d\zeta, \quad j = 0, \pm 1, \pm 2, \dots.
\end{aligned}
\tag{1.7}
$$

Note that $K_{-1} = P_{\lambda_i}$. The set $\{\varphi_{ij};\ 1 \leqslant i \leqslant \nu,\ 1 \leqslant j \leqslant m_i\}$ forms a basis for H_n. Each $x \in H_n$ is uniquely expressed as $x = \sum_{i,j} x_{ij}\varphi_{ij}$. Let T be a bijection, defined as $Tx = \begin{pmatrix} x_{11}\, x_{12}\, \dots\, x_{\nu m_\nu} \end{pmatrix}^{\mathrm{T}}$. Then, L is identified with the upper triangular matrix Λ;

$$TLT^{-1} = \Lambda = \operatorname{diag}\begin{pmatrix} \Lambda_1 & \Lambda_2 & \dots & \Lambda_\nu \end{pmatrix}. \tag{1.8}$$

Let us turn to the operator M in (1.5). Let

$$\{\eta_{ij};\ 1 \leqslant i \leqslant n,\ \ 1 \leqslant j \leqslant \ell_i\}$$

be an orthonormal basis for H_s. Then necessarily $s = \sum_{i=1}^{n} \ell_i \geqslant n$. Every vector $v \in H_s$ is expressed as

$$v = \sum_{i=1}^{n}\sum_{j=1}^{\ell_i} v_{ij}\eta_{ij}, \quad \text{where} \quad v_{ij} = \langle v, \eta_{ij}\rangle_s.$$

Let $\{\mu_i\}_{1=1}^{n}$ be a set of positive numbers such that $0 < \mu_1 < \cdots < \mu_n$, and set

$$Mv = \sum_{i=1}^{n}\sum_{j=1}^{\ell_i} \mu_i v_{ij}\eta_{ij} \tag{1.9}$$

for $v = \sum_{i,j} v_{ij}\eta_{ij}$. It is apparent that (i) $\sigma(M) = \{\mu_i\}_{i=1}^{n}$; and (ii) $(\mu_i - M)\eta_{ij} = 0$, $1 \leqslant i \leqslant n,\ 1 \leqslant j \leqslant \ell_i$. The operator M is self-adjoint, and potive-definite,

$$\langle Mv, v\rangle_s = \sum_{i=1}^{n}\sum_{j=1}^{\ell_i} \mu_i |v_{ij}|^2 \geqslant \mu_1 \|v\|_s^2.$$

Let Q_{μ_i} be the projector in H_s corresponding to the eigenvalue $\mu_i \in \sigma(M)$, say $Q_{\mu_i}v = \sum_{j=1}^{\ell_i} v_{ij}\eta_{ij}$ for $v = \sum_{i,j} v_{ij}\eta_{ij}$. We put an additional condition on M:

$$\sigma(L) \cap \sigma(M) = \varnothing. \tag{1.10}$$

Assuming (1.10), we derive our first result as Proposition 1.1. Since the proof is carried out in exactly the same manner as in [37, 44, 45, 50], it is omitted.

Proposition 1.1. *Suppose that the condition* (1.10) *is satisfied. Then, Sylvester's equation* (1.5) *admits a unique operator solution* $X \in \mathscr{L}(H_n; H_s)$. *The solution* X *is expressed as*

$$\begin{aligned} Xu &= \frac{1}{2\pi \mathrm{i}}\int_C (\lambda - M)^{-1}\Xi W(\lambda - L)^{-1} u\, d\lambda \\ &= \sum_{\lambda \in \sigma(M)} Q_\lambda \Xi W(\lambda - L)^{-1} u \\ &= \sum_{i=1}^{n} Q_{\mu_i}\Xi W(\mu_i - L)^{-1} u, \end{aligned} \tag{1.11}$$

where C denotes a Jordan contour encircling $\sigma(M)$ in its inside, with $\sigma(L)$ outside C. The above first expression is the so called Rosenblum formula [6].

The main results are stated as Theorem 2.1 and Proposition 2.2 in the next section, where a more explicit and concrete expression than ever before of a set of stabilizing actuators g_k in (1.3) is obtained. As we see in the next section, an advantage of considering the operator $X \in \mathscr{L}(H_n; H_s)$ with $s \geqslant n$ is that the bounded inverse $(X^*X)^{-1}$ is ensured under a reasonable assumption on the operator Ξ. A numerical example is also given. Finally, Proposition 2.3 is stated, where our feedback scheme exactly coincides with the standard pole assignment theory [70] in the case where we can choose $N = 1$.

1.2 Main Results

We assume that $\sigma(L) \cap \mathbb{C}_- \neq \varnothing$, so that the semigroup e^{-tL}, $t \geqslant 0$, is *unstable*. We construct suitable actuators $g_k \in H_n$ in (1.3) such that $e^{-t(L-GW)}$ has a preassigned decay rate, say $-\mu_1$ (see (1.9)). The operator $\begin{pmatrix} W & WL & \dots & WL^{n-1} \end{pmatrix}^{\mathrm{T}}$ belongs to $\mathscr{L}(H_n; \mathbb{C}^{nN})$. The observability condition on the pair (W, L) means that the above operator is injective, in other words, $\ker \begin{pmatrix} W & WL & \dots & WL^{n-1} \end{pmatrix}^{\mathrm{T}} = \{0\}$. Throughout the section, the separation condition (1.10) is assumed in Sylvester's equation (1.5). Then, we obtain one of the main results:

Theorem 2.1. *Assume that the conditions*

$$\begin{gathered} \ker \begin{pmatrix} W & WL & \dots & WL^{n-1} \end{pmatrix}^{\mathrm{T}} = \{0\}, \quad \textit{and} \\ \ker Q_{\mu_i} \Xi = \{0\}, \quad 1 \leqslant i \leqslant n \end{gathered} \tag{2.1}$$

are satisfied. Then, $\ker X = \{0\}$.

Proof. Let $Xu = 0$. In view of Proposition 1.1, we see that

$$Q_{\mu_i} \Xi W (\mu_i - L)^{-1} u = 0, \quad 1 \leqslant i \leqslant n.$$

Since $\ker Q_{\mu_i} \Xi = \{0\}$, $1 \leqslant i \leqslant n$, by (2.1), we obtain

$$\begin{gathered} W(\mu_i - L)^{-1} u = 0, \quad 1 \leqslant i \leqslant n, \quad \text{or} \\ \left\langle (\mu_i - L)^{-1} u, w_k \right\rangle_n = 0, \quad 1 \leqslant k \leqslant N, \quad 1 \leqslant i \leqslant n. \end{gathered} \tag{2.2}$$

Set $f_k(\lambda; u) = \left\langle (\lambda - L)^{-1} u, w_k \right\rangle_n$. By recalling that $T(\lambda - L)^{-1} T^{-1} = (\lambda - \Lambda)^{-1}$ (see (1.8)), $f_k(\lambda; u)$ is rewritten as $\left\langle (\lambda - \Lambda)^{-1} T u, \left(T^{-1}\right)^* w_k \right\rangle_{\mathbb{C}^n}$. Each element of the $n \times n$ matrix $(\lambda - \Lambda)^{-1}$ is a rational function of λ; its denominator consists of a polynomial of order n; and the numerator at most of order $n - 1$. This means that each $f_k(\lambda; u)$ is a rational function of λ, the denominator of which is a polynomial of order n, and the numerator of order $n-1$. Since the numerator of f_k has *at least* n distinct zeros μ_i, $1 \leqslant i \leqslant n$, by (2.2), we conclude that

$$f_k(\lambda; u) = \left\langle (\lambda - L)^{-1} u, w_k \right\rangle_n = 0, \quad \lambda \in \rho(L), \quad 1 \leqslant k \leqslant N. \tag{2.3}$$

Let c be a number such that $-c \in \rho(L)$, and set $L_c = L + c$. In view of the identity

$$\begin{aligned}(\lambda - L)^{-1} &= L_c(\lambda - L)^{-1} L_c{}^{-1} \\ &= -L_c{}^{-1} + (\lambda + c)(\lambda - L)^{-1} L_c{}^{-1},\end{aligned}$$

let us introduce a series of rational functions $f_k^l(\lambda; u)$, $l = 0, 1, \ldots$, as

$$f_k^0(\lambda; u) = f_k(\lambda; u), \qquad f_k^{l+1}(\lambda; u) = \frac{f_k^l(\lambda; u)}{\lambda + c}, \quad l = 0, 1, \ldots. \tag{2.4}$$

It is easily seen that

$$f_k^l(\lambda; u) = \left\langle (\lambda - L)^{-1} L_c^{-l} u, w_k \right\rangle_n - \sum_{i=1}^{l} \frac{1}{(\lambda + c)^i} \left\langle L_c^{-(l+1-i)} u, w_k \right\rangle_n \tag{2.5}$$

and

$$f_k^l(\lambda; u) = 0, \quad \lambda \in \rho(L) \setminus \{-c\}, \quad 1 \leqslant k \leqslant N, \quad l \geqslant 0.$$

In view of Laurent's expansion (1.7) of $(\lambda - L)^{-1}$ in a neighborhood of λ_i, we obtain the relation

$$\begin{aligned}0 &= f_k(\lambda; u) \\ &= \sum_{j=1}^{l_i} \frac{\left\langle K_{-j} u, w_k \right\rangle_n}{(\lambda - \lambda_i)^j} + \sum_{j=0}^{\infty} (\lambda - \lambda_i)^j \left\langle K_j u, w_k \right\rangle_n, \quad 1 \leqslant k \leqslant N,\end{aligned}$$

in a neighborhood of λ_i. Calculation of the residue of $f_k(\lambda; u)$ at λ_i implies that

$$\begin{aligned}\left\langle K_{-1} u, w_k \right\rangle_n = \left\langle P_{\lambda_i} u, w_k \right\rangle_n = 0, \quad 1 \leqslant i \leqslant \nu, \quad 1 \leqslant k \leqslant N, \\ \text{or} \quad W P_{\lambda_i} u = 0, \quad 1 \leqslant i \leqslant \nu.\end{aligned} \tag{2.6}$$

As for $f_k^l(\lambda; u)$, $\ell \geqslant 1$, we have a similar expression in a neighborhood of λ_i,

$$\begin{aligned}f_k^l(\lambda; u) = \sum_{j=1}^{l_i} \frac{\left\langle A_{-j} L_c^{-l} u, w_k \right\rangle_n}{(\lambda - \lambda_i)^j} + \sum_{j=0}^{\infty} (\lambda - \lambda_i)^j \left\langle A_j L_c{}^{-l} u, w_k \right\rangle_n \\ - \sum_{i=1}^{l} \frac{1}{(\lambda + c)^i} \left\langle L_c^{-(l+1-i)} u, w_k \right\rangle_n = 0\end{aligned}$$

by (2.5). Note that $K_{-1}L_c^{-l}u = P_{\lambda_i}L_c^{-l}u = L_c^{-l}P_{\lambda_i}u$. Calculation of the residue of $f_k^l(\lambda; u)$ at λ_i similarly implies that

$$\left\langle K_{-1}L_c^{-l}u, w_k \right\rangle_n = \left\langle L_c^{-l}P_{\lambda_i}u, w_k \right\rangle_n = 0, \quad 1 \leqslant i \leqslant \nu, \quad 1 \leqslant k \leqslant N,$$
$$\text{or} \quad WL_c^{-l}P_{\lambda_i}u = 0, \quad 1 \leqslant i \leqslant \nu, \quad l \geqslant 1.$$

Combining these with the above relation (2.6), we see that

$$\begin{pmatrix} W & WL_c^{-1} & \dots & WL_c^{-(n-1)} \end{pmatrix}^{\mathrm{T}} P_{\lambda_i}u = 0, \quad 1 \leqslant i \leqslant \nu. \tag{2.7}$$

It is clear that

$$\ker \begin{pmatrix} W & WL & \dots & WL^{n-1} \end{pmatrix}^{\mathrm{T}} = \ker \begin{pmatrix} W & WL_c & \dots & WL_c^{n-1} \end{pmatrix}^{\mathrm{T}},$$

where $L_c = L + c$. Thus, by the first condition of (2.1), it is easily seen that

$$\ker \begin{pmatrix} W & WL_c^{-1} & \dots & WL_c^{-(n-1)} \end{pmatrix}^{\mathrm{T}} = \ker \begin{pmatrix} W & WL & \dots & WL^{n-1} \end{pmatrix}^{\mathrm{T}} = \{0\}.$$

Thus, (2.7) immediately implies that $P_{\lambda_i}u = 0$ for $1 \leqslant i \leqslant \nu$, and finally that $u = 0$. □

By Theorem 2.1, there is a positive constant such that

$$\|Xu\|_s \geqslant \text{const}\, \|u\|, \quad \forall u \in H_n.$$

The derivation of the above positive lower bound of $\|Xu\|_s$ is due to a specific nature of finite-dimensional spaces. The operator $X^*X \in \mathscr{L}(H_n)$ is self-adjoint, and positive-definite. In fact, by the relation

$$\text{const}\, \|u\|^2 \leqslant \|Xu\|_s^2 = \langle Xu, Xu \rangle_s = \langle X^*Xu, u \rangle_n \leqslant \|X^*Xu\|\, \|u\|,$$

we see that $\|X^*Xu\| \geqslant \text{const}\, \|u\|$. Thus the bounded inverse $(X^*X)^{-1} \in \mathscr{L}(H_n)$ exists. We go back to Sylvester's equation (1.5). Setting $X^*X = \mathscr{X} \in \mathscr{L}(H_n)$ and $X^*MX = \mathscr{M} \in \mathscr{L}(H_n)$, we obtain the relation,

$$\begin{gathered} L - (X^*X)^{-1}X^*MX = -(X^*X)^{-1}X^*\Xi W \quad \text{or} \\ L + \sum_{k=1}^{N} \langle \cdot, w_k \rangle_n \mathscr{X}^{-1}X^*\xi_k = \mathscr{X}^{-1}\mathscr{M}. \end{gathered} \tag{2.8}$$

Both operators $\mathscr{X}$ and $\mathscr{M}$ are self-adjoint, but $\mathscr{X}^{-1}\mathscr{M}$ is not. The following assertion is the second of our main results, and leads to a stabilization result:

Proposition 2.2. *Assume that* (2.1) *is satisfied. Then,* $\sigma(\mathscr{X}^{-1}\mathscr{M})$ *is contained in* $\mathbb{R}^1_+$. *Actually,*

$$\lambda_* = \min \sigma(\mathscr{X}^{-1}\mathscr{M}) \geqslant \mu_1. \tag{2.9}$$

In addition, there is no generalized eigenspace for any $\lambda \in \sigma(\mathscr{X}^{-1}\mathscr{M})$.

Remark: By Proposition 2.2, we obtain a decay estimate

$$\begin{aligned} \left\|\exp\left(-t\left(L+(X^*X)^{-1}X^*\Xi W\right)\right)\right\| &= \left\|e^{-t(\mathscr{X}^{-1}\mathscr{M})}\right\| \\ &\leqslant \text{const}\, e^{-\mu_1 t}, \quad t \geqslant 0. \end{aligned} \tag{2.10}$$

In fact, the last assertion of the proposition ensures that no algebraic growth in time arises in the semigroup, regarding the smallest eigenvalue. Thus, a set of actuators $g_k = -(X^*X)^{-1}X^*\xi_k$, $1 \leqslant k \leqslant N$, in other words, $G = -(X^*X)^{-1}X^*\Xi$ explicitly gives a desired set of actuators in (1.3).

Proof of Proposition 2.2. Since $\mathscr{X}$ is positive-definite, we can find a non-unique bijection $\mathscr{U} \in \mathscr{L}(H_n)$ such that

$$\mathscr{X} = X^*X = \mathscr{U}^*\mathscr{U}, \tag{2.11}$$

the so called Cholesky factorization. Let us define

$$\mathscr{M}' = (\mathscr{U}^*)^{-1}\mathscr{M}\mathscr{U}^{-1} = (\mathscr{U}^{-1})^*\mathscr{M}\mathscr{U}^{-1}.$$

Then, $\mathscr{M}' \in \mathscr{L}(H_n)$ is a self-adjoint operator, enjoying some properties similar to those of $\mathscr{X}^{-1}\mathscr{M}$. In fact, let $\lambda \in \sigma(\mathscr{X}^{-1}\mathscr{M})$, or $(\lambda\mathscr{X} - \mathscr{M})u = 0$ for some $u \neq 0$. Then, since

$$\begin{aligned} 0 = (\lambda\mathscr{U}^*\mathscr{U} - \mathscr{M})u &= \mathscr{U}^*\left(\lambda - (\mathscr{U}^*)^{-1}\mathscr{M}\mathscr{U}^{-1}\right)\mathscr{U}u \\ &= \mathscr{U}^*\left(\lambda - \mathscr{M}'\right)\mathscr{U}u = 0, \end{aligned}$$

we see that λ belongs to $\sigma(\mathscr{M}')$. The converse relation is also correct, which means that

$$\sigma(\mathscr{X}^{-1}\mathscr{M}) = \sigma(\mathscr{M}') \subset \mathbb{R}^1. \tag{2.12}$$

Inequality (2.9) is achieved by applying the well known min-max principle [11] to $\mathscr{M}'$, or more directly by the following observation: Let $\lambda \in \sigma(\mathscr{X}^{-1}\mathscr{M})$, and $(\lambda\mathscr{X} - \mathscr{M})u = 0$ for some $u \neq 0$. Then

$$\lambda\|Xu\|_s^2 = \lambda\langle\mathscr{X}u, u\rangle_n = \langle\mathscr{M}u, u\rangle_n = \langle MXu, Xu\rangle_s \geqslant \mu_1\|Xu\|_s^2,$$

from which (2.9) immediately follows, since $Xu \neq 0$.

Next let us show that there is *no* generalized eigenspace for $\lambda \in \sigma(\mathscr{X}^{-1}\mathscr{M})$. Let $(\lambda - \mathscr{X}^{-1}\mathscr{M})^2u = 0$ for some $u \neq 0$. Setting $v = (\lambda - \mathscr{X}^{-1}\mathscr{M})u$, we calculate

$$\begin{aligned} 0 &= \mathscr{X}(\lambda - \mathscr{X}^{-1}\mathscr{M})^2u = (\lambda\mathscr{X} - \mathscr{M})v \\ &= (\lambda\mathscr{U}^*\mathscr{U} - \mathscr{M})v = \mathscr{U}^*\left(\lambda - (\mathscr{U}^*)^{-1}\mathscr{M}\mathscr{U}^{-1}\right)\mathscr{U}v \\ &= \mathscr{U}^*\left(\lambda - \mathscr{M}'\right)w = 0, \quad w = \mathscr{U}v, \end{aligned}$$

or $(\lambda - \mathscr{M}')w = 0$. On the other hand, since

$$w = \mathscr{U}v = \mathscr{U}(\lambda - \mathscr{X}^{-1}\mathscr{M})u = \mathscr{U}(\lambda - \mathscr{U}^{-1}(\mathscr{U}^*)^{-1}\mathscr{M})u \\ = (\lambda - (\mathscr{U}^*)^{-1}\mathscr{M}\mathscr{U}^{-1})\mathscr{U}u = (\lambda - \mathscr{M}')\mathscr{U}u,$$

we see that

$$0 = (\lambda - \mathscr{M}')w = (\lambda - \mathscr{M}')^2\mathscr{U}u, \quad \mathscr{U}u \neq 0.$$

But, $\mathscr{M}'$ is self-adjoint, so that there is no generalized eigenspace for $\lambda \in \sigma(\mathscr{M}')$. Thus, $\mathscr{U}u$ turns out to be an eigenvector of $\mathscr{M}'$ for λ, and

$$0 = \mathscr{U}^*(\lambda - \mathscr{M}')\mathscr{U}u = \mathscr{U}^*(\lambda - (\mathscr{U}^*)^{-1}\mathscr{M}\mathscr{U}^{-1})\mathscr{U}u \\ = (\lambda\mathscr{U}^*\mathscr{U} - \mathscr{M})u = (\lambda\mathscr{X} - \mathscr{M})u.$$

This means that u is an eigenvector of $\mathscr{X}^{-1}\mathscr{M}$ for λ. □

The following example shows that $\lambda_* = \min\sigma(\mathscr{X}^{-1}\mathscr{M})$ does not generally coincide with the prescribed μ_1.

Example: Let $n = 3$, and set $H_3 = \mathbb{C}^3$, so that L is a 3×3 matrix. Let

$$L = \mathrm{diag}\begin{pmatrix} a & a & b \end{pmatrix},$$

where $a, b \leqslant 0$ and $a \neq b$. Since $n = 3$, $\nu = 2$, $m_1 = 2$, and $m_2 = 1$, we choose $N = 2$, $s = 6$, $H_6 = \mathbb{C}^6$, and $\ell_1 = \ell_2 = \ell_3 = 2$. As for the operator $W \in \mathscr{L}(\mathbb{C}^3; \mathbb{C}^2)$, let us consider the case, for example, where $w_1 = (1\ 0\ 1)^{\mathrm{T}}$ and $w_2 = (0\ 1\ 0)^{\mathrm{T}}$. The operator W is a 2×3 matrix given by $\begin{pmatrix} 1 & 0 & 1 \\ 0 & 1 & 0 \end{pmatrix}$. The pair (W, L) is then observable, and the first condition of (2.1) is satisfied.

To consider Sylvester's equation (1.5), let $\{\eta_{ij};\ 1 \leqslant i \leqslant 3,\ j = 1, 2\}$ be a standard basis for $\mathbb{C}^6$ such that $\eta_{11} = (1\ 0\ 0 \ldots 0)^{\mathrm{T}}$, $\eta_{12} = (0\ 1\ 0 \ldots 0)^{\mathrm{T}}$, $\eta_{21} = (0\ 0\ 1 \ldots 0)^{\mathrm{T}}, \ldots$, and $\eta_{32} = (0 \ldots 0\ 1)^{\mathrm{T}}$. Set

$$M = \mathrm{diag}\begin{pmatrix} \mu_1 & \mu_1 & \mu_2 & \mu_2 & \mu_3 & \mu_3 \end{pmatrix}$$

for $0 < \mu_1 < \mu_2 < \mu_3$. In the operator Ξ given by

$$\Xi u = u_1\xi_1 + u_2\xi_2 \quad \text{for} \quad u = (u_1\ u_2)^{\mathrm{T}} \in \mathbb{C}^2,$$

set $\xi_1 = (1\ 0\ 1\ 0\ 1\ 0)^{\mathrm{T}}$ and $\xi_2 = (0\ 1\ 0\ 1\ 0\ 1)^{\mathrm{T}}$. Then, we see that $\ker Q_{\mu_i}\Xi = \{0\}$, $1 \leqslant i \leqslant 3$, and the second condition of (2.1) is satisfied. The unique solution $X \in \mathscr{L}(\mathbb{C}^3; \mathbb{C}^6)$ to Sylvester's equation (1.5) is a 6×3 matrix described as ($u = (u_{11}\ u_{12}\ u_{21})^{\mathrm{T}} \in \mathbb{C}^3$)

$$Xu = \begin{pmatrix} \langle (\mu_1 - L)^{-1}u, w_1 \rangle_3 \\ \langle (\mu_1 - L)^{-1}u, w_2 \rangle_3 \\ \hline \langle (\mu_2 - L)^{-1}u, w_1 \rangle_3 \\ \langle (\mu_2 - L)^{-1}u, w_2 \rangle_3 \\ \hline \langle (\mu_3 - L)^{-1}u, w_1 \rangle_3 \\ \langle (\mu_3 - L)^{-1}u, w_2 \rangle_3 \end{pmatrix} = \begin{pmatrix} \frac{1}{\mu_1 - a} & 0 & \frac{1}{\mu_1 - b} \\ 0 & \frac{1}{\mu_1 - a} & 0 \\ \frac{1}{\mu_2 - a} & 0 & \frac{1}{\mu_2 - b} \\ 0 & \frac{1}{\mu_2 - a} & 0 \\ \frac{1}{\mu_3 - a} & 0 & \frac{1}{\mu_3 - b} \\ 0 & \frac{1}{\mu_3 - a} & 0 \end{pmatrix} \begin{pmatrix} u_{11} \\ u_{12} \\ u_{21} \end{pmatrix}, \qquad (2.13)$$

where $\langle \cdot, \cdot \rangle_3$ denotes the inner product in $\mathbb{C}^3$. Setting, for computational convenience,

$$\alpha = \left(\frac{1}{\mu_1 - a} \;\; \frac{1}{\mu_2 - a} \;\; \frac{1}{\mu_3 - a} \right)^{\mathrm{T}}, \quad \beta = \left(\frac{1}{\mu_1 - b} \;\; \frac{1}{\mu_2 - b} \;\; \frac{1}{\mu_3 - b} \right)^{\mathrm{T}},$$

and $\quad 1 = (1 \;\; 1 \;\; 1)^{\mathrm{T}}$,

we see that

$$(X^*X)^{-1} = \frac{1}{\gamma} \begin{pmatrix} |\beta|^2 & 0 & -\langle \alpha, \beta \rangle_3 \\ 0 & |\beta|^2 - \langle \alpha, \beta \rangle_3{}^2 / |\alpha|^2 & 0 \\ -\langle \alpha, \beta \rangle_3 & 0 & |\alpha|^2 \end{pmatrix},$$

where $\gamma = |\alpha|^2 |\beta|^2 - \langle \alpha, \beta \rangle_3{}^2$. By noting that $X^*\xi_1 = (\langle \alpha, 1 \rangle_3 \;\; 0 \;\; \langle \beta, 1 \rangle_3)^{\mathrm{T}}$ and $X^*\xi_2 = (0 \;\; \langle \alpha, 1 \rangle_3 \;\; 0)^{\mathrm{T}}$, the matrix $L + (X^*X)^{-1}X^*\Xi W$ is concretely described as

$$\mathrm{diag}\,(a \quad a \quad b) + \frac{1}{\gamma} \times \left(\begin{array}{c|c|c} |\beta|^2 \langle \alpha, 1 \rangle_3 - \langle \alpha, \beta \rangle_3 \langle \beta, 1 \rangle_3 & 0 & |\beta|^2 \langle \alpha, 1 \rangle_3 - \langle \alpha, \beta \rangle_3 \langle \beta, 1 \rangle_3 \\ \hline 0 & \langle \alpha, 1 \rangle_3 \left(|\beta|^2 - \frac{\langle \alpha, \beta \rangle_3{}^2}{|\alpha|^2} \right) & 0 \\ \hline |\alpha|^2 \langle \beta, 1 \rangle_3 - \langle \alpha, \beta \rangle_3 \langle \alpha, 1 \rangle_3 & 0 & |\alpha|^2 \langle \beta, 1 \rangle_3 - \langle \alpha, \beta \rangle_3 \langle \alpha, 1 \rangle_3 \end{array} \right).$$

It is apparent that one of the eigenvalues of this matrix is the $(2,2)$-element:

$$a + \frac{\langle \alpha, 1 \rangle_3}{\gamma} \left(|\beta|^2 - \frac{\langle \alpha, \beta \rangle_3{}^2}{|\alpha|^2} \right) = a + \frac{\langle \alpha, 1 \rangle_3}{|\alpha|^2},$$

and is certainly greater than μ_1. Note that

$$0 < \lambda_* - \mu_1 \leqslant \frac{1}{|\alpha|^2} \left(\frac{\mu_2 - \mu_1}{(\mu_2 - a)^2} + \frac{\mu_3 - \mu_1}{(\mu_3 - a)^2} \right) \to 0, \quad \mu_2, \mu_3 \to \infty.$$

The other eigenvalues are those of the matrix,

$$\frac{1}{\gamma}\begin{pmatrix} |\beta|^2\langle\alpha,1\rangle_3-\langle\alpha,\beta\rangle_3\langle\beta,1\rangle_3+\gamma a & |\beta|^2\langle\alpha,1\rangle_3-\langle\alpha,\beta\rangle_3\langle\beta,1\rangle_3 \\ |\alpha|^2\langle\beta,1\rangle_3-\langle\alpha,\beta\rangle_3\langle\alpha,1\rangle_3 & |\alpha|^2\langle\beta,1\rangle_3-\langle\alpha,\beta\rangle_3\langle\alpha,1\rangle_3+\gamma b \end{pmatrix}. \tag{2.14}$$

To see that these eigenvalues are generally greater than μ_1, let us consider a numerical example: Let $(\mu_1\ \mu_2\ \mu_3)=(2\ 3\ 4)$, $a=0$, and $b=-1$. Then,

$$\alpha=\left(\frac{1}{2}\ \frac{1}{3}\ \frac{1}{4}\right)^{\mathrm{T}},\quad \beta=\left(\frac{1}{3}\ \frac{1}{4}\ \frac{1}{5}\right)^{\mathrm{T}},\quad |\alpha|^2=\frac{61}{144},\quad |\beta|^2=\frac{769}{3600},$$
$$\langle\alpha,\beta\rangle_3=\frac{3}{10},\quad \langle\alpha,1\rangle_3=\frac{13}{12},\quad \langle\beta,1\rangle_3=\frac{47}{60},$$
$$\gamma=|\alpha|^2|\beta|^2-\langle\alpha,\beta\rangle_3{}^2=\frac{253}{518400}.$$

One of the eigenvalues $a+\langle\alpha,1\rangle_3/|\alpha|^2$ is $156/61>2\,(=\mu_1)$. The matrix (2.14) is then $\frac{1}{253}\begin{pmatrix}-1860 & -1860\\ 3540 & 3287\end{pmatrix}$, the eigenvalues of which are denoted as ζ_1 and ζ_2. Then, $\mu_1=2<\zeta_1<156/61<\zeta_2$, and thus $\lambda_*=\zeta_1>2$.

We close this section with the following remark: There is a case where λ_* coincides with μ_1. Following [52], let us consider (1.3) in the space $H_n=\mathbb{C}^n$ (see (1.8)). All operators L, G, and W are then matrices of respective size. Let $\sigma(L)$ consist only of simple eigenvalues, so that $m_i=1$, $1\leqslant i\leqslant n$, and $n=\nu$. Thus we can choose $N=1$, $\ell_i=1$, $1\leqslant i\leqslant n$, and thus $s=n$. The operator in (2.10) is written as $L+(X^*X)^{-1}X^*\Xi W$, where $\Xi u=u\xi$ for $u\in\mathbb{C}^1$, and $W=\langle\cdot,w\rangle_n$, $w=\left(w_1\ w_2\ \dots\ w_n\right)^{\mathrm{T}}\in\mathbb{C}^n$. The observability condition then turns out to be $w_i\neq 0$, $1\leqslant i\leqslant n$. Let us consider Sylvester's equation (1.5) in $H_s=\mathbb{C}^n$. By setting $\xi=\left(1\ 1\ \dots\ 1\right)^{\mathrm{T}}\in\mathbb{C}^n$, the solution X to (1.5) is an $n\times n$ matrix, and has a bounded inverse:

$$X=\Phi\tilde{W}, \tag{2.15}$$

where

$$\Phi=\left(\frac{1}{\mu_i-\lambda_j};\ \begin{matrix} i & \downarrow & 1,\dots,n\\ j & \rightarrow & 1,\dots,n\end{matrix}\right),\quad \text{and}\quad \tilde{W}=\mathrm{diag}\left(w_1\quad w_2\quad \dots\quad w_n\right).$$

Thus, $L+(X^*X)^{-1}X^*\Xi W=L+X^{-1}\xi w^{\mathrm{T}}$. It is shown [52] that, given a set $\{\mu_i\}_{1\leqslant i\leqslant n}$, there is a unique $g\in\mathbb{C}^n$ such that $\sigma\left(L-gw^{\mathrm{T}}\right)=\{\mu_i\}_{1\leqslant i\leqslant n}$, and that g is concretely expressed as

$$g = \begin{pmatrix} g_1 \\ g_2 \\ g_3 \\ \vdots \\ g_n \end{pmatrix} = \frac{1}{\Delta} \begin{pmatrix} \frac{1}{w_1} \Delta_1 f(\lambda_1) \\ -\frac{1}{w_2} \Delta_2 f(\lambda_2) \\ \frac{1}{w_3} \Delta_3 f(\lambda_3) \\ \vdots \\ (-1)^{n-1} \frac{1}{w_n} \Delta_n f(\lambda_n) \end{pmatrix}, \tag{2.16}$$

where

$$f(\lambda) = \prod_{i=1}^{n} (\lambda - \mu_i), \quad \text{and}$$

$$\Delta = \prod_{1 \leqslant i < j \leqslant n} (\lambda_i - \lambda_j), \qquad \Delta_k = \prod_{\substack{1 \leqslant i < j \leqslant n, \\ i, j \neq k}} (\lambda_i - \lambda_j), \quad 1 \leqslant k \leqslant n.$$

The proof will be given later in Section 4.

Proposition 2.3. *Suppose in Proposition 2.2 that $\sigma(L)$ consists only of simple eigenvalues. Set $\xi = (1\ 1\ \ldots\ 1)^{\mathrm{T}}$ as above. Then $X^{-1}\xi = -g$, and thus $\lambda_* = \mu_1$. In fact, we have $\sigma\left(L + (X^*X)^{-1}X^*\Xi W\right) = \{\mu_i\}_{1 \leqslant i \leqslant n}$.*

Proof. The relation, $X^{-1}\xi = -g$ is rewritten as

$$-\Delta \begin{pmatrix} 1 \\ 1 \\ 1 \\ \vdots \\ 1 \end{pmatrix} = \Phi \tilde{W} \begin{pmatrix} \frac{1}{w_1} \Delta_1 f(\lambda_1) \\ -\frac{1}{w_2} \Delta_2 f(\lambda_2) \\ \frac{1}{w_3} \Delta_3 f(\lambda_3) \\ \vdots \\ (-1)^{n-1} \frac{1}{w_n} \Delta_n f(\lambda_n) \end{pmatrix} = \Phi \begin{pmatrix} \Delta_1 f(\lambda_1) \\ -\Delta_2 f(\lambda_2) \\ \Delta_3 f(\lambda_3) \\ \vdots \\ (-1)^{n-1} \Delta_n f(\lambda_n) \end{pmatrix}.$$

In other words, we show that

$$-\sum_{j=1}^{n} \frac{(-1)^{j-1} \Delta_j f(\lambda_j)}{\mu_i - \lambda_j} = \sum_{j=1}^{n} (-1)^{j-1} \Delta_j \overbrace{\prod_{\substack{1 \leqslant \ell \leqslant n, \\ \ell \neq i}} (\lambda_j - \mu_\ell)}^{\left(=\lambda_j^{n-1} + \cdots\right)} \tag{2.17}$$
$$= \Delta, \quad 1 \leqslant i \leqslant n.$$

The left-hand side of (2.17), a polynomial of λ_i, $1 \leqslant i \leqslant n$, is in particular a polynomial of λ_1 of order $n-1$, and the coefficient of λ_1^{n-1} is $\Delta_1 = \prod_{2 \leqslant i < j \leqslant n} (\lambda_i - \lambda_j)$. For $j < k$, let us compare the jth and the kth terms. The following lemma is elementary:

Lemma 2.4 . *Let $1 \leqslant j < k \leqslant n$. In the product Δ_k, a polynomial of $\{\lambda_i\}_{i \neq k}$, set $\lambda_j = \lambda_k$. Then*

$$\Delta_k = (-1)^{k-1+j} \Delta_j.$$

In the left-hand side of (2.17), set $\lambda_j = \lambda_k$. Since the terms other than the jth and the kth terms contain the factor $(\lambda_j - \lambda_k)$, they become to be 0. The kth term is then

$$\begin{aligned}(-1)^{k-1} \Delta_k \prod_{\substack{1 \leqslant \ell \leqslant n, \\ \ell \neq i}} (\lambda_k - \mu_\ell) &= (-1)^{k-1} (-1)^{k-1-j} \Delta_j \prod_{1 \leqslant \ell \leqslant n, `\ell \neq i} (\lambda_k - \mu_\ell) \\ &= -(-1)^{j-1} \Delta_j \prod_{\substack{1 \leqslant \ell \leqslant n, \\ \ell \neq i}} (\lambda_j - \mu_\ell) = -(\text{the } j\text{th term}).\end{aligned}$$

Thus the left-hand side of (2.17) has factors $\lambda_j - \lambda_k$, $j < k$, and is written as $c\Delta$. But, $c\Delta$ is a polynomial of λ_1 of order $n-1$, and the coefficient of λ_1^{n-1} is $c\Delta_1$. This means that $c = 1$, and the proof of relation (2.17) is now complete. □

1.3 Observability: Reduction to Substructures

The first condition of (2.1) is the observability condition on the pair (W, L). The operator $\begin{pmatrix} W & WL & \dots & WL^{n-1} \end{pmatrix}^{\mathrm{T}}$ is rewritten for convenience as

$$\left(WL^k;\ k \downarrow 0, \dots, n-1 \right).$$

Similar expressions for other operators and matrices will be employed hereafter without any confusion. Following [48], we show in this section that the observability condition is reduced to a set of observabillity conditions on subsystems. Let $L_i = L|_{P_{\lambda_i} H_n}$ be the restriction of L onto the invariant subspace $P_{\lambda_i} H_n$, and let $W_i = W|_{P_{\lambda_i} H_n}$. To obtain the reduction, it is convenient to employ matrix representations. The set $\{\varphi_{ij};\ 1 \leqslant i \leqslant \nu,\ 1 \leqslant j \leqslant m_i\}$ introduced in Section 1 forms a basis for H_n. Recall that T is a bijection, $Tu = \hat{u} = \begin{pmatrix} u_{11}\ u_{12}\ \dots\ u_{\nu m_\nu} \end{pmatrix}^{\mathrm{T}}$ for $u = \sum_{i,j} u_{ij} \varphi_{ij} \in H_n$. According to the basis, the operator W is rewritten as

$$\begin{aligned} Wu &= \hat{W} \hat{u} \\ &= \begin{pmatrix} w_{ij}^k; & \begin{matrix} (i,j) & \rightarrow & (1,1), \dots, (\nu, m_\nu) \\ k & \downarrow & 1, \dots, N \end{matrix} \end{pmatrix} \hat{u}, \quad w_{ij}^k = \langle \varphi_{ij}, w_k \rangle_n. \end{aligned} \tag{3.1}$$

Then the operator $W_i \in \mathscr{L}(P_{\lambda_i} H_n; \mathbb{C}^N)$ is clearly

$$W_i u = \hat{W}_i \hat{u} = \begin{pmatrix} w_{ij}^k; & \begin{matrix} j & \rightarrow & 1, \dots, m_i \\ k & \downarrow & 1, \dots, N \end{matrix} \end{pmatrix} \hat{u}, \quad u \in P_{\lambda_i} H_n.$$

The observability condition on (W, L) is, in terms of these symbols, equivalent to

$$\begin{aligned} &\ker\left(\hat{W}\Lambda^k;\ k\downarrow 0,\dots,n-1\right) = \{0\}, \quad \text{or} \\ &\operatorname{rank}\left(\hat{W}\Lambda^k;\ k\downarrow 0,\dots,n-1\right) = n, \end{aligned} \tag{3.2}$$

where $\Lambda = TLT^{-1}$ (see (1.8)). The result in this section is stated as

Proposition 3.1. *In order that the pair (W, L) is observable, it is necessary and sufficient that the pairs (W_i, L_i), $1 \leqslant i \leqslant \nu$, are observable, in other words*

$$\ker\left(W_iL_i^k;\ k\downarrow 0,\dots,m_i-1\right) = \{0\}, \quad 1\leqslant i\leqslant \nu. \tag{3.3}$$

Proof. The proof is elementary. Suppose first that (W, L) is observable. Then it is clear that (3.3) holds. Conversely, suppose (3.3), or equivalently

$$\operatorname{rank}\left(\hat{W}_i\Lambda_i^k;\ k\downarrow 0,\dots,m_i-1\right) = m_i, \quad 1\leqslant i\leqslant \nu.$$

We apply elementary row operations to the matrix $\left(\hat{W}\Lambda^k;\ k\downarrow 0,\dots,n-1\right)$. In order to show (3.2), it is enough to prove that

$$\operatorname{rank}\left(\hat{W}(\Lambda-\lambda_\nu)^k;\ k\downarrow 0,\dots,n-1\right) = n.$$

The matrix just above is written as

$$\left(\begin{array}{cccc|c} \hat{W}_1 & \dots & \hat{W}_{\nu-1} & & \hat{W}_\nu \\ \hat{W}_1(\Lambda_1-\lambda_\nu) & \dots & \hat{W}_{\nu-1}(\Lambda_{\nu-1}-\lambda_\nu) & & \hat{W}_\nu N_\nu \\ \vdots & \vdots & \vdots & & \vdots \\ \hat{W}_1(\Lambda_1-\lambda_\nu)^{m_\nu-1} & \dots & \hat{W}_{\nu-1}(\Lambda_{\nu-1}-\lambda_\nu)^{m_\nu-1} & & \hat{W}_\nu N_\nu^{m_\nu-1} \\ \hline \hat{W}_1(\Lambda_1-\lambda_\nu)^{m_\nu} & \dots & \hat{W}_{\nu-1}(\Lambda_{\nu-1}-\lambda_\nu)^{m_\nu} & & 0 \\ \vdots & \vdots & \vdots & & \vdots \\ \hat{W}_1(\Lambda_1-\lambda_\nu)^{n-1} & \dots & \hat{W}_{\nu-1}(\Lambda_{\nu-1}-\lambda_\nu)^{n-1} & & 0 \end{array}\right). \tag{3.4}$$

The submatrix $\left(\hat{W}_\nu\ \hat{W}_\nu N_\nu \dots \hat{W}_\nu N_\nu^{m_\nu-1}\ 0 \dots 0\right)^{\mathrm{T}}$ on the right side of (3.4) has the full rank $(= m_\nu)$ by the assumption (3.3). Thus (3.2) will be proven, if

$$\begin{aligned} &\operatorname{rank}\begin{pmatrix} \hat{W}_1(\Lambda_1-\lambda_\nu)^{m_\nu} & \dots & \hat{W}_{\nu-1}(\Lambda_{\nu-1}-\lambda_\nu)^{m_\nu} \\ \vdots & \vdots & \vdots \\ \hat{W}_1(\Lambda_1-\lambda_\nu)^{n-1} & \dots & \hat{W}_{\nu-1}(\Lambda_{\nu-1}-\lambda_\nu)^{n-1} \end{pmatrix} \\ &= m_1+\dots+m_{\nu-1}\ (= n'). \end{aligned}$$

By setting $\Lambda' = \mathrm{diag}\left(\Lambda_1 \ \dots \ \Lambda_{\nu-1}\right)$ and $\hat{W}' = \left(\hat{W}_i; i \rightarrow 1, \dots, \nu - 1\right)$, the above relation is equivalent to

$$\begin{aligned}
&\mathrm{rank}\left(\hat{W}'(\Lambda' - \lambda_\nu)^k;\ k \downarrow 0, \dots, n' - 1\right)(\Lambda' - \lambda_\nu)^{m_\nu} \\
&= \mathrm{rank}\left(\hat{W}'(\Lambda' - \lambda_\nu)^k;\ k \downarrow 0, \dots, n' - 1\right) \\
&= \mathrm{rank}\left(\hat{W}'\Lambda'^k;\ k \downarrow 0, \dots, n' - 1\right) = n'.
\end{aligned}$$

The problem is thus reduced to the problem of proving the observability of the pair $(\hat{W}', \Lambda')$. By continueing the reduction procedure -via the assumption (3.3) at each stage, it finally leads to the problem of proving

$$\begin{aligned}
\mathrm{rank}\left(\hat{W}_1\Lambda_1^k;\ k \downarrow 0, \dots, m_1 - 1\right) &= m_1, \quad \text{or} \\
\ker\left(W_1 L_1^k;\ k \downarrow 0, \dots, m_1 - 1\right) &= \{0\}.
\end{aligned}$$

However, this is nothing but our assumption (3.3) when $i = 1$. □

Remark: Let us consider the case where $N_i = 0,\ 1 \leqslant i \leqslant \nu$. This occurs, for example, when the L is a self-adjoint operator. In this case, the relation (3.3) means that

$$\begin{aligned}
\mathrm{rank}\left(\hat{W}_i\Lambda_i^k;\ k \downarrow 0, \dots, m_i - 1\right) &= \mathrm{rank}\left(\lambda_i^k\hat{W}_i;\ k \downarrow 0, \dots, m_i - 1\right) \\
&= \mathrm{rank}\ \hat{W}_i = m_i, \qquad 1 \leqslant i \leqslant \nu.
\end{aligned} \tag{3.5}$$

Thus we need to choose the N greater than or equal to $\max_{1 \leqslant i \leqslant \nu} m_i$ in this case.

This case: $N_i = 0$, $1 \leqslant i \leqslant \nu$ is already discussed in [58], where the result can be viewed as a special case of our Proposition 3.1. Following [58], let us briefly give an alternative proof. The matrix $\left(\hat{W}\Lambda^k;\ k \downarrow 0, \dots, n - 1\right)$ is decomposed into the product of two matrices Φ and Ψ:

$$\begin{aligned}
\left(\hat{W}\Lambda^k;\ k \downarrow 0, \dots, n - 1\right) &= \left(\lambda_i^k\hat{W}_i;\ \begin{matrix} i & \rightarrow & 1, \dots, \nu \\ k & \downarrow & 0, \dots, n-1 \end{matrix}\right) \\
&= \left(\lambda_i^k I_N;\ \begin{matrix} i & \rightarrow & 1, \dots, \nu \\ k & \downarrow & 0, \dots, n-1 \end{matrix}\right) \mathrm{diag}\left(\hat{W}_1 \ \dots \ \hat{W}_\nu\right) \\
&= \Phi\Psi,
\end{aligned}$$

where Φ and Ψ denote the $nN \times \nu N$ and the $\nu N \times n$ matrices, respectively. Suppose first that $\mathrm{rank}\ \hat{W}_i = m_i,\ 1 \leqslant i \leqslant \nu$. Then the rank of Ψ is clearly equal to $m_1 + \cdots + m_\nu = n$. It is easily seen that the rank of Φ is equal to $\nu N\ (\geqslant n)$. Thus, we see that

$$n = \mathrm{rank}\ \Phi + \mathrm{rank}\ \Psi - \nu N \leqslant \mathrm{rank}\ \Phi\Psi \leqslant \min\left(\mathrm{rank}\ \Phi, \mathrm{rank}\ \Psi\right) = n.$$

Conversely, suppose that $\mathrm{rank}\ \Phi\Psi = n$. Then, we see that $n = \mathrm{rank}\ \Phi\ \Psi \leqslant \mathrm{rank}\ \Psi$, so that n column vectors of the $\nu N \times n$ matrix Ψ are linearly independent. But, this means that $\mathrm{rank}\ \hat{W}_i = m_i,\ 1 \leqslant i \leqslant \nu$.

1.4 The Case of a Single Observation

Let us consider the feedback control system (1.3) in the case where $\sigma(L)$ consists only of simple eigenvalues. As long as (W, L) is an obesrvable pair, a single sensor $W = \langle \cdot, w \rangle_n$ is enough for stabilization: We can construct an actuator $G = g$ such that $\sigma(L - GW)$ lies in the right half-plain $\mathbb{C}_+$. In Section 2, it is shown without proof that, given a set of mutually distinct positive numbers $S = \{\mu_i\}_{1 \leqslant i \leqslant n}$, an actuator g achieving the relation, $\sigma(L - \langle \cdot, w \rangle_n g) = S$ is uniquely determined by the formula (2.16): In this section, we show the formula in a little more general situation, where the set S is instead a set of arbitrary numbers.

Let φ_i, $1 \leqslant i \leqslant n$, be the eigenvectors corresponding to the eigenvalues λ_i of L. According to the basis $\{\varphi_i\}_{1 \leqslant i \leqslant n}$ for H_n, the operators L, W, and G are regarded respectively as $\Lambda = \mathrm{diag}\,(\lambda_1 \; \dots \; \lambda_n)$, $\hat{w} = (w_1 \; w_2 \; \dots \; w_n)^{\mathrm{T}} \in \mathbb{C}^n$, $w_i = \langle \varphi_i, w \rangle_n$ (see (3.1)), and $\hat{g} = (g_1 \; g_2 \; \dots \; g_n)^{\mathrm{T}} \in \mathbb{C}^n$. The observability condition is then rewritten as the condition; $w_i \neq 0$, $1 \leqslant i \leqslant n$. Let $S = \{\mu_i\}_{1 \leqslant i \leqslant l}$ be a set of n complex numbers, where some μ_i may coincide. By assuming that $w_i \neq 0$, $1 \leqslant i \leqslant n$, the well known pole allocation theory ensures a $g \in H_n$ or $\hat{g} \in \mathbb{C}^n$ such that $\sigma(L - \langle \cdot, w \rangle_n g) = \sigma(\Lambda - \hat{g}\hat{w}^{\mathrm{T}}) = S$. We show in this section somewhat deeper properties of $\Lambda - \hat{g}\hat{w}^{\mathrm{T}}$: The first property is stated as follows.

Proposition 4.1. *Let (W, L) be an observable pair, or $w_i = \langle \varphi_i, w \rangle_n \neq 0$, $1 \leqslant i \leqslant n$. The actuator $g \in H_n$ or $\hat{g} = (g_1 \; g_2 \; \dots \; g_n)^{\mathrm{T}} \in \mathbb{C}^n$ achieving the relation, $\sigma(L - \langle \cdot, w \rangle_n g) = \sigma(\Lambda - \hat{g}\hat{w}^{\mathrm{T}}) = S$ is uniquely determined by the set S, and given by the formula* (2.16) *or* (4.5) *below.*

Proof. The determinant of the matrix $\lambda - (\Lambda - \hat{g}\hat{w}^{\mathrm{T}})$ is a polynomial of order n, and is calculated as

$$\begin{aligned}
&\lambda^n - \left(\sum_{1 \leqslant i \leqslant n} \lambda_i - \sum_{i=1}^n g_i w_i \right) \lambda^{n-1} \\
&+ \left(\sum_{1 \leqslant i < j \leqslant n} \lambda_i \lambda_j - \sum_{i=1}^n g_i w_i \sum_{1 \leqslant j \leqslant n,\, j \neq i} \lambda_j \right) \lambda^{n-2} - \cdots\cdots \\
&+ (-1)^{n-1} \left(\sum_{1 \leqslant j_1 < \cdots < j_{n-1} \leqslant n} \lambda_{j_1} \cdots \lambda_{j_{n-1}} - \sum_{i=1}^n g_i w_i \sum_{\substack{1 \leqslant j_1 < \cdots < j_{n-2} \leqslant n, \\ j_1, \dots, j_{n-2} \neq i}} \lambda_{j_1} \cdots \lambda_{j_{n-2}} \right) \lambda \\
&+ (-1)^n \left(\lambda_1 \cdots \lambda_n - \sum_{i=1}^n g_i w_i \sum_{\substack{1 \leqslant j_1 < \cdots < j_{n-1} \leqslant n, \\ j_1, \dots, j_{n-1} \neq i}} \lambda_{j_1} \cdots \lambda_{j_{n-1}} \right).
\end{aligned} \tag{4.1}$$

The proof of the formula (4.1) is elementary, and carried out by induction relative to n and an expansion of the determinant according, e.g., to the first row. Since $\det\left(\lambda - (\Lambda - \hat{g}\hat{w}^{\mathrm{T}})\right)$ is equal to $\prod_{i=1}^{n}(\lambda - \mu_i)$, we have to solve the equation:

$$\sum_{i=1}^{n} g_i w_i = \sum_{1\leqslant i\leqslant n} (\lambda_i - \mu_i),$$

$$\sum_{i=1}^{n}\left(\sum_{\substack{1\leqslant j\leqslant n,\\ j\neq i}} \lambda_j\right) g_i w_i = \sum_{1\leqslant i<j\leqslant n} (\lambda_i\lambda_j - \mu_i\mu_j),$$

$$\sum_{i=1}^{n}\left(\sum_{\substack{1\leqslant j,k\leqslant n,\\ j,k\neq i}} \lambda_j\lambda_k\right) g_i w_i = \sum_{1\leqslant i<j<k\leqslant n} (\lambda_i\lambda_j\lambda_k - \mu_i\mu_j\mu_k),$$

$$\cdots \qquad \cdots \qquad\qquad \cdots \qquad \cdots$$

$$\sum_{i=1}^{n}\left(\sum_{\substack{1\leqslant j_1<\cdots<j_{n-2}\leqslant n,\\ j_1,\dots,j_{n-2}\neq i}} \lambda_{j_1}\cdots\lambda_{j_{n-2}}\right) g_i w_i = \sum_{1\leqslant j_1<\cdots<j_{n-1}\leqslant n} (\lambda_{j_1}\cdots\lambda_{j_{n-1}} - \mu_{j_1}\cdots\mu_{j_{n-1}}),$$

$$\sum_{i=1}^{n}\left(\sum_{\substack{1\leqslant j_1<\cdots<j_{n-1}\leqslant n,\\ j_1,\dots,j_{n-1}\neq i}} \lambda_{j_1}\cdots\lambda_{j_{n-1}}\right) g_i w_i = \lambda_1\lambda_2\cdots\lambda_n - \mu_1\mu_2\dots\mu_n,$$

for $(g_1w_1 \;\; g_2w_2 \;\; \dots \;\; g_nw_n)^{\mathrm{T}}$, or in matrix form

$$\begin{pmatrix} 1 & 1 & \dots & 1 \\ \sum_{i\neq 1}\lambda_i & \sum_{i\neq 2}\lambda_i & \dots & \sum_{i\neq n}\lambda_i \\ \sum_{\substack{i<j,\\ i,j\neq 1}}\lambda_i\lambda_j & \sum_{\substack{i<j,\\ i,j\neq 2}}\lambda_i\lambda_j & \dots & \sum_{\substack{i<j,\\ i,j\neq n}}\lambda_i\lambda_j \\ \vdots & \vdots & \vdots & \vdots \\ \prod_{i\neq 1}\lambda_i & \prod_{i\neq 2}\lambda_i & \dots & \prod_{i\neq n}\lambda_i \end{pmatrix} \begin{pmatrix} g_1w_1 \\ g_2w_2 \\ g_3w_3 \\ \vdots \\ g_nw_n \end{pmatrix}$$
$$= \begin{pmatrix} \sum_{1\leqslant i\leqslant n}(\lambda_i - \mu_i) \\ \sum_{1\leqslant i<j\leqslant n}(\lambda_i\lambda_j - \mu_i\mu_j) \\ \sum_{\substack{1\leqslant i<j\\ <k\leqslant n}}(\lambda_i\lambda_j\lambda_k - \mu_i\mu_j\mu_k) \\ \vdots \\ \lambda_1\lambda_2\cdots\lambda_n - \mu_1\mu_2\dots\mu_n \end{pmatrix}. \tag{4.2}$$

The $n \times n$ matrix on the left-hand side is denoted as Φ. Then Φ is invertible, and the inverse Φ^{-1} is calculated as

$$\frac{1}{\Delta}\begin{pmatrix} \Delta_1\lambda_1^{n-1} & -\Delta_1\lambda_1^{n-2} & \Delta_1\lambda_1^{n-3} & \dots & (-1)^{n-1}\Delta_1 \\ -\Delta_2\lambda_2^{n-1} & \Delta_2\lambda_2^{n-2} & -\Delta_2\lambda_2^{n-3} & \dots & (-1)^{n}\Delta_2 \\ \Delta_3\lambda_3^{n-1} & -\Delta_3\lambda_3^{n-2} & \Delta_3\lambda_3^{n-3} & \dots & (-1)^{n-1}\Delta_3 \\ \vdots & \vdots & \vdots & \vdots & \vdots \\ (-1)^{n-1}\Delta_n\lambda_n^{n-1} & -(-1)^{n-1}\Delta_n\lambda_n^{n-2} & (-1)^{n-1}\Delta_n\lambda_n^{n-3} & \dots & \Delta_n \end{pmatrix}, \tag{4.3}$$

where the factors Δ and Δ_k, $1 \leqslant k \leqslant n$, are introduced just after (2.16). To see this, we only have to prove that the matrix on the above right-hand side times Φ is equal to the identity matrix. In fact, the $(1,1)$ element of the product is calculated as

$$\begin{aligned} &\frac{\Delta_1}{\Delta}\Bigg(\lambda_1^{n-1} - \left(\textstyle\sum_{i\neq 1}\lambda_i\right)\lambda_1^{n-2} + \Bigg(\textstyle\sum_{\substack{i<j,\\ i,j\neq 1}}\lambda_i\lambda_j\Bigg)\lambda_1^{n-3} - \Bigg(\textstyle\sum_{\substack{i<j<k,\\ i,j,k\neq 1}}\lambda_i\lambda_j\lambda_k\Bigg)\lambda_1^{n-4} \\ &\qquad + \cdots + (-1)^{n-1}\lambda_2\cdots\lambda_n\Bigg) \\ &= \frac{\Delta_1}{\Delta}\prod_{i=2}^{n}(\lambda_1-\lambda_i) = \frac{\Delta}{\Delta} = 1, \end{aligned}$$

and the $(1,k)$ element, $k \geqslant 2$, becomes

$$\begin{aligned} &\frac{\Delta_1}{\Delta}\Bigg(\lambda_1^{n-1} - \left(\textstyle\sum_{i\neq k}\lambda_i\right)\lambda_1^{n-2} + \Bigg(\textstyle\sum_{\substack{i<j,\\ i,j\neq k}}\lambda_i\lambda_j\Bigg)\lambda_1^{n-3} - \cdots + (-1)^{n-1}\textstyle\prod_{i\neq k}\lambda_i\Bigg) \\ &= \frac{\Delta_1}{\Delta}\prod_{i\neq k}(\lambda_1-\lambda_i) = 0. \end{aligned}$$

The other elements are similarly calculated. Thus eqn. (4.2) has a unique solution. Since $w_i = \langle\varphi_i, w\rangle_n \neq 0$, $1 \leqslant i \leqslant n$, the actuator $\hat{g}$ is thus uniquely determined by the set $S = \{\mu_i\}_{1\leqslant i\leqslant n}$.

To see the actuator $\hat{g}$ more concretely, set $f(\lambda) = \prod_{i=1}^{n}(\lambda-\mu_i)$ (see (2.16)). In view of (4.3), we calculate, for example, as

$$\begin{aligned} \frac{\Delta}{\Delta_1}g_1w_1 &= \lambda_1^{n-1}\sum_i(\lambda_i-\mu_i) - \lambda_1^{n-2}\sum_{i<j}(\lambda_i\lambda_j-\mu_i\mu_j) + \cdots \\ &\quad + (-1)^{n-2}\lambda_1\sum_{1\leqslant j_1<\cdots<j_{n-1}\leqslant n}(\lambda_{j_1}\cdots\lambda_{j_{n-1}} - \mu_{j_1}\cdots\mu_{j_{n-1}}) \\ &\quad + (-1)^{n-1}\Bigg(\prod_{1\leqslant i\leqslant n}\lambda_i - \prod_{1\leqslant i\leqslant n}\mu_i\Bigg). \end{aligned} \tag{4.4}$$

In the expression of the other $g_i w_i$, the polynomial

$$\lambda^{n-1}\sum_i(\lambda_i-\mu_i)-\lambda^{n-2}\sum_{i<j}(\lambda_i\lambda_j-\mu_i\mu_j)$$
$$+\lambda^{n-3}\sum_{i<j<k}(\lambda_i\lambda_j\lambda_k-\mu_i\mu_j\mu_k)-\cdots+(-1)^{n-1}\left(\prod_{1\leqslant i\leqslant n}\lambda_i-\prod_{1\leqslant i\leqslant n}\mu_i\right)$$

similarly appears with λ replaced by λ_i. In view of the identity:

$$\prod_{1\leqslant i\leqslant n}(\lambda-\lambda_i)$$
$$=\lambda^n-\left(\sum_i\lambda_i\right)\lambda^{n-1}+\left(\sum_{i<j}\lambda_i\lambda_j\right)\lambda^{n-2}-\left(\sum_{i<j<k}\lambda_i\lambda_j\lambda_k\right)\lambda^{n-3}$$
$$+\cdots+(-1)^{n-1}\left(\sum_{1\leqslant j_1<\cdots<j_{n-1}\leqslant n}\lambda_{j_1}\cdots\lambda_{j_{n-1}}\right)\lambda+(-1)^n\left(\prod_{1\leqslant i\leqslant n}\lambda_i\right),$$

we see that, for $1\leqslant l\leqslant n$,

$$\lambda_l^n=\left(\sum_i\lambda_i\right)\lambda_l^{n-1}-\left(\sum_{i<j}\lambda_i\lambda_j\right)\lambda_l^{n-2}+\left(\sum_{i<j<k}\lambda_i\lambda_j\lambda_k\right)\lambda_l^{n-3}$$
$$-\cdots+(-1)^{n-2}\left(\sum_{1\leqslant j_1<\cdots<j_{n-1}\leqslant n}\lambda_{j_1}\cdots\lambda_{j_{n-1}}\right)\lambda_l+(-1)^{n-1}\left(\prod_i\lambda_i\right).$$

Thus the right-hand side of (4.4) is rewritten as

$$\lambda_1^n-\left(\sum_i\mu_i\right)\lambda_1^{n-1}+\left(\sum_{i<j}\mu_i\mu_j\right)\lambda_1^{n-2}-\left(\sum_{i<j<k}\mu_i\mu_j\mu_k\right)\lambda_1^{n-3}$$
$$+\cdots+(-1)^{n-1}\left(\sum_{1\leqslant j_1<\cdots<j_{n-1}\leqslant n}\mu_{j_1}\cdots\mu_{j_{n-1}}\right)\lambda_1+(-1)^n\prod_{1\leqslant i\leqslant n}\mu_i$$
$$=\prod_{1\leqslant i\leqslant n}(\lambda_1-\mu_i)=f(\lambda_1),$$

or $g_1=\frac{1}{w_1}\frac{\Delta_1}{\Delta}f(\lambda_1)$. The other g_i have similar expressions. Thus we have

$$\hat{g}=\frac{1}{\Delta}\begin{pmatrix}\frac{\Delta_1}{w_1}f(\lambda_1) & -\frac{\Delta_2}{w_2}f(\lambda_2) & \frac{\Delta_3}{w_3}f(\lambda_3) & \ldots & (-1)^{n-1}\frac{\Delta_n}{w_n}f(\lambda_n)\end{pmatrix}^{\mathrm{T}}. \tag{4.5}$$

This is nothing but (2.16). □

In the case where $\sigma(L)\cap S=\varnothing$, we derive a more detailed result. This occurs, for example, in stabilization problems in which $\sigma(L)$ is contained in $\mathbb{C}_-$ and S

in $\mathbb{C}_+$. Let $S = \{\mu_i\}_{1 \leqslant i \leqslant l}$ be a set of n complex numbers such that $\mu_i \neq \mu_j$ for $i \neq j$; that each μ_i has multiplicity m_i; and that $\sum_{1 \leqslant i \leqslant l} m_i = n$.

Proposition 4.2. *Assume that $\sigma(L) \cap S = \varnothing$ in Proposition 4.1. As long as $m_i \geqslant 2$, each eigenvalue μ_i of $L - \langle \cdot, w \rangle_n g$ admits a generalized eigenspace. More precisely, set*

$$W_{\mu_i}^{(k)} = \left\{ p \in H_n;\ \left(\mu_i - L + \langle \cdot, w \rangle_n g\right)^k p = 0 \right\}, \quad k \geqslant 1.$$

Then,

$$\dim W_{\mu_i}^{(k)} = \begin{cases} k, & 1 \leqslant k \leqslant m_i, \\ m_i, & m_i < k. \end{cases} \tag{4.6}$$

Set $p_{ij} = (L - \mu_i)^{-j} g$, $j \geqslant 1$. Then, the system $\{p_{ij}\}_{1 \leqslant j \leqslant k}$ forms a basis for $W_{\mu_i}^{(k)}$, $1 \leqslant k \leqslant m_i$, and the system $\{p_{ij};\ 1 \leqslant i \leqslant l,\ 1 \leqslant j \leqslant m_i\}$ forms a basis for H_n.

When $m_i > 1$, set $\boldsymbol{p}_i = \left(p_{i1}\ \ldots\ p_{im_i}\right)$. For any $p = \sum_{1 \leqslant j \leqslant m_i} c_{ij} p_{ij} = \boldsymbol{p}_i \boldsymbol{c}_i \in W_{\mu_i}^{(m_i)}$, $\boldsymbol{c}_i$ being $\left(c_{i1}\ \ldots\ c_{im_i}\right)^{\mathrm{T}}$, we then have a matrix represenation of the operator $L - \langle \cdot, w \rangle_n g$:

$$\left(L - \langle \cdot, w \rangle_n g\right) \boldsymbol{p}_i \boldsymbol{c}_i = \boldsymbol{p}_i M_i \boldsymbol{c}_i, \quad M_i = \mu_i + N_i, \quad N_i^{m_i} = 0, \tag{4.7}$$

where M_i denotes an $m_i \times m_i$ matrix.

Proof. Assuming, for example, that $m_1 > 1$, let us first examine $W_{\mu_1}^{(1)}$. Let $\left(\mu_1 - L + \langle \cdot, w \rangle_n g\right) p = (\mu_1 - L) p + \langle p, w \rangle_n g = 0$. Then, since $\mu_1 \notin \sigma(L)$, we have $p = \langle p, w \rangle_n (L - \mu_1)^{-1} g$ as a necessary condition. We show that $p_{11} = (L - \mu_1)^{-1} g$ certainly belongs to $W_{\mu_1}^{(1)}$. Recall that, for $\mu \in \rho(L)$,

$$(L - \mu)^{-1} g = \sum_{1 \leqslant i \leqslant n} \frac{g_i}{\lambda_i - \mu} \varphi_i, \quad g = \sum_{1 \leqslant i \leqslant n} g_i \varphi_i.$$

In view of (4.5) then, we calculate as

$$\langle p_{11}, w \rangle_n = \frac{1}{\Delta} \sum_{1 \leqslant i \leqslant n} \frac{(-1)^{i-1} \Delta_i f(\lambda_i)}{\lambda_i - \mu_1} = \frac{1}{\Delta} \sum_{1 \leqslant i \leqslant n} (-1)^{i-1} \Delta_i g(\lambda_i),$$

where we have set $f(\lambda) = (\lambda - \mu_1) g(\lambda)$. The last term is a polynomial of $\{\lambda_i\}_{1 \leqslant i \leqslant n}$. Let $1 \leqslant k < l \leqslant n$. Note that Δ_i, $i \neq k, l$, contains a factor $\lambda_k - \lambda_l$ in the product. In view of Lemma 2.4, we see, by setting $\lambda_k = \lambda_l$, that

$$\begin{aligned} \sum_{1 \leqslant i \leqslant n} (-1)^{i-1} \Delta_i g(\lambda_i) &= (-1)^{k-1} \Delta_k g(\lambda_k) + (-1)^{l-1} \Delta_l g(\lambda_l) \\ &= (-1)^{k-1} \Delta_k g(\lambda_k) + (-1)^{l-1+(l-1+k)} \Delta_k g(\lambda_k) = 0. \end{aligned}$$

Thus, $\sum_{i=1}^{n}(-1)^{i-1}\Delta_i g(\lambda_i) = c\prod_{k<l}(\lambda_k - \lambda_l) = c\Delta$, both sides of which are polynomials of λ_1 of order $n-1$. Comparing then the coefficients of λ_1^{n-1} of the both sides, we have $c=1$, in other words, $\langle p_{11}, w\rangle_n = 1$. We have shown that $\dim W_{\mu_1}^{(1)} = 1$, and that

$$\left(\mu_1 - L + \langle\cdot, w\rangle_n g\right)p_{11} = (\mu_1 - L)p_{11} + \langle p_{11}, w\rangle_n g = -g + g = 0.$$

Assuming inductively that the system $\{p_{1j}\}_{1\leqslant j\leqslant k}$ forms a basis for $W_{\mu_1}^{(k)}$, $1\leqslant k\leqslant m_1-1$, let us examine $W_{\mu_1}^{(m_1)}$. For any $p\in W_{\mu_1}^{(m_1)}$, we note that $\left(\mu_1 - L + \langle\cdot, w\rangle_n g\right)p$ belongs to $W_{\mu_1}^{(m_1-1)}$. Thus,

$$\left(\mu_1 - L + \langle\cdot, w\rangle_n g\right)p = (\mu_1 - L)p + \langle p, w\rangle_n g = \sum_{1\leqslant j\leqslant m_1-1} \alpha_j p_{1j},$$

from which p has an expression:

$$\begin{aligned} p &= \langle p, w\rangle_n (L-\mu_1)^{-1}g - \sum_{1\leqslant j\leqslant m_1-1} \alpha_j (L-\mu_1)^{-1}p_{1j} \\ &= \langle p, w\rangle_n p_{11} - \sum_{1\leqslant j\leqslant m_1-2} \alpha_j p_{1(j+1)} - \alpha_{m_1-1}(L-\mu_1)^{-m_1}g. \end{aligned} \tag{4.8}$$

We show that the last vector $p_{1m_1} = (L-\mu_1)^{-m_1}g$ belongs to $W_{\mu_1}^{(m_1)}$. In the relation

$$\left(\mu_1 - L + \langle\cdot, w\rangle_n g\right)p_{1m_1} = (\mu_1 - L)p_{1m_1} + \langle p_{1m_1}, w\rangle_n g,$$

set $f(\lambda) = (\lambda - \mu_1)^{m_1} g(\lambda)$, where $g(\lambda) = \prod_{2\leqslant i\leqslant l}(\lambda - \mu_i)^{m_i}$. Then,

$$\langle p_{1m_1}, w\rangle_n = \frac{1}{\Delta}\sum_{1\leqslant i\leqslant n}(-1)^{i-1}\frac{\Delta_i f(\lambda_i)}{(\lambda_i - \mu_1)^{m_1}} = \frac{1}{\Delta}\sum_{1\leqslant i\leqslant n}(-1)^{i-1}\Delta_i g(\lambda_i).$$

Here, $g(\lambda_1)$ is a polynomial of λ_1 of order $n-m_1$, and $\Delta_2, \dots, \Delta_n$ of order $n-2$. Let $1\leqslant k<l\leqslant n$. In the polynomial $\sum_{i=1}^{n}(-1)^{i-1}\Delta_i g(\lambda_i)$, set $\lambda_k = \lambda_l$. Then, we similarly find -via Lemma 2.4- that $\sum_{i=1}^{n}(-1)^{i-1}\Delta_i g(\lambda_i) = c\Delta$. The left-hand side is a polynomial of λ_1 of order $n-2$, but Δ is of order $n-1$. Thus we have $c=0$, and $\langle p_{1m_1}, w\rangle_n = 0$. This means that $\left(\mu_1 - L + \langle\cdot, w\rangle_n g\right)p_{1m_1} = -(L-\mu_1)p_{1m_1} = -p_{1(m_1-1)}$, and thus $\left(\mu_1 - L + \langle\cdot, w\rangle_n g\right)^{m_1}p_{1m_1} = 0$. We have inductively shown that

$$\left(\mu_1 - L + \langle\cdot, w\rangle_n g\right)p_{1i} = -(L-\mu_1)p_{1i} = -p_{1(i-1)}, \quad 2\leqslant i\leqslant m_1.$$

Next we show that $W_{\mu_1}^{(m_1+1)} = W_{\mu_1}^{(m_1)}$. Let p be in $W_{\mu_1}^{(m_1+1)}$. Then, as in (4.8),

$$p = \langle p, w\rangle_n p_{11} - \sum_{1\leqslant j\leqslant m_1-1} \alpha_j p_{1(j+1)} - \alpha_{m_1}(L-\mu_1)^{-(m_1+1)}g, \tag{4.9}$$

the first and the second terms of which belong to $\in W_{\mu_1}^{(m_1)}$. If $\alpha_{m_1} \neq 0$, we must have

$$p^* = (L-\mu_1)^{-(m_1+1)} g \in W_{\mu_1}^{(m_1+1)}.$$

We show that this leads to a contradiction. In fact, since

$$\begin{aligned}(\mu_1 - L + \langle \cdot, w \rangle_n g) p^* &= (\mu_1 - L) p^* + \langle p^*, w \rangle_n g \\ &= -p_{1m_1} + \langle p^*, w \rangle_n g \in W_{\mu_1}^{(m_1)},\end{aligned}$$

we see that $\langle p^*, w \rangle_n g$ belongs to $W_{\mu_1}^{(m_1)}$ in this case. Setting $f(\lambda) = (\lambda - \mu_1)^{m_1} g(\lambda)$ as before, we calculate as

$$\begin{aligned}\langle p^*, w \rangle_n &= \frac{1}{\Delta} \sum_{1 \leqslant i \leqslant n} (-1)^{i-1} \frac{\Delta_i f(\lambda_i)}{(\lambda_i - \mu_1)^{m_1+1}} \\ &= \frac{1}{\Delta} \sum_{1 \leqslant i \leqslant n} (-1)^{i-1} \frac{\Delta_i g(\lambda_i)}{\lambda_i - \mu_1} \\ &= \frac{1}{\Delta} \frac{1}{\prod_{i \geqslant 1} (\lambda_i - \mu_1)} \sum_{1 \leqslant i \leqslant n} (-1)^{i-1} \cdot \prod_{j \neq i} (\lambda_j - \mu_1) \cdot \Delta_i g(\lambda_i).\end{aligned} \tag{4.10}$$

It will be shown just below that $\langle p^*, w \rangle_n \neq 0$. Once it is proven, g *must* belong to $W_{\mu_1}^{(m_1)}$, and thus $g = \sum_{1 \leqslant i \leqslant n} h_i \varphi_i$ is expressed as $g = \sum_{1 \leqslant j \leqslant m_1} \gamma_j p_{1j}$. Here we note that $g_i \neq 0$, $1 \leqslant i \leqslant n$, by our assumption, $\sigma(L) \cap S = \varnothing$. Since $g_i \neq 0$, $1 \leqslant i \leqslant n$, the expression leads to the relation,

$$\begin{pmatrix} \frac{1}{\lambda_1 - \mu_1} & \frac{1}{(\lambda_1 - \mu_1)^2} & \cdots & \frac{1}{(\lambda_1 - \mu_1)^{m_1}} \\ \frac{1}{\lambda_2 - \mu_1} & \frac{1}{(\lambda_2 - \mu_1)^2} & \cdots & \frac{1}{(\lambda_2 - \mu_1)^{m_1}} \\ \vdots & \vdots & \cdots & \vdots \\ \frac{1}{\lambda_n - \mu_1} & \frac{1}{(\lambda_n - \mu_1)^2} & \cdots & \frac{1}{(\lambda_n - \mu_1)^{m_1}} \end{pmatrix} \begin{pmatrix} \gamma_1 \\ \gamma_2 \\ \vdots \\ \gamma_{m_1} \end{pmatrix} = \begin{pmatrix} 1 \\ 1 \\ \vdots \\ 1 \end{pmatrix}. \tag{4.11}$$

However, this relation forms a so-called *overdetermined system*, and has *no* set of solutions $\{\gamma_j\}_{1 \leqslant j \leqslant m_1}$ satisfying (4.11), which is a contradiction: In fact, assuming that (4.11) admits a solution $\begin{pmatrix} \gamma_1 & \dots & \gamma_{m_1} \end{pmatrix}^{\mathrm{T}}$, consider a polynomial $p(\lambda) = \gamma_{m_1} \lambda^{m_1} + \cdots + \gamma_2 \lambda^2 + \gamma_1 \lambda - 1$ of order at most m_1. Then $h(\lambda)$ has $n (> m_1)$ distinct zeros $(\lambda_i - \mu_1)^{-1}$, $1 \leqslant i \leqslant n$, and thus $h(\lambda)$ is identically equal to 0. But, it is a contradiction. We have shown that $\alpha_{m_1} = 0$, and that the vector p in (4.9) belongs to $W_{\mu_1}^{(m_1)}$. Thus, $W_{\mu_1}^{(m_1+1)} = W_{\mu_1}^{(m_1)}$.

By assuming that $m_1 > 1$, the system $\{p_{1j}\}_{1 \leqslant j \leqslant m_1}$ forms a basis for $W_{\mu_1}^{(m_1)}$. Let us show the relation (4.7) when $i = 1$. For any $p = \sum_{1 \leqslant i \leqslant m_1} c_{1i} p_{1i} = \boldsymbol{p}_1 \boldsymbol{c}_1$, we calculate as

$$\begin{aligned}(L-\langle\cdot,w\rangle_n g)p &= \sum_{1\leqslant i\leqslant m_1} c_{1i}(L-\langle\cdot,w\rangle_n g)p_{1i}\\ &= c_{11}\mu_1 p_{11} + \sum_{2\leqslant i\leqslant m_1} c_{1i}(\mu_1 p_{1i} + p_{1(i-1)})\\ &= \sum_{1\leqslant i\leqslant m_1-1} (\mu_1 c_{1i} + c_{1(i+1)})p_{1i} + \mu_1 c_{1m_1} p_{1m_1}\\ &= (p_{11}\ \dots\ p_{1m_1})M_1\boldsymbol{c}_1 = \boldsymbol{p}_1 M_1 \boldsymbol{c}_1,\end{aligned}$$

where M_1 denotes an $m_1\times m_1$ matrix given by

$$M_1 = \begin{pmatrix} \mu_1 & 1 & 0 & \dots & 0\\ 0 & \mu_1 & 1 & \dots & 0\\ 0 & 0 & \mu_1 & \dots & 0\\ \vdots & \vdots & \vdots & \ddots & \vdots\\ 0 & 0 & 0 & \dots & \mu_1\end{pmatrix} = \mu_1 + N_1, \quad N_1^{m_1} = 0,$$

and means a matrix representation of the operator $L-\langle\cdot,w\rangle_n g$.

Proof of $\langle p^*,\hat{w}\rangle_n \neq 0$:

In (4.10), we slightly modify the polynomial $g(\lambda)$ with zeros μ_i, $2\leqslant i\leqslant l$, to $g_\varepsilon(\lambda)$, so that $g_\varepsilon(\lambda)$ has $n-m_1$ distinct zeros: For a small $\varepsilon\neq 0$ and $2\leqslant i\leqslant l$, change the set of m_i multiple of μ_i to $\{\mu_i+\varepsilon^k\}_{0\leqslant k\leqslant m_i-1}$, and relabel them as $\{\zeta_i\}_{1\leqslant i\leqslant n-m_1}$, where $\zeta_i=\zeta(\varepsilon)$. Set then

$$g_\varepsilon(\lambda) = \prod_{1\leqslant i\leqslant n-m_1} (\lambda-\zeta_i).$$

It is clear that $\zeta_i\neq\zeta_j$ for $i\neq j$ as long as ε is small enough, and that $\lim_{\varepsilon\to 0} g_\varepsilon(\lambda) = g(\lambda)$. In (4.10), we consider a slightly modified polynomial

$$\sum_{1\leqslant i\leqslant n} (-1)^{i-1}\cdot\prod_{j\neq i}(\lambda_j-\mu_1)\cdot\Delta_i g_\varepsilon(\lambda_i) = G, \tag{4.12}$$

which is a polynomial of $\{\lambda_i\}_{i\geqslant 1}$ and μ_1. In (4.12), set $\lambda_k=\lambda_l$ for $1\leqslant k<l\leqslant n$. Then, $\Delta_i=0$ for $i\neq k,l$. By Lemma 2.4, the sum of the kth and the lth terms becomes

$$\begin{aligned}&\frac{(-1)^{k-1}\cdot\prod_{j\neq k}(\lambda_j-\mu_1)\cdot\Delta_k g_\varepsilon(\lambda_k) + (-1)^{l-1}\cdot\prod_{j\neq l}(\lambda_j-\mu_1)\cdot\Delta_l g_\varepsilon(\lambda_l)}{\prod_{j\geqslant 1}(\lambda_j-\mu_1)}\\ &= (-1)^{k-1}\frac{\Delta_k g_\varepsilon(\lambda_k)}{\lambda_k-\mu_1} + (-1)^{l-1}\frac{\Delta_l g_\varepsilon(\lambda_l)}{\lambda_l-\mu_1}\\ &= (-1)^{k-1}\frac{\Delta_k g_\varepsilon(\lambda_k)}{\lambda_k-\mu_1} + (-1)^{l-1}(-1)^{l-1+k}\frac{\Delta_k g_\varepsilon(\lambda_k)}{\lambda_k-\mu_1} = 0.\end{aligned}$$

Thus, G contains the factor Δ, a polynomial of λ_1 of order $n-1$.

On the other hand, G is also a polynomial of μ_1 at most of order $n-1$. We show that G has the factor $g_\varepsilon(\mu_1)$, a polynomial of μ_1 of order $n-m_1$. Let $g_\varepsilon(\lambda) = (\lambda - \zeta_1)h_\varepsilon(\lambda)$, h_ε being of order $n-m_1-1$. By setting $\mu_1 = \zeta_1$ in (4.12), G is calculated as

$$\begin{aligned}&\sum_{1\leqslant i\leqslant n}(-1)^{i-1}\cdot\prod_{j\neq i}(\lambda_j-\mu_1)\cdot\Delta_i(\lambda_i-\mu_1)h_\varepsilon(\lambda_i)\\ &=\prod_{j\geqslant 1}(\lambda_j-\mu_1)\sum_{1\leqslant i\leqslant n}(-1)^{i-1}\Delta_i h_\varepsilon(\lambda_i)=0.\end{aligned}$$

In fact, the above right-hand side is a polynomial of λ_1 at most of order $n-2$, and by Lemma 2.4, contains the factor Δ. Thus we see that G contains the factor $\mu_1 - \zeta_1$. A similar factorization of $g_\varepsilon(\lambda)$ shows that G contains the factor $\prod_{1\leqslant i\leqslant n-m_1}(\mu_1-\zeta_i) = g_\varepsilon(\mu_1) \neq 0$. Thus we have an expression

$$G=\sum_{1\leqslant i\leqslant n}(-1)^{i-1}\cdot\prod_{j\neq i}(\lambda_j-\mu_1)\cdot\Delta_i g_\varepsilon(\lambda_i)=c\,\Delta\, g_\varepsilon(\mu_1). \tag{4.13$_1$}$$

Both sides of (4.13$_1$) are polynomials of λ_1 of order $n-1$. Comparing the coefficients of λ_1^{n-1}, we have a relation of polynomials with reduced order:

$$\sum_{2\leqslant i\leqslant n}(-1)^{i-1}\cdot{\prod_{j\neq i}}'(\lambda_j-\mu_1)\cdot\Delta_i' g_\varepsilon(\lambda_i)=c\,g_\varepsilon(\mu_1)\Delta_1. \tag{4.13$_2$}$$

Here, $\prod'$ and Δ_i' mean the products in λ_j, $j \geqslant 2$, e.g., $\Delta_i' = \prod_{\substack{2\leqslant p<q\leqslant n,\\ p,q\neq i}}(\lambda_p-\lambda_q)$. Both sides of (4.13$_2$) are polynomials of λ_2 of order $n-2$. Similarly comparing the coefficients of λ_2^{n-2}, we have

$$\sum_{3\leqslant i\leqslant n}(-1)^{i-1}\cdot{\prod_{j\neq i}}''(\lambda_j-\mu_1)\cdot\Delta_i'' g_\varepsilon(\lambda_i)=c\,g_\varepsilon(\mu_1)\Delta_2'. \tag{4.13$_3$}$$

Here, $\prod''$ and Δ_i'' mean the products in λ_j, $j \geqslant 3$. Both sides of (4.13$_3$) are polynomials of λ_3 of order $n-3$. Continuing the same procedure, we have

$$\sum_{m_1\leqslant i\leqslant n}(-1)^{i-1}\cdot{\prod_{j\neq i}}^*(\lambda_j-\mu_1)\cdot\Delta_i^* g_\varepsilon(\lambda_i)=c\,g_\varepsilon(\mu_1)\prod_{m_1\leqslant k<l\leqslant n}(\lambda_k-\lambda_l). \tag{4.13$_4$}$$

Here, $\prod^*$ and Δ_i^* mean the products in λ_j, $j \geqslant m_1$. At this stage, we note that

(i) $g_\varepsilon(\mu_1)$: a polynomial of μ_1 of order $n-m_1$;

(ii) $\Delta_i^* = \prod_{\substack{m_1\leqslant k<l,\\ k,l\neq i}}(\lambda_k-\lambda_l)$, $i \geqslant m_1$; and

(iii) $\prod_{j\neq i}^*(\lambda_j-\mu_1) = \prod_{\substack{m_1\leqslant j,\\ j\neq i}}(\lambda_j-\mu_1)$: a polynomial of μ_1 of order $n-m_1$.

Thus, both sides of (4.13_4) are polynomials of μ_1 of order $n-m_1$. Comparing the coefficients of $\mu_1^{n-m_1}$, we have

$$(-1)^{n-m_1} \sum_{m_1 \leqslant i \leqslant n} (-1)^{i-1} \Delta_i^* g_\varepsilon(\lambda_i) = c \prod_{m_1 \leqslant k < l \leqslant n} (\lambda_k - \lambda_l). \qquad (4.13_5)$$

Let us examine the left-hand side of (4.13_5) which is a polynomial of λ_j, $m_1 \leqslant j \leqslant n$. Set $\lambda_k = \lambda_l$ for $m_1 \leqslant k < l \leqslant n$. Then, $\Delta_i^* = 0$ for $i \neq k,\ l$. As for the kth and the lth terms, we similarly have the relation: $\Delta_l^* = (-1)^{l-1+k} \Delta_k^*$ (see Lemma 2.4). The sum of the kth and the lth terms then becomes

$$\begin{aligned} &(-1)^{k-1} \Delta_k^* g_\varepsilon(\lambda_k) + (-1)^{l-1} \Delta_l^* g_\varepsilon(\lambda_l) \\ &= (-1)^{k-1} \Delta_k^* g_\varepsilon(\lambda_k) + (-1)^{l-1+(l-1+k)} \Delta_k^* g_\varepsilon(\lambda_k) = 0. \end{aligned}$$

Thus the left-hand side of (4.13_5) contains a factor $\prod_{m_1 \leqslant k < l \leqslant n} (\lambda_k - \lambda_l)$:

$$(-1)^{n-m_1} \sum_{m_1 \leqslant i \leqslant n} (-1)^{i-1} \Delta_i^* g_\varepsilon(\lambda_i) = c' \prod_{m_1 \leqslant k < l \leqslant n} (\lambda_k - \lambda_l).$$

The above both sides are polynomials of λ_{m_1} of order $n-m_1$. Comparing the coefficients of $\lambda_{m_1}^{n-m_1}$, we have

$$(-1)^{n-m_1} (-1)^{m_1-1} \Delta_{m_1}^* = c' \Delta_{m_1}^* \quad \text{or} \quad c' = (-1)^{n-1}.$$

Thus, going back to (4.13_5), we see that $c = (-1)^{n-1}$. Passage to the limit regarding ε $(\to 0)$ in (4.13_1) shows that

$$\sum_{1 \leqslant i \leqslant n} (-1)^{i-1} \cdot \prod_{j \neq i} (\lambda_j - \mu_1) \cdot \Delta_i g(\lambda_i) = (-1)^{n-1} \Delta g(\mu_1). \qquad (4.14)$$

Finally, $\langle p^*, \hat{w} \rangle_n$ in (4.10) is rewritten as

$$\begin{aligned} \langle p^*, \hat{w} \rangle_n &= \frac{1}{\Delta} \frac{1}{\prod_{i \geqslant 1} (\lambda_i - \mu_1)} \sum_{i=1}^{n} (-1)^{i-1} \cdot \prod_{j \neq i} (\lambda_j - \mu_1) \cdot \Delta_i g(\lambda_i) \\ &= (-1)^{n-1} \frac{g(\mu_1)}{\prod_{i \geqslant 1} (\lambda_i - \mu_1)} \neq 0. \end{aligned}$$

This completes the proof of Proposition 4.2. □

Chapter 2

Preliminary results: Basic theory of elliptic operators

2.1 Introduction

Following the finite-dimensional stabilization problems in the preceding chapter, we are mainly interested in control problems governed by partial differential equations. In this chapter, some preliminary results on elliptic problems necessary for the following chapters are discussed. In this monograph, a uniformly elliptic operator $\mathscr{L}$ of order 2 equipped with a boundary operator τ is studied. Let Ω be a bounded domain in $\mathbb{R}^m$ with the boundary Γ which consists of a finite number of smooth components of $(m-1)$-dimension. The pair $(\mathscr{L}, \tau)$ is a standard one, and described as

$$\begin{aligned} \mathscr{L}u &= -\sum_{i,j=1}^{m} \frac{\partial}{\partial x_i}\left(a_{ij}(x)\frac{\partial u}{\partial x_j}\right) + \sum_{i=1}^{m} b_i(x)\frac{\partial u}{\partial x_i} + c(x)u, \\ \tau u &= \alpha(\xi)u + (1-\alpha(\xi))\frac{\partial u}{\partial \nu}, \end{aligned} \tag{1.1}$$

where $a_{ij}(x) = a_{ji}(x)$ for $1 \leqslant i, j \leqslant m$ and $x = (x_1, \ldots, x_m) \in \overline{\Omega}$; for some positive δ

$$\sum_{i,j=1}^{m} a_{ij}(x)\xi_i\xi_j \geqslant \delta|\xi|^2, \quad \forall \xi = (\xi_1, \ldots, \xi_m) \in \mathbb{R}^m, \quad \forall x \in \overline{\Omega};$$

and

$$0 \leqslant \alpha(\xi) \leqslant 1, \qquad \frac{\partial u}{\partial \nu} = \sum_{i,j=1}^{m} a_{ij}(\xi)\nu_i(\xi) \left.\frac{\partial u}{\partial x_j}\right|_{\Gamma},$$

$\nu(\xi) = (\nu_1(\xi), \ldots, \nu_m(\xi))$ being the unit outer normal at each point $\xi \in \Gamma$. The last term, $\left.\frac{\partial u}{\partial x_j}\right|_{\Gamma}$ is understood as the *trace* $\gamma\left(\frac{\partial u}{\partial x_j}\right)$ of the function $\frac{\partial u}{\partial x_j}$ on Γ (see Section 2 below for the trace operator γ). As for the regularity of the coefficients, it is enough to assume that $a_{ij}(\cdot)$, $b_i(\cdot)$, $c(\cdot)$, and $\alpha(\cdot)$ belong to $C^2(\overline{\Omega})$, $C^2(\overline{\Omega})$, $C^{\omega}(\overline{\Omega})$, and $C^{2+\omega}(\Gamma)$, respectively, where ω, $0 < \omega < 1$, denote respective constants. Throughout the chapter (and the whole monograph, too), all arguments will be based on the $L^2(\Omega)$-framework.

While the pair $(\mathscr{L}, \tau)$ in (1.1) is standard and looks simple, it contains general spectral properties, so that it would work as a prototype in various applications. We are interested in basic properties of the operator L which is derived from $\mathscr{L}|_{\ker \tau}$. As long as the author knows, there are two approaches: One is a typical approach based on functional analysis, in the case where $\alpha(\xi) \equiv 1$ (the Dirichret boundary) or $0 \leqslant \alpha(\xi) < 1$ (the Robin boundary). The other is based both on the classical C^{α}-theory and functional analysis. The former approach is very well known, and nowadays there are rich theories established, based on functional analysis and distribution theory. The latter approach is useful when the problem setting cannot be properly described in the former framework, such as the case where the Dirichlet boundary is continuously connected with the Robin boundary: It is based on the classical fundamental solution $U(t,x,y)$ associated with the corresponding parabolic equation. In this boundary operator τ, it seems difficult to characterize the domain of the fractional powers of the right shift of L. Both approaches have merits and demerits. Thus, we briefly introduce these approaches in the following sections. Since these theories are broad and rich, the readers may refer to other monographs and textbooks on functional analytic approach on pde and its applications, e.g., [1, 3, 15, 20, 24, 32, 35, 71]. It is preferable that the readers are familiar with Lebesgue integrals and elementary functional analysis.

2.2 Brief Survey of Sobolev Spaces

As in Section 1, let Ω be a bounded domain in $\mathbb{R}^m$. For $1 \leqslant p < \infty$, the set of all measurable functions u such that $\int_{\Omega} |u(x)|^p \, dx < \infty$ forms a Banach space, equipped with norm $\left(\int_{\Omega} |u(x)|^p \, dx\right)^{1/p}$. The space is denoted as $L^p(\Omega)$. When $p = 2$, the space $L^2(\Omega)$ is especially a Hilbert space equipped with inner product $\langle \cdot, \cdot \rangle$ and norm $\|\cdot\|$:

$$\langle u, v \rangle = \int_{\Omega} u(x)\,\overline{v(x)}\, dx, \qquad \|u\| = \langle u, u \rangle^{1/2}.$$

When u is in $L^1_{\mathrm{loc}}(\Omega)$, that is, u is locally integrable in Ω, u defines a *distribution* by the formula:

$$u[\varphi] = \int_\Omega u(x)\varphi(x)\,dx, \quad \forall \varphi \in \mathscr{D}(\Omega), \tag{2.1}$$

where $\mathscr{D}(\Omega)$ means the space of infinitely differentiable functions φ in Ω with compact supports, $\mathrm{supp}\,[\varphi] \subsetneq \Omega$, and the symbol $u[\cdot]$ a linear functional on $\mathscr{D}(\Omega)$. There are, of course, more general distributions (or functionals) $u[\varphi]$ other than the above $u[\varphi]$. A distribution $u[\varphi]$ is generally defined as a linear functional which is continuous in the topology of $\mathscr{D}(\Omega)$, that is, $u[\varphi_n] \to 0$ as $\varphi_n \to 0$ in $\mathscr{D}(\Omega)$: The convergence of $\varphi_n \to 0$ in $\mathscr{D}(\Omega)$ means that $\mathrm{supp}\,[\varphi_n]$ of the sequence $\{\varphi_n\} \subset \mathscr{D}(\Omega)$ are contained in a fixed compact set in Ω, and that all partial derivatives $D^\alpha \varphi_n$ uniformly converge to 0 in the above compact set. Here,

$$D^\alpha = \frac{\partial^{|\alpha|}}{\partial x_1^{\alpha_1} \cdots \partial x_m^{\alpha_m}}, \quad |\alpha| = \alpha_1 + \cdots + \alpha_m.$$

For every distribution u in Ω, we define $D^\alpha u$ as

$$(D^\alpha u)[\varphi] = (-1)^{|\alpha|} u[D^\alpha \varphi], \quad u \in \mathscr{D}(\Omega), \tag{2.2}$$

the right-hand side of which is also a distribution in Ω. Thus, every distribution is infinitely differentiable in the above sense. When u is especially a function in $L^1_{\mathrm{loc}}(\Omega)$, then

$$(D^\alpha u)[\varphi] = (-1)^{|\alpha|} u[D^\alpha \varphi] = (-1)^{|\alpha|} \int_\Omega u(x) D^\alpha \varphi(x)\,dx.$$

Roughly speaking, $D^\alpha u$ is viewed as a generalization of the concept of integration by parts. Since Ω is bounded, it is clear that $L^2(\Omega)$ is contained in $L^1(\Omega)$.

For an integer $n \geqslant 1$, let us consider a function $u \in L^2(\Omega)$ such that all derivatives $D^\alpha u$ (in the sense of distribution) up to the nth order belong to $L^2(\Omega)$. More precisely it means the following: For each α, $|\alpha| \leqslant n$, there is a function $u_\alpha \in L^2(\Omega)$ such that[1]

$$(D^\alpha u)[\varphi] = (-1)^{|\alpha|} \int_\Omega u(x) D^\alpha \varphi(x)\,dx = \int_\Omega u_\alpha(x)\varphi(x)\,dx, \quad \forall \varphi \in \mathscr{D}(\Omega).$$

[1]As a simple example in the interval $I = (-2, 2) \subset \mathbb{R}^1$, let us consider a function $u(x) = 1 - |x|$ for $|x| \leqslant 1$ and $= 0$ for $|x| > 1$. For an arbitrary $\varphi \in \mathscr{D}(I)$, integration by parts shows that

$$(u')[\varphi] = -\int_I u\varphi'\,dx = \int_I v\varphi\,dx,$$

where $v(x) = 1$ for $-1 < x < 0$; $= -1$ for $0 < x < 1$; and $= 0$ otherwise. Thus, $u' = v$ in the sense of distribution.

The set of all such functions u forms a Hilbert space, which is called the *Sobolev space* of order n, and denoted as $H^n(\Omega)$. There is no necessity to distinguish the difference between $D^\alpha u$ and u_α. A standard inner product and norm in $H^n(\Omega)$ are defined as

$$\langle u, v\rangle_{H^n(\Omega)} = \sum_{|\alpha|\leqslant n} \langle D^\alpha u, D^\alpha v\rangle, \quad \text{and} \quad \|u\|_{H^n(\Omega)} = \left(\langle u, u\rangle_{H^n(\Omega)}\right)^{1/2},$$

respectively. Let $C_0^n(\Omega)$ be the set of functions u in Ω such that $\operatorname{supp}[u]$ are contained in Ω, and all derivatives of u up to the nth order are continuous functions. It is clear that functions in $C_0^n(\Omega)$ belong to $H^n(\Omega)$. The completion of $\mathscr{D}(\Omega)$ in the toplogy of $H^n(\Omega)$ is denoted as $H_0^n(\Omega)$, which is also a Hilbert space. Thus, for every function $u \in H_0^n(\Omega)$, there is a sequence $\{u_n\} \subset \mathscr{D}(\Omega)$ such that $u_n \to u$ in $H^n(\Omega)$ [2]. Roughly speaking, the boundary value of $u \in H_0^n(\Omega)$, called the *trace* of u on Γ, is equal to 0. Sobolev spaces are similarly defined for other open (bounded or unbounded) sets Ω such as $\mathbb{R}^m$.

Partition of unity :

To study the boundary value problems associated with the pair $(\mathscr{L}, \tau)$, we introduce a local coordinate system for each point $x_0 \in \Gamma$: Let W_{x_0} be a neighborhood of x_0. Every $x \in W_{x_0} \cap \Gamma$ is expressed in terms of an auxiliary parameter $y' = (y_1, \ldots, y_{m-1})$ around $0 \in \mathbb{R}^{m-1}$ as

$$x_i = x_i(y_1, \ldots, y_{m-1}), \quad 1 \leqslant i \leqslant m, \quad \text{or } x = x(y'), \tag{2.3}$$

where the regularities of $x_i(\cdot)$ in y' are tacitly assumed. It is further assumed that

$$\operatorname{rank}\begin{pmatrix} \dfrac{\partial x_i}{\partial y_j}; & \begin{matrix} i & \downarrow & 1, \ldots, m \\ j & \to & 1, \ldots, m-1 \end{matrix} \end{pmatrix} = m-1$$

in a neighborhood of $y' = 0$. Thus the $m-1$ vectors $\left(\frac{\partial x_1}{\partial y_j}, \ldots, \frac{\partial x_m}{\partial y_j}\right)$, $1 \leqslant j \leqslant m-1$, belonging to the tangent plane to Γ form a linearly independent system. For each $x \in W_{x_0} \cap \Gamma$, let $(\nu_A)_i = \sum_j a_{ij}(x)\nu_j(x)$, $1 \leqslant i \leqslant m$, and define a set of functions of $y = (y', y_m)$ in the ball $V = \{y \in \mathbb{R}^m;\ |y| < \delta\}$ with a suitable $\delta > 0$ as

$$x_i = -(\nu_A)_i y_m + x_i(y_1, \ldots, y_{m-1}), \quad 1 \leqslant i \leqslant m, \tag{2.4}$$

which is denoted simply as $x = \Phi(y)$. Then, $\Phi(y', 0) \in W_{x_0} \cap \Gamma$, and Φ maps V and $V|_{y_m > 0}$ onto W_{x_0} and $W_{x_0} \cap \Omega$, respectively. Since the Jacobian $J \neq 0$ at $y = (y', 0)$, the function $x = \Phi(y)$ is a bijection between V and W_{x_0} by choosing narrower V and W_{x_0} if necessary. For a smooth function $u(x)$, $x \in \overline{\Omega}$, we see that $\frac{\partial}{\partial \nu} u(x)\Big|_\Gamma = -\frac{\partial}{\partial y_m} u(\Phi(y))\Big|_{y_m=0}$. For a smaller $0 < \delta' < \delta$, set $\tilde{V} = \{y \in \mathbb{R}^m;\ |y| < \delta'\} \subset V$, and $\tilde{W}_{x_0} = \Phi(\tilde{V}) \subsetneq W_{x_0}$.

[2]Let $\rho_\varepsilon *$ be a *Friedrichs' mollifier* [35]. For $u \in C_0^n(\Omega)$ and a small $\varepsilon > 0$, it is well known that the function $(\rho_\varepsilon * u)(x) = \int \rho_\varepsilon(x-y)u(y)\,dy$ belongs to $\mathscr{D}(\Omega)$, and converges to u in the $H^n(\Omega)$-topology as $\varepsilon \to 0$. Thus, $C_0^n(\Omega) \subset H_0^n(\Omega)$.

For an open covering $\bigcup_{x\in\Gamma}\tilde{W}_x$ of Γ, we can choose a finite number of $x_k \in \Gamma$, $1 \leqslant k \leqslant n$, such that $\Gamma \subset \bigcup_{1\leqslant k\leqslant n}\tilde{W}_{x_k}$ $\left(\subset \bigcup_{1\leqslant k\leqslant n} W_{x_k}\right)$. For each k, the above bijection and its domain are denoted, respectively, as $\Phi_k(\cdot)$ and $V_k = \{y;\ |y| < \delta_k\}$. Let ω be the function in $\mathscr{D}(\mathbb{R}^m_y)$ defined as

$$\omega(y) = \begin{cases} \exp\left(-\dfrac{1}{1-|y|^2}\right), & |y| < 1, \\ 0, & |y| \geqslant 1. \end{cases}$$

Then, $\operatorname{supp}[\omega] = \{y;\ |y| \leqslant 1\}$. Set for $1 \leqslant k \leqslant n$

$$\omega_k(x) = \begin{cases} \omega\left(\dfrac{1}{\delta_k'}\Phi_k^{-1}(x)\right), & x \in \tilde{W}_{x_k}, \\ 0, & x \notin \tilde{W}_{x_k}. \end{cases}$$

Each ω_k is a smooth function in $\mathbb{R}^m$, and has a compact support $\operatorname{supp}[\omega_k] = \overline{\tilde{W}_k} \subsetneq W_k$. The regularity depends on the map Φ_k (we assume enough regularity on Φ_k so that the following arguments are carried out). It is clear that $\frac{\partial}{\partial \nu}\omega_k(x)\Big|_\Gamma = 0$. Set

$$\alpha_k(x) = \begin{cases} \dfrac{\omega_k(x)}{\sum_{i=1}^n \omega_i(x)}, & x \in \bigcup_{1\leqslant i\leqslant n}\tilde{W}_i, \\ 0, & \text{otherwise}. \end{cases} \tag{2.5}$$

Then, we see that (i) $\sum_{k=1}^n \alpha_k(x) = 1$ in a neighborhood of Γ, and $= 0$ otherwise; (ii) $0 \leqslant \alpha_k(x) \leqslant 1$; and $\frac{\partial}{\partial \nu}\alpha_k(x)\Big|_\Gamma = 0$. The set $\{\alpha_k\}_{1\leqslant k\leqslant n}$ is called a *partition of unity* subordinate to the open covering $\bigcup_k W_k$ of Γ. Any function u in Ω is decomposed as

$$u(x) = \sum_{k=1}^n \alpha_k(x)u(x) + \left(1 - \sum_{k=1}^n \alpha_k(x)\right)u(x). \tag{2.6}$$

The study of the behavior of $u(x)$ in a neighborhood of Γ is thus reduced to the study of $\alpha_k u$, or $\alpha_k(\Phi_k(y))u(\Phi_k(y))$, $y \in V_k|_{y_m>0}$, since the support of $\alpha_k u$ is in $V_k|_{y_m>0}$.

Trace operator γ:

Let $\mathbb{R}^m_+ = \{y = (y', y_m);\ y_m > 0\}$, and let u be a function in $H^1(\mathbb{R}^m_+)$. We can show that

(i) For a.e. $y' \in \mathbb{R}^{m-1}$,

$$\lim_{y_m\to+0} u(y', y_m) = u(y', +0) = \varphi(y') \tag{2.7}$$

exists.

(ii) The function φ belongs to $L^2(\mathbb{R}^{m-1}_{y'})$, and

$$\|\varphi\|_{L^2(\mathbb{R}^{m-1})} \leqslant \text{const } \|u\|_{H^1(\mathbb{R}^m_+)} \,. \tag{2.8}$$

In fact, let α be a smooth function of y_m which is equal to 1 in a neighborhood of $y_m = 0$ and equal to 0 for $y_m \geqslant 1$. We only have to show (2.7) and (2.8) for u replaced by $\alpha(y_m)u(y', y_m) \in H^1(\mathbb{R}^m_+)$. Thus, supp $[\alpha u]$ lies in $0 < y_m \leqslant 1$. First note that

$$\iint \left| \frac{\partial(\alpha u)}{\partial y_m} \right|^2 dy' dy_m < \infty.$$

By Fubini's theorem, we see that for a.e. y'

$$\int_0^1 \left| \frac{\partial(\alpha u)}{\partial y_m} \right| dy_m \leqslant \left(\int_0^1 \left| \frac{\partial(\alpha u)}{\partial y_m} \right|^2 dy_m \right)^{1/2} < \infty.$$

Since both αu and $(\alpha u)_{y_m}$ are locally integrable, Nikodym's theorem ensures that (αu) is absolutely continuous in $y_m \in (0, 1]$ for a.e. y'. Thus, we do not have to distinguish $(\alpha u)_{y_m}$ in the sense of distribution from an ordinary partial derivative $(\alpha u)_{y_m}$. Thus, for a.e. y' and a sufficiently small $\varepsilon > 0$

$$u(y', \varepsilon) = -\int_\varepsilon^1 (\alpha u)_{y_m}(y', y_m)\, dy_m,$$

which implies that

$$\varphi(y') = u(y', +0) = -\int_0^1 (\alpha u)_{y_m}(y', y_m)\, dy_m$$

exists. By Schwarz's inequality,

$$|\varphi(y')|^2 \leqslant \int_0^1 |(\alpha u)_{y_m}(y', y_m)|^2\, dy_m$$

for a.e. y'. Integrating the both sides in y', we find that

$$\int_{\mathbb{R}^{m-1}} |\varphi(y')|^2\, dy' \leqslant \int_{\mathbb{R}^m_+} |(\alpha u)_{y_m}(y', y_m)|^2\, dy \leqslant \text{const } \|u\|^2_{H^1(\mathbb{R}^m_+)} \,.$$

Let us examine the behavior of functions in $H^1(\Omega)$ near the boundary Γ. Let u be in $H^1(\Omega)$, and consider the partition of unity stated above. By the decomposition (2.6), each $\alpha_k u \in H^1(\Omega \cap W_k)$ is regarded as a function in $H^1(V_k|_{y_m>0})$ (after transformation of the coordinates by $x = \Phi_k(y)$). Since the support of $\alpha_k u$ lies in $V_k|_{y_m>0}$, we may regard $\alpha_k u$ belongs to $H^1(\mathbb{R}^m_+)$, $1 \leqslant k \leqslant n$. By (2.7) and (2.8), there is a function $(\alpha_k u)|_{y_m=+0} \in L^2(\mathbb{R}^{m-1})$, the support of which lies in $|y'| < \delta_k$. We summerize the above result as follows:

For an arbitrary $u \in H^1(\Omega)$, we define the boundary value φ of u as the limit along a normal of Γ, which is denoted as $\varphi = \gamma u$. Then,

$$\|\gamma u\|_{L^2(\Gamma)} \leqslant \text{const}\, \|u\|_{H^1(\Omega)}, \quad \forall u \in H^1(\Omega). \tag{2.9}$$

The above operator γ is called the *trace operator*, and γu the trace of u on Γ. For a function $u \in H_0^1(\Omega)$, it is apparent that the trace γu is equal to 0. When u belongs to $H^2(\Omega)$, then $\dfrac{\partial u}{\partial \nu} \in L^2(\Gamma)$ is well defined as the trace $\sum_{i,j=1}^{m} a_{ij}(\xi)\nu_i(\xi)\gamma\left(\dfrac{\partial u}{\partial x_j}\right)$. However, the estimate (2.9) is not the best one in the following two meanings: By introducing interpolation spaces lying in Sobolev spaces $H^n(\Omega)$ of discrete order $n = 1, 2, \ldots$, the estimate (2.9) is sharpened:

$$\|\gamma u\|_{L^2(\Gamma)} \leqslant \text{const}\, \|u\|_{H^s(\Omega)}, \quad \forall u \in H^s(\Omega), \quad s > \frac{1}{2}. \tag{2.9$_1$}$$

The definition and detailed properties of the interpolation spaces $H^s(\Omega)$ for real $s \in \mathbb{R}^1$ are found in [32]. When $0 < s < 1$, for example, $H^s(\Omega) = [H^1(\Omega), L^2(\Omega)]_{1-s}$ (see the end of this section for the definition of the bracket).

The argument on the trace operator is based on the properties of $u(y', y_m) \in H^1(\mathbb{R}^m_+)$ and its trace $\varphi(y') = u(y', +0)$. The function $u(y', y_m)$ has an extention to a function in $\mathbb{R}^m$ by setting $u(y', y_m) = u(y', -y_m)$ for $y_m < 0$ (see (3.33), Subsection 3.4). The extention, denoted by the same symbol u, clearly belongs to $H^1(\mathbb{R}^m)$, and

$$\|u\|_{H^1(\mathbb{R}^m)} = \sqrt{2}\, \|u\|_{H^1(\mathbb{R}^m_+)} = \|(1+|\xi|)\hat{u}(\xi)\|_{L^2(\mathbb{R}^m_\xi)},$$

where $\hat{u}(\xi) = \hat{u}(\xi', \xi_m)$ denotes the Fourier transform $\int_{\mathbb{R}^m} e^{-2\pi i \xi \cdot y} u(y)\, dy$ of $u(y)$ in the L^2-space. In addition, $\mathscr{D}(\mathbb{R}^m)$ is dense in $H^1(\mathbb{R}^m)$. Thus an approach based on the Fourier transform is available. Then, for an arbitrary small $\varepsilon > 0$, we can derive the estimate

$$\|\varphi(y')\| \leqslant \varepsilon\, \|u(y', y_m)\|_{H^1(\mathbb{R}^m)} + C(\varepsilon)\, \|u(y', y_m)\|_{L^2(\mathbb{R}^m)}, \quad \forall u \in H^1(\mathbb{R}^m), \tag{2.9$_2$}$$

where $\varphi(y') = \gamma u$, and $C(\varepsilon) > 0$ denotes a constant depending on ε. The above estimate is first proven for functions in $\mathscr{D}(\mathbb{R}^m)$, and then by passage to the limit regarding $u \in H^1(\mathbb{R}^m)$. Going back to our functions in $H^1(\Omega)$, we finally obtain, for an arbitrary $\varepsilon > 0$,

$$\|\gamma u\|_{L^2(\Gamma)} \leqslant \varepsilon\, \|u\|_{H^1(\Omega)} + C(\varepsilon)\, \|u\|_{L^2(\Omega)}, \quad \forall u \in H^1(\Omega), \tag{2.9$_3$}$$

Following [35], let us observe briefly how the estimate (2.9_2) is obtained. Let u be in $\mathscr{D}(\mathbb{R}^m)$, and set $\varphi(y') = u(y', +0)$. Then,

$$\hat{u}(\xi', \xi_m) = \int_{\mathbb{R}^m} e^{-2\pi i(\xi' \cdot y' + \xi_m y_m)} u(y', y_m)\, dy' dy_m, \quad \text{and}$$

$$\hat{\varphi}(\xi') = \int_{-\infty}^{\infty} \hat{u}(\xi', \xi_m)\, d\xi_m, \quad \xi = (\xi', \xi_m), \quad \xi' = (\xi_1, \ldots, \xi_{m-1}).$$

By Hölder's inequality, we calculate as

$$\begin{aligned}
&|\hat{\varphi}(\xi')| \\
&\leqslant \int_{-\infty}^{\infty} (1 + |\xi'| + |\xi_m|)^{1/4} |\hat{u}(\xi', \xi_m)|^{1/2} \, (1 + |\xi'| + |\xi_m|)^{1/2} |\hat{u}(\xi', \xi_m)|^{1/2} \\
&\qquad \times (1 + |\xi'| + |\xi_m|)^{-3/4}\, d\xi_m \\
&\leqslant \Phi_1(\xi')^{1/4}\, \Phi_2(\xi')^{1/4} \frac{2}{(1 + |\xi'|)^{1/4}},
\end{aligned}$$

where

$$\begin{aligned}
\Phi_1(\xi') &= \int_{-\infty}^{\infty} (1 + |\xi'| + |\xi_m|)\, |\hat{u}(\xi', \xi_m)|^2\, d\xi_m, \\
\Phi_2(\xi') &= \int_{-\infty}^{\infty} (1 + |\xi'| + |\xi_m|)^2\, |\hat{u}(\xi', \xi_m)|^2\, d\xi_m.
\end{aligned}$$

By Schwarz's inequallity,

$$\begin{aligned}
\int_{\mathbb{R}^{m-1}} |\hat{\varphi}(\xi')|^2\, d\xi' &\leqslant \int_{\mathbb{R}^{m-1}} (1 + |\xi'|)^{1/2} |\hat{\varphi}(\xi')|^2\, d\xi' \\
&\leqslant 4 \left(\int_{\mathbb{R}^{m-1}} \Phi_1(\xi')\, d\xi' \right)^{1/2} \left(\int_{\mathbb{R}^{m-1}} \Phi_2(\xi')\, d\xi' \right)^{1/2}.
\end{aligned}$$

In other words,

$$\begin{aligned}
\|\hat{\varphi}\|^2 \leqslant 4 &\left(\int_{\mathbb{R}^m} (1 + |\xi'| + |\xi_m|)\, |\hat{u}(\xi)|^2\, d\xi \right)^{1/2} \\
&\times \left(\int_{\mathbb{R}^m} (1 + |\xi'| + |\xi_m|)^2\, |\hat{u}(\xi)|^2\, d\xi \right)^{1/2}.
\end{aligned}$$

The space of all functions $u \in L^2(\mathbb{R}^m_x)$ such that $(1 + |\xi|)^s \hat{u}(\xi)$ belong to $L^2(\mathbb{R}^m_\xi)$ is denoted as H_s, the norm of which is defined by $\left(\int_{\mathbb{R}^m} (1 + |\xi|)^{2s} |\hat{u}(\xi)|^2\, d\xi \right)^{1/2}$. Apparently, $H_s = H^s(\mathbb{R}^m)$ for $s \geqslant 0$. The above inequality is then rewrirren as

$$\|\hat{\varphi}\|^2 \leqslant \text{const}\, \|u\|_{H_{1/2}}\, \|u\|_{H_1}.$$

For an arbitrary $\varepsilon > 0$, we choose a $C_\varepsilon > 0$ such that $1 + |\xi| \leqslant \varepsilon^2 (1 + |\xi|)^2 + C_\varepsilon^2$ for $\forall \xi \in \mathbb{R}^m$. Then,

$$\|u\|_{H_{1/2}}^2 \leqslant \varepsilon^2\, \|u\|_{H_1}^2 + C_\varepsilon^2\, \|u\|_{H_0}^2$$

Thus, (2.9_2) holds for $u \in \mathscr{D}(\mathbb{R}^m)$. Finally, passage to the limit regarding $u \in H^1(\mathbb{R}^m)$ shows that (2.9_2) also holds for $u \in H^1(\mathbb{R}^m)$.

Conversely, given functions f and g on Γ, we can define a non-unique operator of prolongation R such that

$$u = R(f,g); \quad \gamma u = f, \quad \gamma\left(\frac{\partial u}{\partial \nu}\right) = g. \tag{2.10}$$

Here, the regularities of f, g, and u will be characterized in the following. Let $\{\tilde{W}_{x_k}\}_{1 \leqslant k \leqslant n}$ be an open covering of Γ stated before, and let $W_{x_k} \supset \tilde{W}_{x_k}$, such that (i) the map $x = \Phi_k(y)$ is a bijection between $V_k = \{y \in \mathbb{R}^m;\ |y| < \delta_k\}$ and W_{x_k}, (ii) $\Phi_k(y',0) \in W_{x_k} \cap \Gamma$, and (iii) $V_k|_{y_m>0}$ is mapped onto $W_{x_k} \cap \Omega$. Let $\{\alpha_k\}_{1 \leqslant k \leqslant n}$ be the partition of unity subordinate to the open covering $\bigcup_{1 \leqslant k \leqslant n} W_k$, such that $\text{supp}[\alpha_k] = \overline{\tilde{W}_k} \subsetneq W_{x_k}$ (see (2.5)). For $f,\ g \in C^2(\Gamma)$ and $1 \leqslant k \leqslant n$, let u_k be the function defined in V_k as

$$u_k(y) = u_k(y', y_m) = f(y') - y_m g(y').$$

Then, $u_k(y',0) = f(y')$ and $-\dfrac{\partial}{\partial y_m} u_k(y',0) = g(y')$. Set

$$u(x) = \sum_{k=1}^{n} \alpha_k(x) u_k\big(\Phi_k^{-1}(x)\big).$$

The function $u_k\big(\Phi_k^{-1}(x)\big)$ is not defined for $x \notin W_k$. However, since $\text{supp}[\alpha_k] \subsetneq W_k$, we may regard it as a function of class C^2 defined in $\mathbb{R}^m$. Then it is clear that the prolongation u satisfies (2.10).

We remark that the operator R is extended to a bounded operator from $H^{3/2}(\Gamma) \times H^{1/2}(\Gamma)$ to $H^2(\Omega)$.

On the interpolation space $[X,Y]_\theta$.

We introduced a fractional Sobolev space $H^s(\Omega) = [H^1(\Omega), L^2(\Omega)]_{1-s}$ in (2.9_1) without definition. The square bracket $[X,Y]_\theta$, $0 \leqslant \theta \leqslant 1$, is defined in a standard manner for two Hilbert spaces X and Y such that $X \subset Y$ and X is densely embedded in Y with continuous injection. The space $[X,Y]_\theta$ lies between X and Y, and is defined in the following manner: Let $\langle\cdot,\cdot\rangle_X$ and $\langle\cdot,\cdot\rangle_Y$ be the inner products in X and Y, respectively. If $\langle u, v\rangle_X$ is an anti-linear form which is continuous in the topology of Y, the Riesz representation theorem [71] ensures a unique $Su \in Y$ such that $\langle u, v\rangle_X = \langle Su, v\rangle_Y$ for $\forall v \in X$. The set of all these $u \in X$ is denoted as $\mathscr{D}(S)$. Then, S is an unbounded operator in Y. The domain $\mathscr{D}(S) \subset X$ is dense in X, and thus in Y, too. This fact is similar to the fact that the adjoint of a densely defined closed operator has dense domain. In fact, let $G_S = \{(u, Su);\ u \in \mathscr{D}(S)\}$ be the the *graph* of S in the product space $X \times Y$. Apparently G_S is a closed subspace. By definition, we see that

$$G_S = \{(v, -v) \in X \times Y\}^\perp.$$

If $\mathscr{D}(S)$ were not dense in X, there should be an $h\,(\neq 0) \in X$ such that $h \perp \mathscr{D}(S)$. Then, $(h,0)$ is an element of $G_S{}^{\perp}$, and thus belongs to the subspace $\{(v,-v);\ v \in X\}$, from which we find that $h = 0$. But, this is a contradiction.

The operator S in Y is self-adjoint and positive-definite by definition. Thus, S generates a unique *spectral resolution of the identity*, $\{E_\lambda;\ \lambda \in \mathbb{R}^1\}$ such that

$$Su = \int_a^\infty \lambda\, dE_\lambda u, \quad \text{for } u \in \mathscr{D}(S),$$

where $a > 0$ is the lower bound of $\sigma(S)$ (see, e.g., [71] for detailed properties on the family of the operators $\{E_\lambda\}$). The domain $\mathscr{D}(S)$ consists of all elements u such that $\int_a^\infty \lambda^2 d\,\langle E_\lambda u,\, u\rangle_Y < \infty$. Since S is positive-definite, we define the operator $\Lambda = S^{1/2}$. The domain $\mathscr{D}(\Lambda)$ consists of all u such that $\int_a^\infty \lambda\, d\,\langle E_\lambda u,\, u\rangle_Y < \infty$. The domain $\mathscr{D}(S)$ is dense in $\mathscr{D}(\Lambda)$. For $u \in \mathscr{D}(S)\,(\subset X)$, note that

$$\langle \Lambda u,\, \Lambda u\rangle_Y = \langle Su,\, u\rangle_Y = \langle u,\, u\rangle_X, \quad \text{or } \|\Lambda u\|_Y = \|u\|_X.$$

Since $\mathscr{D}(S)$ is dense both in $\mathscr{D}(\Lambda)$ and in X, we find that $\mathscr{D}(\Lambda) = X$.

With these preparations, the interpolation space $[X,Y]_\theta$ is defined as

$$[X,Y]_\theta = \mathscr{D}(\Lambda^{1-\theta}), \quad 0 \leqslant \theta \leqslant 1.$$

Thus, $[X,Y]_0 = \mathscr{D}(\Lambda) = X$, and $[X,Y]_1 = Y$. The norm of $[X,Y]_\theta$ is the graph norm of $\Lambda^{1-\theta}$, i.e., $\left(\|u\|_Y^2 + \|\Lambda^{1-\theta}u\|_Y^2\right)^{1/2}$.

2.3 Elliptic Boundary Valule Problems

2.3.1 The Dirichlet boundary

Let us first consider the pair $(\mathscr{L},\, \tau)$ in (1.1) in the case where $b_i(x) = 0,\ 1 \leqslant i \leqslant m$, and $\alpha(\xi) \equiv 1$ (the Dirichlet boundary), so that $\mathscr{L}$ is formally self-adjoint. Associated with $\mathscr{L}$ is the Hermitian form $B(u,v)$ in $H_0^1(\Omega)$:

$$B(u,v) = \sum_{i,j=1}^{m} \left\langle a_{ij}(x)\,\frac{\partial u}{\partial x_j},\, \frac{\partial v}{\partial x_i}\right\rangle + \langle c(x)u,\, v\rangle, \quad u,\, v \in H_0^1(\Omega). \tag{3.1}$$

If a constant $c > 0$ is chosen large enough, we see that

$$m_c\,\|u\|^2 \leqslant B(u,u) + c\,\|u\|^2 \leqslant M_c\,\|u\|^2, \quad \forall u \in H_0^1(\Omega),$$

where $m_c > 0$ and $M_c > 0$ are constants depending on c. Thus the positive-definite Hermitian form $B_c(u,v) = B(u,v) + c\,\langle u,\, v\rangle$ defines another inner product and the corresponding norm which is equivalent to $\|\cdot\|_{H_0^1(\Omega)}$ (see Section 2). This space, which is equal to $H_0^1(\Omega)$ algebraically and topologically, is denoted as H. Let f be an arbitrary function in $L^2(\Omega)$. If there is a solution u in $H^2(\Omega)$ to the boundary value problem

$$(c+\mathscr{L})u = f \quad \text{in } \Omega, \quad \tau u = u|_\Gamma = 0 \quad \text{on } \Gamma, \tag{3.2}$$

we derive via integration by parts the relation

$$B_c(u,v) = \langle f, v \rangle, \quad \forall v \in H_0^1(\Omega) = H. \tag{3.3}$$

Conversely, the right-hand side defines an anti-linear form in H. By the Riesz representation theorem, there is a unique $u \in H$, such that (3.3) holds. Set $u = G_c f$. The operator G_c is called *Green's operator*, and satisfies the estimate

$$\|G_c f\| \leqslant \|G_c f\|_{H_0^1(\Omega)} \leqslant \text{const}\,\|f\|, \quad \forall f \in L^2(\Omega). \tag{3.4}$$

Thus, $G_c \in \mathscr{L}(L^2(\Omega))$, where $\mathscr{L}(L^2(\Omega))$ denotes the space of all bounded linear operators in $L^2(\Omega)$. We note that the boundary value $\gamma u = \gamma(G_c f)$ is equal to 0 by (2.9). The following result is immediate:

Theorem 3.1. *Green's operator G_c is a compact and positive Hermitian operator in $L^2(\Omega)$.*

Proof. Boundedness of G_c is already shown. For any f and g in $H = H_0^1(\Omega)$, we calculate as

$$B_c(G_c f, g) = \langle f, g \rangle = \overline{\langle g, f \rangle} = \overline{B_c(G_c g, f)} = B_c(f, G_c g).$$

Then, by (3.3),

$$\begin{aligned} \langle G_c f, g \rangle &= B_c(G_c^2 f, g) = B_c(G_c f, G_c g) \\ &= B_c(f, G_c^2 g) = \langle f, G_c g \rangle . \end{aligned}$$

Since $H = H_0^1(\Omega)$ is dense in $L^2(\Omega)$, and B_c bounded, we see that

$$\langle G_c f, g \rangle = \langle f, G_c g \rangle, \quad \forall f, \forall g \in L^2(\Omega).$$

As shown just above, note that $\langle G_c f, f \rangle = B_c(G_c f, G_c f) \geqslant 0$ first for $f \in H$, and then for $f \in L^2(\Omega)$ by continuity. Equality holds if and only if $G_c f = 0$, or $f = 0$.

To show compactness of G_c, we use Rellich's theorem [1, 20, 35]. It asserts a compactness property of sequences in $H_0^1(\Omega)$: *Given any bounded sequence, $\{f_n\} \subset H_0^1(\Omega)$, we can extract a subsequence $\{f_{n_k}\}$ such that it converges in the topology of $L^2(\Omega)$* [3] . Note that the result is limited to the case where Ω is bounded. Now, let the sequence $\{f_n\}$ be a bounded set in $L^2(\Omega)$. By (3.4),

[3] A generalized version of this result to fractional Sobolev spaces $H^s(\Omega)$ is found in [32].

$\{G_c f_n\}$ is a bounded set in $H_0^1(\Omega)$. By Rellich's theorem, there is a subsequence $\{G_c f_{n_k}\}$ such that it converges in $L^2(\Omega)$. □

The solution $u = G_c f$ in (3.3) satisfies the relation: $\mathscr{L}_c u = f$ in the sense of distribution in Ω, in other words,

$$(\mathscr{L}_c u)[\varphi] = \langle f, \varphi \rangle, \quad \forall \varphi \in \mathscr{D}(\Omega), \quad \mathscr{L}_c = \mathscr{L} + c.$$

For a given $\lambda \in \mathbb{C}$, the problem

$$(\lambda - \mathscr{L})u = f \quad \text{in } \Omega, \quad \tau u = u|_\Gamma = 0 \quad \text{on } \Gamma \tag{3.5}$$

turns out to be the problem of seeking a $u \in L^2(\Omega)$ of the equation:

$$((\lambda + c)G_c - 1)\, u = G_c f.$$

The Hilbert-Schmidt theory [3, 20, 35, 71] is applied to the operator G_c: Then, the following results are well known:

(i) The spectrum $\sigma(G_c)$ consists only of eigenvalues μ_i in $\mathbb{R}^1_+$; is bounded from above; and 0 is the only accumulation point, where $\mu_i \neq \mu_j$ for $i \neq j$, and $\mu_i \downarrow 0$.

(ii) Each eigenspace corresponding to μ_i is finite-dimensional with dimension m_i.

(iii) For each μ_i, let φ_{ij}, $1 \leqslant j \leqslant m_i$, be the normalized eigenfunction, that is, $(\mu_i - G_c)\varphi_{ij} = 0$, and $\|\varphi_{ij}\| = 1$. The set $\{\varphi_{ij}\}$ forms a orthonormal basis for $L^2(\Omega)$. Thus, any $u \in L^2(\Omega)$ is expressed as a Fourier series:

$$u = \sum_{i=1}^{\infty} \sum_{j=1}^{m_i} \langle u, \varphi_{ij} \rangle \varphi_{ij}.$$

Set $\dfrac{1}{\lambda_i + c} = \mu_i \downarrow 0$. Then, $\lambda_1 < \cdots < \lambda_i < \cdots \to \infty$; $(\lambda_i - \mathscr{L})\varphi_{ij} = 0$ in the sense of distribution in Ω; and the unique solution $u \in H^1(\Omega)$ to (3.5) for $\lambda \notin \lambda_i$, $1 \leqslant i < \infty$, is expressed as

$$u = \sum_{i,j} \frac{1}{\lambda - \lambda_i} \langle f, \varphi_{ij} \rangle \varphi_{ij}.$$

We can prove that the solution u actually belongs to $H^2(\Omega)$. The proof consists of two parts: One is on interior regularity which is independent of boundary conditions, such that the assumption $f \in H^l_{\mathrm{loc}}(\Omega)$ implies that $u \in H^{l+2}_{\mathrm{loc}}(\Omega)$, and the other on regularity near the boundary Γ. Since it requires much preparation, the readers may refer to, e.g., [20, 35] for the detailed proof. We remark that the same regularity result holds for a general $(\mathscr{L}, \tau)$ in (1.1). Now let

$$\begin{aligned} Lu &= \mathscr{L}u, \quad u \in \mathscr{D}(L), \\ \mathscr{D}(L) &= \left\{u \in H^2(\Omega);\ \tau u = 0 \text{ on } \Gamma\right\} = H^2(\Omega) \cap H_0^1(\Omega). \end{aligned} \tag{3.6}$$

Then the above solution to (3.5) belongs to $\mathscr{D}(L)$, and is written as $(\lambda - L)^{-1} f$. There is a sector $\overline{\Sigma} = \{\lambda - b \in \mathbb{C};\ \theta_0 \leqslant |\arg\lambda| \leqslant \pi\}$, $0 < \forall\theta_0 < \pi/2$, $\exists b \in \mathbb{R}^1$ such that

$$\left\|(\lambda - L)^{-1}\right\|_{\mathscr{L}(L^2(\Omega))} \leqslant \frac{\text{const}}{1 + |\lambda|}, \quad \lambda \in \overline{\Sigma}\ (\subset \rho(L)). \tag{3.7}$$

The domain $\mathscr{D}(L)$ is regarded as a subspace of $H^2(\Omega)$. By the preceding arguments, the operator $L_c = L + c$ is (i) linear; (ii) continuously maps $\mathscr{D}(L)$ *onto* $L^2(\Omega)$; and (iii) is one-to-one. Thus, the inverse map is also continuos, that is,

$$\|u\|_{H^2(\Omega)} \leqslant \text{const}\ \|L_c u\|_{L^2(\Omega)}, \quad \forall u \in \mathscr{D}(L). \tag{3.8}$$

Note that L_c is self-adjoint and positive-definite, since $\langle L_c u, u\rangle = B_c(u,u) \geqslant m_c \|u\|^2$ for $u \in \mathscr{D}(L)$. Thus, fractional powers of L_c is well defined. Heinz's inequality [26] and (3.8) imply that $\mathscr{D}(L_c^\omega)$ is contained in $H^{2\omega}(\Omega)$, and that

$$\|u\|_{H^{2\omega}(\Omega)} \leqslant \text{const}\ \|L_c^\omega u\|, \quad \forall u \in \mathscr{D}(L_c^\omega), \quad 0 \leqslant \omega \leqslant 1. \tag{3.9}$$

By the moment inequality[4]

$$\|L_c^\omega u\| \leqslant \text{const}\ \|u\|^{1-\omega}\,\|L_c u\|^\omega, \quad u \in \mathscr{D}(L),$$

we see that

$$\left\|L_c^\omega(\lambda - L)^{-1}\right\|_{\mathscr{L}(L^2(\Omega))} \leqslant \frac{\text{const}}{1 + |\lambda|^{1-\omega}}, \quad \lambda \in \Sigma\ (\subset \rho(L)). \tag{3.10}$$

A characterization of $\mathscr{D}(L_c^\omega)$ in terms of fractional Sobolev spaces is obtainable. While unnecessary for a moment, some comments on the characterization will be stated at the end of this section for later use.

[4]The moment inequality is obtained not only for self-adjoint L_c but also for *sectorial operators* (see (4.1) later). Let $0 < \alpha < 1$. Since $\sigma(L_c) \subset \mathbb{C}_+$, we define $L_c^{-\alpha} = \frac{-1}{2\pi i}\int_{\partial\overline{\Sigma}+c} \lambda^{-\alpha}(\lambda - L_c)^{-1}\,d\lambda$, where $|\arg\lambda| < \pi$. We contract the contour $\partial\overline{\Sigma} + c$ into the negative real axis to obtain

$$\begin{aligned} L_c^{-\alpha}u &= \tfrac{\sin\pi\alpha}{\pi}\int_0^\infty \lambda^{-\alpha}(\lambda + L_c)^{-1}u\,d\lambda \\ &= \tfrac{\sin\pi\alpha}{\pi}\left(\int_0^N \lambda^{-\alpha}L_c(\lambda + L_c)^{-1}L_c^{-1}u\,d\lambda + \int_N^\infty \lambda^{-\alpha}(\lambda + L_c)^{-1}u\,d\lambda\right), \quad N > 0. \end{aligned}$$

Then, $\|L_c^{-\alpha}u\| \leqslant c(\alpha)\left(N^{1-\alpha}\|L_c^{-1}u\| + N^{-\alpha}\|u\|\right)$. Minimizing the last term with respect to $N > 0$, we obtain $\|L_c^{-\alpha}u\| \leqslant c(\alpha)\|L_c^{-1}u\|^\alpha\|u\|^{1-\alpha}$ for $\forall u$, from which the above moment inequality is derived.

Let us consider a general $\mathscr{L}$ with $b_i(x) \neq 0$ in (1.1). One attempt to obtain Green's operator is to decompose $\mathscr{L}$ formally into

$$\mathscr{L} = \frac{1}{2}(\mathscr{L} + \mathscr{L}^*) + \mathrm{i}\frac{1}{2\mathrm{i}}(\mathscr{L} - \mathscr{L}^*) = \mathscr{L}_1 + \mathrm{i}\mathscr{L}_2,$$

$$\text{where} \quad \mathscr{L}^* v = -\sum_{i,j=1}^{m} \frac{\partial}{\partial x_i}\left(a_{ij}(x)\frac{\partial v}{\partial x_j}\right) - \operatorname{div}(\boldsymbol{b}(x)v) + c(x)v, \quad \boldsymbol{b}(x) = (b_1(x), \ldots, b_m(x)).$$

The operators $\mathscr{L}_1$ and $\mathscr{L}_2$ are formally self-adjoint. Let $c > 0$ be chosen large enough. The equation,

$$(c + \mathscr{L})u = f \quad \text{in } \Omega, \quad \tau u = u|_\Gamma = 0 \quad \text{on } \Gamma, \tag{3.2$'$}$$

for a given $f \in L^2(\Omega)$ leads to the problem of seeking a $u \in H_0^1(\Omega)$ satisfying

$$\tilde{B}_c(u, v) + \mathrm{i}\langle \mathscr{L}_2 u, v\rangle = \langle f, v\rangle, \quad \forall v \in H_0^1(\Omega), \tag{3.3$'$}$$

where $\tilde{B}_c(\cdot,\cdot)$ is a positive-definite Hermitian form associated with $\mathscr{L}_1|_{\ker\tau}$, and defines the same topology as in $H_0^1(\Omega)$. As before, let H be the space $H_0^1(\Omega)$ equipped with inner product $\langle \cdot,\cdot\rangle_H = \tilde{B}_c(\cdot,\cdot)$ and norm $\|\cdot\|_H$. By noting that $|\langle \mathscr{L}_2 u, u\rangle| \leqslant \text{const}\,\|u\|_H^2$ for $\forall u \in H$, there is a unique Hermitian operator $E \in \mathscr{L}(H)$ such that

$$\langle \mathscr{L}_2 u, v\rangle = \langle Eu, v\rangle_H, \quad \forall u,\ \forall v \in H.$$

Thus, (3.3$'$) is rewritten as

$$\langle (1 + \mathrm{i}E)u, v\rangle_H = \langle f, v\rangle = \langle Cf, v\rangle_H, \quad \forall v \in H, \quad C \in \mathscr{L}(L^2(\Omega); H).$$

Thus, we see that

$$u = G_c f = (1 + \mathrm{i}E)^{-1} Cf, \quad G_c \in \mathscr{L}(L^2(\Omega); H). \tag{3.11}$$

It is clear that G_c is compact as an operator in $\mathscr{L}(L^2(\Omega))$. For a given $\lambda \in \mathbb{C}$, the problem

$$(\lambda - \mathscr{L})u = f \quad \text{in } \Omega, \quad \tau u = u|_\Gamma = 0 \quad \text{on } \Gamma \tag{3.5$'$}$$

is similarly reduced to the problem of solving the equation: $((\lambda + c)G_c - 1)u = G_c f$ in $L^2(\Omega)$. According to the Riesz-Schauder theory [35], the problem has a unique solution if and only if the homogeneous equation $((\lambda + c)G_c - 1)u = 0$ admits no solution other than the trivial solution 0. As before, the solution $u \in H_0^1(\Omega)$ to (3.5$'$) actually belongs to $H^2(\Omega)$. Let the operator L be defined by (3.6) with $\mathscr{L}$ replaced by a general (non self-adjoint) one. By the compactness property of G_c, the spectrum $\sigma(L)$ of L consists only of eigenvalues λ_i, $i \geqslant 1$, of finite multiplicities, for which ∞ is the only accumulation point. However,

$\sigma(L)$ does not generally lie in $\mathbb{R}^1$. Instead, since Ω is a bounded domain, it is known that $\sigma(L)$ lies in some parabola with the x-axis as the line of symmetry. Generally, each eigenvalue may admit a generalized eigenspace. It is not hard to show that there is a sector $\overline{\Sigma}$ such that $\overline{\Sigma}$ is contained in $\rho(L)$, and the decay estimate (3.7) holds for the resolvent $(\lambda - L)^{-1}$. The angle θ_0, $0 < \theta_0 < \pi/2$, of $\overline{\Sigma}$ can be chosen arbitrary, $b \in \mathbb{R}^1$ being chosen suitably.

There is another simple approach to (3.5′): Let

$$\mathscr{L}u = \widetilde{\mathscr{L}}u + Du, \quad \text{where}$$

$$\widetilde{\mathscr{L}}u = -\sum_{i,j=1}^{m} \frac{\partial}{\partial x_i}\left(a_{ij}(x)\frac{\partial u}{\partial x_j}\right) + c(x)u, \quad Du = \sum_{i=1}^{m} b_i(x)\frac{\partial u}{\partial x_i}, \tag{3.12}$$

and set $\tilde{L}u = \widetilde{\mathscr{L}}u$, $u \in \mathscr{D}(\tilde{L}) = H^2(\Omega) \cap H_0^1(\Omega)$. As we have seen, $\tilde{L}$ is self-adjoint, and the resolvent $(\lambda - \tilde{L})^{-1}$ exists in a sector $\overline{\Sigma}$ satisfying the decay estimate (3.7). Eqn. (3.5′) is rewritten as

$$(\lambda - \tilde{L} - D)u = \left(1 - D(\lambda - \tilde{L})^{-1}\right)(\lambda - \tilde{L})u = f.$$

Thus, as long as $\left\|D(\lambda - \tilde{L})^{-1}\right\|_{\mathscr{L}(L^2(\Omega))}$ is smaller than 1, the solution is expressed as

$$u = (\lambda - \tilde{L})^{-1}\left(1 - D(\lambda - \tilde{L})^{-1}\right)^{-1} f \in \mathscr{D}(\tilde{L}). \tag{3.13}$$

Applying (3.9) and (3.10) with $\omega = 1/2$ to D and choosing an $R > 0$ large enough, we estimate as

$$\left\|D\tilde{L}_c^{-1/2}\tilde{L}_c^{1/2}(\lambda - \tilde{L})^{-1}\right\|_{\mathscr{L}(L^2(\Omega))} \leqslant \frac{\text{const}}{1 + |\lambda|^{1/2}} < \frac{1}{2}, \quad \lambda \in \overline{\Sigma} \cap \{|\lambda| > R\}.$$

Thus, by choosing b in $\overline{\Sigma}$ large enough if necessary, (3.13) is justified, and the decay (3.7) holds for $\lambda \in \overline{\Sigma}$.

2.3.2 The Robin boundary

In the case where $0 \leqslant \alpha(\xi) < 1$ in (1.1), we may write the boundary operator τ as $\tau u = \dfrac{\partial u}{\partial \nu} + \sigma(\xi)u$. The coefficient $\sigma(\xi)$ may be replaced by any function of class C^1 on Γ. In this setting, similar results hold for the pair $(\mathscr{L}, \tau)$ with technical changes.

Let us begin with the case where $b_i(x) = 0$, $1 \leqslant i \leqslant m$, so that $\mathscr{L}$ is formally self-adjoint. Associated with $(\mathscr{L}, \tau)$ is the Hermitian form $B(u,v)$ in $H^1(\Omega)$:

$$\begin{aligned} B(u,v) = &\sum_{i,j=1}^{m} \left\langle a_{ij}(x)\frac{\partial u}{\partial x_j}, \frac{\partial v}{\partial x_i}\right\rangle + \langle \sigma(\xi)u, v\rangle_\Gamma \\ &+ \langle c(x)u, v\rangle, \quad u, v \in H^1(\Omega). \end{aligned} \tag{3.14}$$

In (3.14), $\langle \cdot, \cdot \rangle_\Gamma$ denotes the inner product in $L^2(\Gamma)$. By (2.9$_2$), for any small $\varepsilon > 0$,

$$B(u,u) \geqslant \text{const} \sum_{i=1}^{m} \left\| \frac{\partial u}{\partial x_i} \right\|^2 - \sup |\sigma(\xi)| \left(\varepsilon \|u\|^2_{H^1(\Omega)} + C(\varepsilon) \|u\|^2 \right) + \inf c(x) \|u\|^2 .$$

By choosing a constant $c > 0$ large enough, $B_c(u,v) = B(u,v) + c \langle u, v \rangle$ defines another inner product and norm which is equivalent to $\|\cdot\|_{H^1(\Omega)}$. As in the Dirichlet case, there is Green's operator $G_c \in \mathscr{L}(L^2(\Omega); H^1(\Omega))$ such that

$$u = G_c f; \quad B_c(u,v) = \langle f, v \rangle, \quad \forall v \in H^1(\Omega). \tag{3.15}$$

The solution $u = G_c f$ thus satisfies the equation, $\mathscr{L}_c u = f$ in the sense of distribution in Ω. Theorem 3.1 similarly holds for the operator $G_c \in \mathscr{L}(L^2(\Omega))$ in this problem. For a given $\lambda \in \mathbb{C}$, the problem

$$(\lambda - \mathscr{L})u = f \quad \text{in } \Omega, \quad \tau u = \frac{\partial u}{\partial \nu} + \sigma(\xi) u = 0 \quad \text{on } \Gamma \tag{3.16}$$

turns out to be the problem of seeking a solution $u \in L^2(\Omega)$ to the equation: $((\lambda + c)G_c - 1) u = G_c f$. The Hilbert-Schmidt theory is again applied to the problem (3.16). Thus, as in the Dirichlet case, there is a set of eigenpairs $\{\lambda_i, \varphi_{ij}\}$, $i \geqslant 1$, $1 \leqslant j \leqslant m_i (< \infty)$, such that (i) $\lambda_1 < \cdots < \lambda_i < \cdots \to \infty$, (ii) $(\lambda_i - \mathscr{L})\varphi_{ij} = 0$, and (iii) the set $\{\varphi_{ij}\}$ forms an orthonormal basis for $L^2(\Omega)$.

In light of the assumption that $\sigma(\xi)$ in the Hermitian form $B(\cdot,\cdot)$ is in $C^1(\Gamma)$, it is shown that the u is actually a function in $H^2(\Omega)$: In fact, a further regularity of u owes much to the property of $B(\cdot,\cdot)$. The proof is, however, omitted as in the case of the Dirichlet boundary (the detailed proof is found, e.g., in [20, 35]). Thus, by Green's formula, we calculate as

$$\begin{aligned} \langle \mathscr{L}_c u, v \rangle &= - \left\langle \frac{\partial u}{\partial \nu}, v \right\rangle_\Gamma + \sum_{i,j=1}^{m} \left\langle a_{ij}(x) \frac{\partial u}{\partial x_j}, \frac{\partial v}{\partial x_i} \right\rangle + \langle (c(x) + c) u, v \rangle \\ &= - \left\langle \frac{\partial u}{\partial \nu} + \sigma(\xi) u, v \right\rangle_\Gamma + \langle f, v \rangle, \quad \forall v \in H^1(\Omega), \end{aligned}$$

which means that

$$\langle \tau u, v \rangle_\Gamma = \left\langle \frac{\partial u}{\partial \nu} + \sigma(\xi) u, v \right\rangle_\Gamma = 0, \quad \forall v \in H^1(\Omega).$$

Since the set $\{\gamma v;\ v \in H^1(\Omega)\}$ is dense in $L^2(\Gamma)$, we conclude that $\tau u = 0$ on Γ.

Now let

$$\begin{aligned} &Lu = \mathscr{L} u, \quad u \in \mathscr{D}(L), \\ &\mathscr{D}(L) = \left\{ u \in H^2(\Omega);\ \tau u = \frac{\partial u}{\partial \nu} + \sigma(\xi) u = 0 \ \text{on } \Gamma \right\}. \end{aligned} \tag{3.17}$$

Then the solution $u = ((\lambda + c)G_c - 1)^{-1} G_c f$ to (3.16) belongs to $\mathscr{D}(L)$, and is written as $(\lambda - L)^{-1} f$. There is a sector $\overline{\Sigma} = \{\lambda - b \in \mathbb{C}; \theta_0 \leqslant |\arg \lambda| \leqslant \pi\}$, $0 < \forall \theta_0 < \pi/2$, $\exists b \in \mathbb{R}^1$ such that

$$\left\| (\lambda - L)^{-1} \right\|_{\mathscr{L}(L^2(\Omega))} \leqslant \frac{\text{const}}{1 + |\lambda|}, \quad \lambda \in \Sigma \ (\subset \rho(L)). \tag{3.18}$$

Since the map, $u \in H^2(\Omega) \to \tau u \in L^2(\Gamma)$ is continuous, the domain $\mathscr{D}(L)$ is a subspace of $H^2(\Omega)$. Thus,

$$\|u\|_{H^2(\Omega)} \leqslant \text{const} \, \|L_c u\|_{L^2(\Omega)}, \quad \forall u \in \mathscr{D}(L) \tag{3.19}$$

(compare it with (3.8)). A decay estimate similar to (3.10) also holds for the resolvent $(\lambda - L)^{-1}$.

Let us consider (3.16) for a general $\mathscr{L}$ with $b_i(x) \neq 0$. In view of (3.12), rewrite $\mathscr{L}$ as $\widetilde{\mathscr{L}} + D$, and set $\tilde{L} u = \widetilde{\mathscr{L}} u$, $u \in \mathscr{D}(\tilde{L}) = \mathscr{D}(L)$, where $\widetilde{\mathscr{L}}$ is formally self-adjoint. Choosing an $R > 0$ large enough, we have the estimate,

$$\left\| D(\lambda - \tilde{L})^{-1} \right\|_{\mathscr{L}(L^2(\Omega))} < \frac{1}{2}, \quad \lambda \in \overline{\Sigma} \cap \{|\lambda| > R\}.$$

For such λ, the problem (3.16) admits a unique solution

$$\begin{aligned} u &= (\lambda - L)^{-1} f \\ &= (\lambda - \tilde{L})^{-1} \left(1 - D(\lambda - \tilde{L})^{-1}\right)^{-1} f \in \mathscr{D}(\tilde{L}) = \mathscr{D}(L). \end{aligned}$$

Thus the resolvent $(\lambda - L)^{-1}$ satisfies the decay estimate

$$\left\| (\lambda - L)^{-1} \right\|_{\mathscr{L}(L^2(\Omega))} \leqslant \frac{\text{const}}{1 + |\lambda|}, \quad \lambda \in \overline{\Sigma}, \tag{3.20}$$

by replacing b in $\overline{\Sigma}$ by a greater constant.

2.3.3 The case of a general boundary

Let us consider the pair $(\mathscr{L}, \tau)$ in the general case where $0 \leqslant \alpha(\xi) \leqslant 1$ in (1.1). In this boundary, the Dirichlet boundary is locally continuously connected with the Robin boundary. Thus, the preceding formulation based on functional analysis faces serious difficulties. Instead, there is an alternative approach which is partially based on a classical theory and partially on functional analysis. The approach is based on constructing a *fundamental solution* $U(t,x,y)$ to the initial-boundary value problem of parabolic type for $u = u(t,x)$,

$$\begin{cases} \dfrac{\partial u}{\partial t} + \mathscr{L} u = f(t,x) & \text{in } \mathbb{R}^1_+ \times \Omega, \\ \tau u = g(t,\xi), & \text{on } \mathbb{R}^1_+ \times \Gamma, \\ u(0,x) = u_0(x), & \text{in } \Omega. \end{cases} \tag{3.21}$$

The functions f and g in (3.21) mean external forces or a kind of controls. The fundamental solution $U(t,x,y)$ is first constructed in the whole space $\mathbb{R}^m$. Then, detailed, complicated, and somewhat tedious calculation by successive approximations -via a partition of unity subordinate to $\overline{\Omega}$ (not Γ) shows that one can construct a unique fundamental solution $U(t,x,y)$, $t>0$, $x, y\in\overline{\Omega}$ with the following properties:

(i) $\left(\dfrac{\partial}{\partial t}+\mathscr{L}_x\right)U(t,x,y)=0, \quad \tau_\xi U(t,\xi,y)=0,$

where the subindex x to $\mathscr{L}$, for example, means to apply $\mathscr{L}$ to $U(t,x,y)$ as a function of x, and the subsequent subindices τ_ξ, etc. will be self-explanatory.

(ii) $\left(\dfrac{\partial}{\partial t}+\mathscr{L}_y^*\right)U(t,x,y)=0, \quad \tau_\xi^* U(t,x,\xi)=0.$

Here, the pair $(\mathscr{L}^*, \tau^*)$ denotes the formal adjoint of $(\mathscr{L}, \tau)$ defined by

$$\begin{aligned}\mathscr{L}^*\varphi &= -\sum_{i,j=1}^m \frac{\partial}{\partial x_i}\left(a_{ij}(x)\frac{\partial\varphi}{\partial x_j}\right)-\operatorname{div}(\boldsymbol{b}(x)\varphi)+c(x)\varphi,\\ \tau^*\varphi &= \alpha(\xi)\varphi+(1-\alpha(\xi))\left(\frac{\partial\varphi}{\partial\nu}+(\boldsymbol{b}(\xi)\cdot\boldsymbol{\nu}(\xi))\varphi\right).\end{aligned} \tag{3.22}$$

(iii) $U(t,x,y)>0$ for $t>0$ and $x, y\in\overline{\Omega}\setminus\Gamma_1$, where $\Gamma_1=\{\xi\in\Gamma;\ \alpha(\xi)=1\}$.

(iv) For each $t>0$, the operator U_t defined by

$$U_t u_0=\int_\Omega U(t,x,y)u_0(y)\,dy, \quad u_0\in L^2(\Omega)$$

belongs to $\mathscr{L}(L^2(\Omega))$. The function $u=U_t u_0$ satisfies (3.21) with $f=g=0$, and $\|U_t u_0-u_0\|\to 0$ as $t\downarrow 0$ (see (4.7) below).

(v) In addition to the above properties, there is a constant $C>0$ such that, for $t>0$ and $x, y\in\overline{\Omega}$,

$$U(t,x,y)\leqslant \frac{C}{t^{m/2}}\exp\left(Ct-\frac{|x-y|^2}{Ct}\right),$$

$$\left|\frac{\partial}{\partial x_i}U(t,x,y)\right|\leqslant \frac{C}{t^{(m+1)/2}}\exp\left(Ct-\frac{|x-y|^2}{Ct}\right), \quad \text{and}$$

$$\left.\begin{array}{r}\left|\dfrac{\partial^2}{\partial x_i\partial x_j}U(t,x,y)\right|,\\ \left|\dfrac{\partial}{\partial t}U(t,x,y)\right|\end{array}\right\}\leqslant \frac{C}{t^{m/2+1}}\exp\left(Ct-\frac{|x-y|^2}{Ct}\right).$$

The above operator U_t, $t > 0$, is nothing but the analytic semigroup e^{-tL} which appears later. The readers may refer to [23, 24] for the detailed proof of construction of $U(t,x,y)$ and its properties. If $u(t,x)$ is a genuine solution to the problem (3.21), then $u(t,x)$ is expressed as

$$\begin{aligned} u(t,x) &= \int_\Omega U(t,x,y)u_0(y)\,dy + \int_0^t ds \int_\Omega U(t-s,x,y) f(s,y)\,dy \\ &+ \int_0^t ds \int_\Gamma \left((1 - \boldsymbol{b}(\xi)\cdot\boldsymbol{\nu}(\xi))U(t-s,x,\xi) - \frac{\partial}{\partial \nu_\xi} U(t-s,x,\xi) \right) g(s,\xi)\,d\Gamma. \end{aligned} \tag{3.23}$$

There are a variety of regularity assumptions on $u_0(x)$, $f(t,x)$, and $g(t,\xi)$ to ensure that the right-hand side of (3.23) actually gives a unique genuine solution to (3.21). Such assumptions are, for example, that (i) $u_0(x)$ is in $L^2(\Omega)$; (ii) $f(t,x)$ is uniformly Hölder continuous in t, i.e., $\sup_{\overline{\Omega}} |f(t,x) - f(s,x)| \leqslant \text{const}\,|t-s|^\gamma$, $0 < \gamma < 1$; and (iii) $g_t(t,\xi)$ and $g_{\xi_i\xi_j}(t,\xi)$ are uniformly Hölder continuous in t for each local coordinate (see [23, 24]).

In the general boundary condition, the elliptic theory owes much to the property of the fundamental solution $U(t,x,y)$. Let us define the operator L derived from the pair $(\mathscr{L}, \tau)$. Set

$$\begin{aligned} \hat{L}u &= \mathscr{L}u, \quad u \in \mathscr{D}(\hat{L}), \\ \mathscr{D}(\hat{L}) &= \left\{ u \in C^2(\Omega) \cap C^1(\overline{\Omega});\ \mathscr{L}u \in L^2(\Omega),\ \tau u = 0 \right\}. \end{aligned} \tag{3.24}$$

There is a closure of $\hat{L}$ in $L^2(\Omega)$: For the existence, it is necessary and sufficient that the implication:

$$u_n \in \mathscr{D}(\hat{L}) \to 0, \quad \mathscr{L}u_n \to v \text{ in } L^2(\Omega) \quad \Rightarrow \quad v = 0$$

holds. To examine this, let $c > 0$ be chosen large enough, and set $\mathscr{L}_c u = (\mathscr{L} + c)u$. Then we calculate, for $u \in \mathscr{D}(\hat{L})$, as

$$\begin{aligned} \langle \mathscr{L}_c u, u \rangle &= \sum_{i,j=1}^m \left\langle a_{ij}(x) \frac{\partial u}{\partial x_j}, \frac{\partial u}{\partial x_i} \right\rangle + \sum_{i=1}^m \left\langle b_i(x) \frac{\partial u}{\partial x_i}, u \right\rangle + \langle (c(x)+c)u, u \rangle \\ &+ \left\langle \frac{\alpha(\xi)}{1-\alpha(\xi)} u, u \right\rangle_{\Gamma \setminus \Gamma_1}, \quad \Gamma_1 = \{\xi \in \Gamma;\ \alpha(\xi) = 1\} \neq \varnothing. \end{aligned}$$

Thus,

$$\text{Re}\,\langle \mathscr{L}_c u, u \rangle \geqslant \text{const}\,\|u\|^2_{H^1(\Omega)}, \quad u \in \mathscr{D}(\hat{L}).$$

By the assumption, we find that $u_n \to 0$ in $H^1(\Omega)$. For an arbitrary $\varphi \in \mathscr{D}(\Omega)$, Green's formula implies that

$$\begin{aligned} \langle \mathscr{L}_c u_n, \varphi \rangle &= \sum_{i,j=1}^m \left\langle a_{ij}(x) \frac{\partial u_n}{\partial x_j}, \frac{\partial \varphi}{\partial x_i} \right\rangle + \sum_{i=1}^m \left\langle b_i(x) \frac{\partial u_n}{\partial x_i}, \varphi \right\rangle \\ &+ \langle (c(x)+c)u_n, \varphi \rangle \to 0, \quad n \to \infty. \end{aligned}$$

We see that $\langle v, \varphi \rangle = 0$ for $\forall \varphi \in \mathscr{D}(\Omega)$. Thus, $v = 0$, and $\hat{L}$ is closable in $L^2(\Omega)$.

The closure of $\hat{L}$ in $L^2(\Omega)$ is denoted by L. The domain $\mathscr{D}(L)$ consists of $u \in L^2(\Omega)$ with the property that (i) there is a sequence $\{u_n\} \subset \mathscr{D}(\hat{L})$ such that $u_n \to u$ and (ii) $\hat{L}u_n$ converges as $n \to \infty$. Unlike in the Dirichlet boundary or the Robin boundary, it seems difficult to define L in the form (3.6) or (3.17). Let a constant c be chosen large enough, and consider the boundary value problem

$$(c + \mathscr{L})u = f \quad \text{in } \Omega, \quad \tau u = 0 \quad \text{on } \Gamma. \tag{3.25}$$

For a given $f \in C^\omega(\overline{\Omega})$, $\omega > 0$, the problem admits a unique solution $u \in \mathscr{D}(\hat{L})$ [24, Theorem 19.2]. The solution u to (3.25) with $f \in C^\omega(\overline{\Omega})$ is expressed as

$$u(x) = \int_\Omega G(x,y) f(y)\, dy, \quad \text{where}$$
$$G(x,y) = \int_0^\infty e^{-ct} U(t,x,y)\, dt, \quad (x,y) \in \overline{\Omega} \times \overline{\Omega},\ x \neq y.$$

A similar result holds for L^* [24, Theorem 19.2*].

When $\mathscr{L}$ is formally self-adjoint, the corresponding L is a self-adjoint operator with compact resolvent. Thus, as in the preceding subsections, there is a set of eigenpairs $\{\lambda_i, \varphi_{ij}\}$, $i \geqslant 1$, $1 \leqslant j \leqslant m_i\,(< \infty)$, such that

(i) $\sigma(L) = \{\lambda_i\}_{i \geqslant 1}$, $\lambda_1 < \cdots < \lambda_i < \cdots \to \infty$;

(ii) $(\lambda_i - L)\varphi_{ij} = 0$, that is, $(\lambda_i - \mathscr{L})\varphi_{ij} = 0$, $\tau\varphi_{ij} = 0$; and

(iii) the set $\{\varphi_{ij}\}$ forms an orthonormal basis for $L^2(\Omega)$.

For $\lambda \notin \sigma(L)$, the resolvent $(\lambda - L)^{-1}$ is expressed as

$$(\lambda - L)^{-1} u = \sum_{i=1}^{\infty} \sum_{j=1}^{m_i} \frac{\langle u, \varphi_{ij} \rangle}{\lambda - \lambda_i} \varphi_{ij}, \quad u \in L^2(\Omega).$$

Thus, as in (3.7), there is a sector $\overline{\Sigma} = \{\lambda - b \in \mathbb{C};\ \theta_0 \leqslant |\arg \lambda| \leqslant \pi\}$, $0 < \forall\theta_0 < \pi/2$, $\exists b \in \mathbb{R}^1$ such that the decay estimate

$$\left\| (\lambda - L)^{-1} \right\|_{\mathscr{L}(L^2(\Omega))} \leqslant \frac{\text{const}}{1 + |\lambda|}, \quad \lambda \in \overline{\Sigma}\ (\subset \rho(L)) \tag{3.26}$$

holds. Since $L_c = L + c$ is positive-definite, fractional powers L_c^ω, $\omega \in \mathbb{R}^1$, are well defined. Let us introduce a Hilbert space $H^1_\alpha(\Omega)$ defined as

$$H^1_\alpha(\Omega) = \left\{ u \in H^1(\Omega);\ u = 0 \text{ on } \Gamma_1,\ \left(\frac{\alpha(\xi)}{1 - \alpha(\xi)} \right)^{1/2} u \in L^2(\Gamma \setminus \Gamma_1) \right\}. \tag{3.27}$$

The inner product is, by choosing a constant $c > 0$ large enough, defined as

$$\langle u, v\rangle_{H^1_\alpha(\Omega)} = \sum_{i,j=1}^{m} \left\langle a_{ij}(x)\frac{\partial u}{\partial x_j}, \frac{\partial v}{\partial x_i}\right\rangle + \langle (c(x)+c)u, v\rangle$$
$$+\left\langle \frac{\alpha(\xi)}{1-\alpha(\xi)}u, v\right\rangle_{\Gamma\setminus\Gamma_1}, \qquad u, v \in H^1_\alpha(\Omega).$$

It is easy to see that the set $\{\varphi_{ij}\}$ forms an orthogonal but not normalized basis for $H^1_\alpha(\Omega)$, and

$$\mathscr{D}(L_c^{1/2}) = H^1_\alpha(\Omega). \tag{3.28}$$

In the case where $\mathscr{L}$ is not a formally self-adjoint operator, we decompose $\mathscr{L}$ as $\mathscr{L}u = \widetilde{\mathscr{L}}u + Du$ as in (3.12), and let $\tilde{L}$ be the self-adjoint operator which is obtained as the closure of $\widetilde{\mathscr{L}}|_{\ker\tau}$ in $L^2(\Omega)$. In view of (3.28), the operator $D\tilde{L}_c^{-1/2}$ is bounded. Thus, by choosing an $R > 0$ large enough, the resolvent

$$(\lambda - L)^{-1} = (\lambda - \tilde{L})^{-1}\left(1 - D\tilde{L}_c^{-1/2}\tilde{L}_c^{1/2}(\lambda - \tilde{L})^{-1}\right)^{-1}$$

exists for $\lambda \in \overline{\Sigma} \cap \{|\lambda| > R\}$, and satisfies the decay estimate

$$\|(\lambda - L)^{-1}\|_{\mathscr{L}(L^2(\Omega))} \leqslant \frac{\text{const}}{1+|\lambda|}, \quad \lambda \in \overline{\Sigma} \cap \{|\lambda| > R\}. \tag{3.29}$$

It is clear that $(\lambda - L)^{-1}$ is compact. We note that, if $c > 0$ is chosen large enough,

$$\mathrm{Re}\,\langle L_c u, u\rangle \geqslant \text{const}\,\|u\|^2_{H^1(\Omega)}, \quad \text{and thus}$$
$$\|L_c u\| \geqslant \text{const}\,\|u\|_{H^1(\Omega)}, \quad u \in \mathscr{D}(L).$$

Thus, we see that $\mathscr{D}(L) = \mathscr{D}(\tilde{L})$ both algebraically and topologically. Via a generalization of Heinz's inequality [25], we see that

$$\mathscr{D}(L_c^{\omega/2}) = \mathscr{D}(\tilde{L}_c^{\omega/2}) \subset H^\omega(\Omega), \quad 0 \leqslant \omega \leqslant 1. \tag{3.30}$$

Thus,

$$\|(\lambda - L)^{-1}\|_{\mathscr{L}(L^2(\Omega); H^1(\Omega))} \leqslant \frac{\text{const}}{1+|\lambda|^{1/2}}, \quad \lambda \in \overline{\Sigma} \cap \{|\lambda| > R\}. \tag{3.31}$$

For the pair $(\mathscr{L}^*, \tau^*)$ in (3.22), we also define the operator L^* as the closure of the closable operator $\widehat{L^*}$ (see the statement following (3.24)), which is actually the adjoint of L. Then, we similarly obtain

$$\mathrm{Re}\,\langle L_c^* u, u\rangle \geqslant \text{const}\,\|u\|^2_{H^1(\Omega)}, \quad \text{and thus}$$
$$\|L_c^* u\| \geqslant \text{const}\,\|u\|_{H^1(\Omega)}, \quad u \in \mathscr{D}(L^*).$$

The relation (3.30) also holds with L_c replaced by L_c^*.

2.3.4 On the domain of fractional powers L_c^θ with Robin boundary

Let us consider again the preceding operator L with Robin boundary: $\tau = \partial/\partial\nu + \sigma(\xi)$. By choosing $c > 0$ large enough, $\sigma(L_c)$ lies in $\mathbb{C}_+$ (see (3.20)). Thus, fractional powers L_c^θ of L_c is well defined. Later in Section 6, Chapter 4 and Section 3, Chapter 6, the characterization of $\mathscr{D}(L_c^\theta)$, $\theta < 3/4$, is usefully applied in studying boundary control systems with boundary input.

Let $\zeta(x)$, $x \in \mathbb{R}^m$, denote the distance from x to the boundary Γ. It is assumed that $\sigma(\xi)$ appearing in τ has a suitable smooth extension to $\overline{\Omega}$. The following characterization of $\mathscr{D}(L_c^\theta)$ is well known (see [19, 21])[5] :

(i) $\mathscr{D}(L_c^\theta) = H^{2\theta}(\Omega), \quad 0 \leqslant \theta < \frac{3}{4}$;

(ii) $\mathscr{D}(L_c^{3/4}) = \left\{ u \in H^{3/2}(\Omega);\ \int_\Omega \frac{1}{\zeta(x)} |\tau_\Omega u|^2\, dx < \infty \right\}$, where τ_Ω is a first order differential operator defined by $\tau_\Omega u = \dfrac{\partial u}{\partial \zeta} + \sigma(x) u$; and

(iii) $\mathscr{D}(L_c^\theta) = \{ u \in H^{2\theta}(\Omega);\ \tau u = 0 \text{ on } \Gamma \}, \quad \frac{3}{4} < \theta \leqslant 1$.

Here, we examine the above relation (i), following [19]:

$$\mathscr{D}(L_c^\theta) = H^{2\theta}(\Omega), \quad 0 \leqslant \theta < \frac{3}{4}. \tag{3.32}$$

Just a sketch of the proof is briefly illustrated. The domain $\mathscr{D}(L)$ is given by (3.17). The identity map; $u \mapsto u$ is continuous from $\mathscr{D}(L)$ to $H^2(\Omega)$, and from $L^2(\Omega)$ to $L^2(\Omega)$, too. Thus, the map is continuous from $[\mathscr{D}(L_c), L^2(\Omega)]_{1-\theta} = \mathscr{D}(L_c^\theta)$ to $[H^2(\Omega), L^2(\Omega)]_{1-\theta} = H^{2\theta}(\Omega)$, $0 \leqslant \theta \leqslant 1$: The former relation is due to m-accretiveness of L_c. Thus, if u belongs to $\mathscr{D}(L_c^\theta)$, then u also belongs to $H^{2\theta}(\Omega)$, that is, $\mathscr{D}(L_c^\theta) \subset H^{2\theta}(\Omega)$.

To show the converse relation for $0 \leqslant \theta < \frac{3}{4}$, we need several steps. Let $\{\alpha_k\}_{1 \leqslant k \leqslant n}$ be a partition of unity subordinate to the open covering $\bigcup_k W_k$ of Γ such that $\operatorname{supp}[\alpha_k] \subsetneq W_k$ (see Section 2). Let $\beta \in C_0^\infty(\Omega) = \mathscr{D}(\Omega)$ be a function such that $\operatorname{supp}[\beta] \supset \Omega \setminus \bigcup_k W_k$, and $0 \leqslant \beta \leqslant 1$, and let $u = \beta u + (1-\beta)u$. The

[5] In the case of the Dirichlet boundary, a similar but more restrictive relation holds:

(i) $\mathscr{D}(L_c^\theta) = H^{2\theta}(\Omega), \quad 0 \leqslant \theta < \frac{1}{4}$;

(ii) $\mathscr{D}(L_c^{1/4}) = \left\{ u \in H^{1/2}(\Omega);\ \int_\Omega \frac{1}{\zeta(x)} |u|^2\, dx < \infty \right\}$; and

(iii) $\mathscr{D}(L_c^\theta) = \{ u \in H^{2\theta}(\Omega);\ u|_\Gamma = 0 \text{ on } \Gamma \}, \quad \frac{1}{4} < \theta \leqslant 1$.

Since the upper bound of the power θ is small in (i), however, this relation seems not very useful in boundary control systems with the Dirichlet boundary.

map; $u \mapsto \beta u$ is continuous from $H^2(\Omega)$ to $\mathscr{D}(L)$, and from $L^2(\Omega)$ to $L^2(\Omega)$. Thus, the map is continuous from $H^{2\theta}(\Omega)$ to $\mathscr{D}(L_c^\theta)$. Thus, if u belongs to $H^{2\theta}(\Omega)$, then βu belongs to $\mathscr{D}(L_c^\theta)$, $0 \leqslant \theta \leqslant 1$.

Let us consider $(1-\beta)u$ for $u \in H^{2\theta}(\Omega)$, $0 \leqslant \theta < \frac{3}{4}$. It is clear that $(1-\beta)u \in H^{2\theta}(\Omega)$. In view of $(1-\beta)u = \sum_{k=1}^n (1-\beta)\alpha_k u$ in a neighborhood of Γ, we may limit ourselves to the property, say, of $(1-\beta)\alpha_1 u$ in $\Omega \cap W_1 = W_1^+$. The function $(1-\beta)u$ belongs to $H^{2\theta}(W_1^+)$. Thus, $(1-\beta)u\big|_{x=\Phi_1(y)}$ belongs to $H^{2\theta}(V_1^+)$, where $V_1^+ = V_1 \cap \{y_m > 0\}$. The map;

$$u \xrightarrow[\text{outside supp}[\alpha_1(y)]]{\times\sqrt{\alpha_1(y)}\text{ and extension by 0}} \sqrt{\alpha_1}\, u$$

is continuous from $H^2(V_1^+)$ to $H^2(\mathbb{R}_+^m)$; from $L^2(V_1^+)$ to $L^2(\mathbb{R}_+^m)$; and thus from $H^{2\theta}(V_1^+)$ to $H^{2\theta}(\mathbb{R}_+^m)$. This shows that $v(y) = e^{-\tilde{\sigma}(y')y_m}(1-\beta)\sqrt{\alpha_1}\,u$ belongs to $H^{2\theta}(\mathbb{R}_+^m)$, where $\tilde{\sigma}(y') = \sigma(\xi)\big|_{\xi=\Phi(y',0)}$.

For functions $u = u(y', y_m)$ in $\mathbb{R}_+^m$, we define two maps of prolongation ν and λ as

$$(\nu u)(y', y_m) = \begin{cases} u(y', y_m), & y_m > 0, \\ u(y', -y_m), & y_m < 0, \end{cases} \quad \text{and}$$
$$(\lambda u)(y', y_m) = \begin{cases} u(y', y_m), & y_m > 0, \\ -u(y', -y_m), & y_m < 0, \end{cases} \tag{3.33}$$

respectively. Then,

$$\nu;\quad \underbrace{[H^1(\mathbb{R}_+^m), L^2(\mathbb{R}_+^m)]_\omega}_{=H^{1-\omega}(\mathbb{R}_+^m)} \xrightarrow{\text{continuously}} \underbrace{[H^1(\mathbb{R}^m), L^2(\mathbb{R}^m)]_\omega}_{=H^{1-\omega}(\mathbb{R}^m)},$$
$$\lambda;\quad \underbrace{[H_0^1(\mathbb{R}_+^m), L^2(\mathbb{R}_+^m)]_\omega}_{=H_0^{1-\omega}(\mathbb{R}_+^m)} \xrightarrow{\text{continuously}} \underbrace{[H^1(\mathbb{R}^m), L^2(\mathbb{R}^m)]_\omega}_{=H^{1-\omega}(\mathbb{R}^m)}. \tag{3.34}$$

Note that $H_0^{1-\omega}(\mathbb{R}_+^m) = H^{1-\omega}(\mathbb{R}_+^m)$ for $\frac{1}{2} < \omega \leqslant 1$ (see [32]). For $u \in H^{2\omega}(\mathbb{R}_+^m)$, $\frac{1}{2} < \omega < \frac{3}{4}$, we see that

$$\frac{\partial}{\partial y_i}(\nu u) = \nu\left(\frac{\partial u}{\partial y_i}\right) \in H^{2\omega-1}(\mathbb{R}^m), \quad 1 \leqslant i \leqslant m-1,$$

$$\frac{\partial}{\partial y_m}(\nu u) = \lambda\left(\frac{\partial u}{\partial y_m}\right) \in H^{2\omega-1}(\mathbb{R}^m),$$

which implies that $\nu u \in H^{2\omega}(\mathbb{R}^m)$. Thus, ν is a continous map from $H^{2\omega}(\mathbb{R}^m_+)$ to $H^{2\omega}(\mathbb{R}^m)$ for $\frac{1}{2} < \omega < \frac{3}{4}$.[6]

For functions u in $\mathbb{R}^m$, let us consider the map π of restriction:

$$(\pi u)(y', y_m) = \frac{1}{2}\left(u(y', y_m) + u(y', -y_m)\right)\Big|_{y_m>0}. \tag{3.35}$$

Then, π is a continuous map from $H^2(\mathbb{R}^m)$ to $\left\{u \in H^2(\mathbb{R}^m_+);\ \frac{\partial u}{\partial y_m}\Big|_{y_m=0} = 0\right\}$; from $L^2(\mathbb{R}^m)$ to $L^2(\mathbb{R}^m_+)$; and thus

$$\pi;\quad H^{2\omega}(\mathbb{R}^m) \xrightarrow{\text{continuously}} \left[\left\{u \in H^2(\mathbb{R}^m_+);\ \frac{\partial u}{\partial y_m}\Big|_{y_m=0} = 0\right\}, L^2(\mathbb{R}^m_+)\right]_{1-\omega},$$
$$0 \leqslant \omega \leqslant 1. \tag{3.36}$$

Setting $\theta = \frac{3}{4} - \varepsilon$, $0 < \varepsilon < \frac{1}{4}$, let us go back to $v(y) = e^{-\tilde{\sigma}(y')y_m}(1-\beta)\sqrt{\alpha_1}\,u \in H^{2\theta}(\mathbb{R}^m_+)$. Since $v = \pi(\nu v)$ by definition, we see that

$$\begin{aligned} v(y) &= e^{-\tilde{\sigma}(y')y_m}(1-\beta)\sqrt{\alpha_1}\,u \\ &\in \left[\left\{u \in H^2(\mathbb{R}^m_+);\ \frac{\partial u}{\partial y_m}\Big|_{y_m=0} = 0\right\}, L^2(\mathbb{R}^m_+)\right]_{1/4+\varepsilon}. \end{aligned}$$

Considering the map $u(y) \mapsto e^{\tilde{\sigma}(y')y_m}u(y)\Big|_{V_1^+}$ for functions u in $\mathbb{R}^m_+$, we see that

$$(1-\beta)\sqrt{\alpha_1}\,u \in \left[\left\{u \in H^2(V_1^+);\ \frac{\partial u}{\partial y_m} + \tilde{\sigma}(y')u\Big|_{y_m=0} = 0\right\}, L^2(V_1^+)\right]_{1/4+\varepsilon},$$

or by changing the coordinate y to x,

$$(1-\beta)\sqrt{\alpha_1}\,u \in \left[\left\{u \in H^2(W_1^+);\ \frac{\partial u}{\partial \nu} + \sigma(\xi)u\Big|_{\Gamma} = 0\right\}, L^2(W_1^+)\right]_{1/4+\varepsilon}.$$

For functions u in W_1^+, the map;

$$u \xrightarrow[\text{outside supp}\,[\alpha_1(x)]]{\times\sqrt{\alpha_1(x)}\ \text{and extension by 0}} \sqrt{\alpha_1}\,u$$

[6] We have a further continuity property of ν:

$$\nu;\quad \left[\left\{u \in H^2(\mathbb{R}^m_+);\ \frac{\partial u}{\partial y_m}\Big|_{y_m=0} = 0\right\}, L^2(\mathbb{R}^m_+)\right]_{1-\omega} \xrightarrow{\text{continuously}} H^{2\omega}(\mathbb{R}^m),\quad 0 \leqslant \omega \leqslant 1.$$

is continuous from $\left\{u \in H^2(W_1^+);\ \frac{\partial u}{\partial \nu} + \sigma(\xi)u\Big|_{\Gamma} = 0\right\}$ to $\mathscr{D}(L)$; from $L^2(W_1^+)$ to $L^2(\Omega)$; and thus from $\left[\left\{u \in H^2(W_1^+);\ \frac{\partial u}{\partial \nu} + \sigma(\xi)u\Big|_{\Gamma} = 0\right\},\ L^2(W_1^+)\right]_{1/4+\varepsilon}$ to $[\mathscr{D}(L), L^2(\Omega)]_{1/4+\varepsilon} = \mathscr{D}(L_c^{3/4-\varepsilon})$. We have shown that $(1-\beta)u$ belongs to $\mathscr{D}(L_c^{3/4-\varepsilon})$. Since both βu and $(1-\beta)u$ belongs to $\mathscr{D}(L_c^{3/4-\varepsilon})$, we conclude that every $u \in H^{2\theta}(\Omega)$ belongs to $\mathscr{D}(L_c^{\theta})$ for $\theta = \frac{3}{4} - \varepsilon$, $0 < \varepsilon < \frac{1}{4}$, and (3.32) is proven. □

2.4 Analytic Semigroup

We have shown that, whatever the boundary operator τ may be, the resolvent $(\lambda - L)^{-1}$ of the operator L in Section 3 satisfies the decay estimate

$$\|(\lambda - L)^{-1}\|_{\mathscr{L}(L^2(\Omega))} \leqslant \frac{\text{const}}{1+|\lambda|}, \quad \lambda \in \overline{\Sigma}, \tag{4.1}$$

by choosing a suitable sector $\overline{\Sigma} = \{\lambda - b \in \mathbb{C};\ \theta_0 \leqslant |\arg\lambda| \leqslant \pi\}$, $0 < \forall\theta_0 < \pi/2$, $\exists b \in \mathbb{R}^1$. An operator L whose resolvent satisfies (4.1) in some sector $\overline{\Sigma}$ with angle more than π is called a *sectorial operator*. It is well known that a sectorial operator generates an analytic semigroup. We briefly review some properties of a sectorial operator in this section. The class of these operators are apparently narrower than those generating C_0-semigroups, since, for example, the operator in wave equations with no damping term has the spectrum lying in the imaginary axis in the complex plane, and thus is not sectorial. Let

$$e^{-tL} = \frac{-1}{2\pi \mathrm{i}} \int_{\partial\overline{\Sigma}} e^{-t\lambda}(\lambda - L)^{-1}\, d\lambda, \quad t > 0, \tag{4.2}$$

where the integral along $\partial\overline{\Sigma}$ is oriented according to increasing $\operatorname{Im}\lambda$. The right-hand side converges in the topology of operator norm. It is standard to see that $\{e^{-tL}\}_{t>0}$ enjoys the semigroup property:

$$e^{-(t+s)L} = e^{-tL}e^{-sL}, \quad t, s > 0.$$

There are constants $M \geqslant 1$ and $\omega \in \mathbb{R}^1$ such that

$$\|e^{-tL}\|_{\mathscr{L}(L^2(\Omega))} \leqslant Me^{\omega t}, \quad t > 0. \tag{4.3}$$

In fact, the estimate is clear for $t \geqslant 1$. A rough estimate shows that ω is smaller than or equal to b, and there is an infimum ω^* of such ω (see a remark below on general well-posed Cauchy problems). To ensure boundedness of the left-hand side for $t \in (0, 1]$, we may change the contour $\partial\overline{\Sigma}$ to $\partial\overline{\Sigma} - 1/t$ in (4.2).

Elementary calculation then shows that $\left\|e^{-tL}\right\|_{\mathscr{L}(L^2(\Omega))}$ is bounded on $(0, 1]$ (see Figure 1 below).

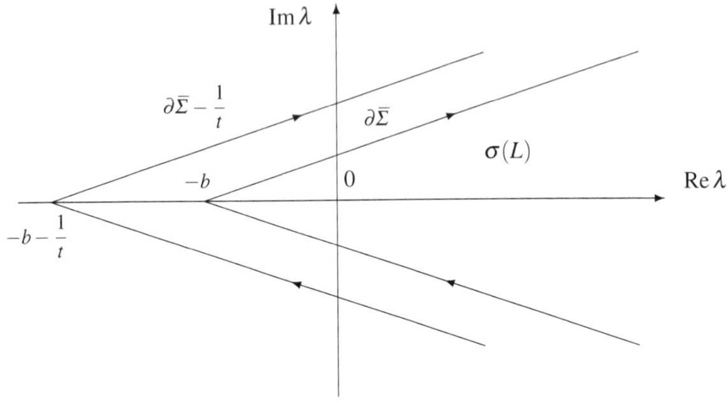

Figure 1

Thus, (4.3) holds. By (4.2), e^{-tL} is infinitely differentiable in $t > 0$, and the range belongs to $\bigcup_{k\geqslant 1} \mathscr{D}(L^k)$. Direct calculation shows that

$$\left\|Le^{-tL}\right\|_{\mathscr{L}(L^2(\Omega))} \leqslant \frac{M'}{t} e^{\omega t}, \quad t > 0. \tag{4.4}$$

Thus, by the moment inequality (see Subsection 3.1),

$$\left\|L_c^{\alpha} e^{-tL}\right\|_{\mathscr{L}(L^2(\Omega))} \leqslant \frac{M''}{t^{\alpha}} e^{\omega t}, \quad t > 0. \tag{4.4$'$}$$

Let $c > 0$ be large enough, so that $-c \in \rho(L)$. For $u \in \mathscr{D}(L)$, let $u = L_c^{-1} v$. Then, we calculate as

$$\begin{aligned} e^{-tL} u &= \frac{-1}{2\pi \mathrm{i}} \int_{\partial \bar{\Sigma}} e^{-t\lambda} (\lambda - L)^{-1} L_c^{-1} v \, d\lambda \\ &= \frac{-1}{2\pi \mathrm{i}} \int_{\partial \bar{\Sigma}} \frac{e^{-t\lambda}}{\lambda + c} \left((\lambda - L)^{-1} + L_c^{-1} \right) v \, d\lambda \\ &= \frac{-1}{2\pi \mathrm{i}} \int_{\partial \bar{\Sigma}} \frac{e^{-t\lambda}}{\lambda + c} (\lambda - L)^{-1} v \, d\lambda \\ &\to \frac{-1}{2\pi \mathrm{i}} \int_{\partial \bar{\Sigma}} \frac{1}{\lambda + c} (\lambda - L)^{-1} v \, d\lambda, \quad t \downarrow 0. \end{aligned}$$

The last integral is equal to the residue of $-(\lambda + c)^{-1}(\lambda - L)^{-1} v$ at $\lambda = -c$, that is, $L_c^{-1} v = u$. For a general $u \in L^2(\Omega)$, we can apply the Banach-Steinhaus theorem [26, 71] to e^{-tL}, since the operator norm of e^{-tL} is bounded on $(0, 1]$. We then conclude that

$$e^{-tL} u \to u, \quad t \downarrow 0, \quad \forall u \in L^2(\Omega)$$

strongly.

In the specific case where L is self-adjoint, we can apply the Fourier series expansion of $(\lambda - L)^{-1}$ to (4.2) to derive the expression:

$$
\begin{aligned}
e^{-tL}u &= \frac{-1}{2\pi \mathrm{i}} \int_{\partial\Sigma} e^{-t\lambda} \sum_{i=1}^{\infty} \sum_{j=1}^{m_i} \frac{\langle u, \varphi_{ij} \rangle}{\lambda - \lambda_i} \varphi_{ij} \, d\lambda \\
&= \sum_{i,j} \frac{-1}{2\pi \mathrm{i}} \int_{\partial\Sigma} \frac{e^{-t\lambda}}{\lambda - \lambda_i} d\lambda \, \langle u, \varphi_{ij} \rangle \varphi_{ij} \\
&= \sum_{i,j} e^{-\lambda_i t} \langle u, \varphi_{ij} \rangle \varphi_{ij}, \quad t > 0.
\end{aligned}
\tag{4.5}
$$

The Cauchy problem

$$
\frac{du}{dt} + Lu = 0, \quad t > 0, \quad u(0) = u_0 \in L^2(\Omega) \tag{4.6}
$$

admits a unique solution $u(t) = e^{-tL}u_0$. The solution $u(t)$ is also expressed in terms of the fundamental solution as

$$
u(t) = e^{-tL}u_0 = \int_{\Omega} U(t,x,y) u_0(y) \, dy. \tag{4.7}
$$

Generally, the Cauchy problem of the differential equation:

$$
\frac{du}{dt} + Lu = 0, \quad t > 0, \quad u(0) = u_0
$$

in a Banach space is called *well posed* on $(0, \infty)$, if (i) there is a unique solution $u(t)$ to the problem for each $u_0 \in \mathscr{D}(L)$, and (ii) $u(t)$ continuously depends on u_0 in the topology of the space for each $t > 0$. The semigroup generated by the problem is called a C_0*-semigroup*. The problem (4.6) in $L^2(\Omega)$ is thus well posed on $(0, \infty)$. When e^{-tL} is a C_0-semigroup, the function $\log \|e^{-tL}\|$ is subadditive in $t > 0$. Thus we see that [3, 16, 26, 71],

$$
\omega^* = \lim_{t \to \infty} \frac{\log \|e^{-tL}\|}{t} = \inf_{t > 0} \frac{\log \|e^{-tL}\|}{t}.
$$

The number ω^*, called the *type of the semigroup*, gives the infimum of ω in the estimate (4.3).

Let us go back to the sectorial operator L. Let F be a linear operator with domain $\mathscr{D}(F)$ which is subordinate to L_c^α, $0 \leqslant \alpha < 1$, that is, $\mathscr{D}(F) \supset \mathscr{D}(L_c^\alpha)$ and $\|Fu\| \leqslant \text{const} \, \|L_c^\alpha u\|$ for $\forall u \in \mathscr{D}(L_c^\alpha)$. When L is perturbed by F, the resultant operator $L + F$ is also sectorial. In fact, $FL_c^{-\alpha}$ is bounded, and by choosing an R large enough, the relation

$$
(\lambda - L - F)^{-1} = (\lambda - L)^{-1} \left(1 - FL_c^{-\alpha} L_c^\alpha (\lambda - L)^{-1}\right)^{-1}
$$

holds for λ in $\overline{\Sigma} \cap \{|\lambda| > R\}$. Thus, $-(L+F)$ is an infinitesimal generattor of an analytic semigroup $e^{-t(L+F)}$, $t > 0$. We will often use the result in Chapters 3 through 7. In the case where $-L$ is instead an infinitesimal generator of a C_0-semigroup, $-(L+F)$ also becomes a generator of a C_0-semigroup, as long as F is bounded [26].

Other properties of e^{-tL} are found, e.g., in the monographs, [3, 16, 26, 35, 71] and the references therein. We mainly consider the above sectorial L in this monograph. In Chapter 7, however, we study control systems with a more general L not satisfying the assumption (4.1), so that $-L$ is the infinitesimal generator of a so called *eventually differentiable* semigroup.

Chapter 3

Stabilization of linear systems of infinite dimension: Static feedback

3.1 Introduction

In this chapter, we study stabilization problems of systems governed by linear differential equations of infinite dimension. Let H be a Hilbert space equipped with inner product $\langle \cdot, \cdot \rangle$ and norm $\|\cdot\|$. The symbol $\|\cdot\|$ will be also used for the $\mathscr{L}(H)$-norm. Our control system has state u, output $y = Wu \in \mathbb{C}^N$, and input $f \in \mathbb{C}^N$, and is described by a linear differential equation in H as

$$\frac{du}{dt} + Lu = Gf, \quad y = Wu, \quad u(0) = u_0 \in H. \tag{1.1}$$

Here, as in Chapter 1,

$$\begin{aligned} Gf &= \sum_{k=1}^{N} f_k g_k \quad \text{for } f = \begin{pmatrix} f_1 & \dots & f_N \end{pmatrix}^{\mathrm{T}} \in \mathbb{C}^N, \\ Wu &= \begin{pmatrix} \langle u, w_1 \rangle & \dots & \langle u, w_N \rangle \end{pmatrix}^{\mathrm{T}} \quad \text{for } u \in H, \end{aligned} \tag{1.2}$$

$(\dots)^{\mathrm{T}}$ being the transpose of vectors or matrices. The vectors $w_k \in H$ denote given weights of the observation (output); and $g_k \in H$ are actuators. In (1.1),

L is a linear closed operator with dense domain $\mathscr{D}(L)$ such that the resolvent $(\lambda - L)^{-1}$ is compact, and satisfies a decay estimate

$$\left\|(\lambda - L)^{-1}\right\| \leqslant \frac{\text{const}}{1+|\lambda|}, \quad \lambda \in \overline{\Sigma}, \tag{1.3}$$

where $\overline{\Sigma}$ denotes some sector described by $\overline{\Sigma} = \{\lambda - b;\ \theta_0 \leqslant |\arg \lambda| \leqslant \pi\}$, $0 < \theta_0 < \pi/2$, $b \in \mathbb{R}^1$. Setting $f = Wu$ or $f_k(t) = \langle u, w_k \rangle$, $1 \leqslant k \leqslant N$, we have a closed loop feedback control system:

$$\frac{du}{dt} + Lu = GWu = \sum_{k=1}^{N} \langle u, w_k \rangle g_k, \quad t > 0, \quad u(0) = u_0. \tag{1.4}$$

In (1.4), the output Wu is directly fed back into the equation. This scheme is the so called *static feedback* scheme [58]. While the scheme looks simple, it has many difficulties in engineering implementations. Recently, in [27] and the references therein, the stabilization of one-dimensional heat equations by this scheme asks to solve specific boundary value problems of wave equations in a triangle: The idea looks interesting, but the support of the designed sensor lies all over the interval, and thus the construction of the scheme faces a serious difficulty in engineering implementation. In [4, 5, 55], a feedback control law for a class of Navier-Stokes equations is constructed by solving an algebraic Riccati equation: There is, however, *no* guarantee on the narrow support of the sensors designed through the solution of the Riccati equation.

Later in Chapters 4 and 5 we study stabilization problems with the *dynamic feedback* scheme which permits the case of boundary observation and boundary feedback. The static scheme developed in this chapter has enough meaning in the sense that it constitutes a part of stabilization studies with the dynamic feedback scheme. The results of this chapter are mainly based on those in [46, 48, 52, 58]. In the case where the system generates a *contraction semigroup*, the decay (1.3) is no more expected. The contraction semigroup is the one such that it satisfies an estimate; $\|e^{-tL}\| \leqslant 1$ for $\geqslant 0$. A typical example is a pure wave equation with no damping term. The feedback scheme for such systems is specifically determined by the law: $G = -W^*$ in (1.4), so that the energy does not increase. Like a decomposition of H into a direct sum of invariant subspaces, $H = H_1 \oplus H_2$ stated next page, there is a decomposition—the so called Nagy-Foias decomposition—of H into the direct sum of invariant subspaces: $H = H_u \oplus H_{cnu}$, where H_u is called a unitary subspace and H_{cnu} a completely nonunitary subspace. Both spaces are infinite-dimensional, so that the stabilization problem for such systems requires an approach fairly different from ours (see [30, 65]). The approach is interesting from a mathematical viewpoint, but not stated in this monograph.

By our assumption (1.3), $-L$ is an infinitesimal generator of an analytic semigroup e^{-tL}, $t > 0$. By the compactness property of the resolvent $(\lambda - L)^{-1}$, the spectrum $\sigma(L)$ of L consists only of eigenvalues. It is assumed throughout the chapter that $\sigma(L) \cap \overline{\mathbb{C}_-} \neq \varnothing$, so that (1.1) with $f_k(t) = 0$, $1 \leqslant k \leqslant N$, is unstable. Given a prescribed $\mu > 0$, it is assumed that $\sigma(L)$ is divided into the union of two disjoint sets σ_1 and σ_2; $\sigma(L) = \sigma_1 \cup \sigma_2$, $\sigma_1 \cap \sigma_2 = \varnothing$, where σ_1 contains a finite number of unstable eigenvalues of L and σ_2 is contained in $\mathbb{C}_+$ such that

$$\sigma_1 \subset \{\lambda \in \mathbb{C};\ \mathrm{Re}\,\lambda < \mu\}, \quad \sigma_2 \subset \{\lambda \in \mathbb{C};\ \mathrm{Re}\,\lambda > \mu\}. \tag{1.5}$$

The projector associated with σ_1 is denoted as P with $\dim PH = n < \infty$:

$$P = \frac{1}{2\pi \mathrm{i}} \int_C (\lambda - L)^{-1}\, d\lambda,$$

where C denotes a Jordan contour encircling σ_1 in its inside, with σ_2 outside C. Then, H is decomposed into the direct sum of two invariant subspaces: $H = H_1 \oplus H_2$, where $H_1 = PH$ and $H_2 = (1-P)H$. Let $L_1 = L|_{H_1}$ be the restriction of L onto the finite dimensional subspace H_1, and $L_2 = L|_{\mathscr{D}(L)\cap H_2}$. We see that $e^{-tL_1} = e^{-tL}|_{H_1}$, and $e^{-tL_2} = e^{-tL}|_{H_2}$. The latter semigroup is stable, that is,

$$\left\| e^{-tL_2} \right\| \leqslant \mathrm{const}\, e^{-\mu t}, \quad t \geqslant 0. \tag{1.6}$$

More is true: Let $\mu_* = \min_{\lambda \in \sigma_2} \mathrm{Re}\,\lambda > \mu$. Then,

$$\left\| e^{-tL_2} \right\| \leqslant \mathrm{const}\,(1 + t^m) e^{-\mu_* t}, \quad t \geqslant 0, \tag{1.6$_1$}$$

where $m \geqslant 0$ denotes an integer. In fact, the algebraic growth $(1+t^m)$, $m \geqslant 1$, arises when the dimension of the generalized eigenspace of $\lambda \in \sigma_2$ ($\mathrm{Re}\,\lambda = \mu_*$) is greater than 1. Let $\sigma_1 = \{\lambda_i\}_{1 \leqslant i \leqslant \nu}$, $\lambda_i \neq \lambda_j$ for $i \neq j$. Since each λ_i is an eigenvalue of L_1, there is a set of generalized eigenpairs $\{\lambda_i,\ \varphi_{ij}\}$ with the following properties:

(i) $\sigma(L_1) = \sigma_1$, $\quad \lambda_i \neq \lambda_j$ for $i \neq j$;

(ii) $L_1 \varphi_{ij} = \lambda_i \varphi_{ij} + \sum_{k<j} \alpha^i_{jk} \varphi_{ik}$, $\quad 1 \leqslant i \leqslant \nu$, $1 \leqslant j \leqslant m_i$; and

(iii) the set $\{\varphi_{ij};\ 1 \leqslant i \leqslant \nu,\ 1 \leqslant j \leqslant m_i\}$ forms a basis for PH.

Let P_{λ_i} be the projector in H corresponding to the eigenvalue λ_i. Then, $P = P_{\lambda_1} + \cdots + P_{\lambda_\nu}$, and $P_{\lambda_i} u = \sum_{j=1}^{m_i} u_{ij} \varphi_{ij}$ for $u \in H$. The restriction of L (and thus of L_1) onto the invariant subspace $P_{\lambda_i} H$ is, in the basis $\{\varphi_{i1}, \ldots, \varphi_{im_i}\}$, is represented by the $m_i \times m_i$ upper triangular matrix Λ_i, where

$$\Lambda_i|_{(j,k)} = \begin{cases} \alpha^i_{kj}, & j < k, \\ \lambda_i, & j = k, \\ 0, & j > k. \end{cases} \tag{1.7}$$

The matrix representation of L_1 is then an $n \times n$ matrix $\Lambda = \operatorname{diag}(\Lambda_1 \ \ldots \ \lambda_\nu)$. This algebraic structure is of fairly general nature, and often appears in later chapters.

The structure of the adjoint operator L_1^* is similar to that of L_1: There is a set of generalized eigenpairs $\{\overline{\lambda_i},\ \psi_{ij}\}$ with the following properties:

(i) $\sigma(L_1^*) = \overline{\sigma_1}$;

(ii) $L_1^* \psi_{ij} = \overline{\lambda_i} \psi_{ij} + \sum_{k<j} \beta^i_{jk} \psi_{ik}, \quad 1 \leqslant i \leqslant \nu,\ 1 \leqslant j \leqslant m_i$; and

(iii) the set $\{\psi_{ij};\ 1 \leqslant i \leqslant \nu,\ 1 \leqslant j \leqslant m_i\}$ forms a basis for P^*H.

The restriction of L_1^* onto the invariant subspace $P^*_{\lambda_i} H$ is, in the basis $\{\psi_{i1}, \ldots, \psi_{im_i}\}$, is represented by the $m_i \times m_i$ upper triangular matrix $\tilde{\Lambda}_i$, where

$$\tilde{\Lambda}_i\big|_{(j,k)} = \begin{cases} \beta^i_{kj}, & j<k, \\ \overline{\lambda_i}, & j=k, \\ 0, & j>k. \end{cases} \tag{1.8}$$

The above results are derived from the general theory of compact operators: Let $c>0$ be chosen large enough such that $-c$ is in $\rho(L)$, and set $L_c = L+c$. Then the results are reduced to those of the compact operator L_c^{-1} and its adjoint operator $(L_c^{-1})^*$. The generalized eigenspace of L corresponding to λ_i is

$$\bigcup_{1 \leqslant k < \infty} W^{(k)}_{\lambda_i}, \quad W^{(k)}_{\lambda_i} = \left\{ u \in \mathscr{D}(L^k);\ (\lambda_i - L)^k u = 0 \right\}.$$

By compactness of the resolvent, there is an integer $n \geqslant 1$ such that $W^{(n)}_{\lambda_i} = W^{(n+1)}_{\lambda_i}$, so that the above space is of finite dimension. The smallest such integer n, say l_i, is called the *ascent* of $\lambda_i - L$ [66]. Thus, the above generalized eigenspace is $W^{(l_i)}_{\lambda_i}$ with $\dim W^{(l_i)}_{\lambda_i} < \infty$. It is easily seen that

$$W^{(k)}_{\lambda_i} = \left\{ u \in H;\ \left(\frac{1}{\lambda_i + c} - L_c^{-1} \right)^k u = 0 \right\}, \quad k \geqslant 1,$$

the right-hand side of which are the generalized eigenspaces of the compact operator L_c^{-1} corresponding to the eigenvalue $\dfrac{1}{\lambda_i + c}$. For a non-zero eigenvalue $\dfrac{1}{\lambda_i + c}$ of L_c^{-1}, the operators $\dfrac{1}{\lambda_i + c} - L_c^{-1}$ and $\overline{\left(\dfrac{1}{\lambda_i + c}\right)} - (L_c^{-1})^*$ have the same ascent (see page 282 of [66]).

What is the relationship between the two matrices Λ_i and $\tilde{\Lambda}_i$? For each i, set

$$\Pi_{\lambda_i} = \left(\langle \varphi_{ij}, \psi_{il} \rangle;\ \begin{matrix} j & \rightarrow & 1, \ldots, m_i \\ l & \downarrow & 1, \ldots, m_i \end{matrix} \right). \tag{1.9}$$

It is easy to see that the inverse $\Pi_{\lambda_i}^{-1}$ exists. In fact, consider a linear combination of the column vectors of Π_{λ_i} satisfying

$$0 = \sum_{1 \leqslant j \leqslant m_i} c_j \left\langle \varphi_{ij}, \psi_{il} \right\rangle = \left\langle \sum_{1 \leqslant j \leqslant m_i} c_j \varphi_{ij}, \psi_{il} \right\rangle, \quad 1 \leqslant l \leqslant m_i.$$

This means that $\sum_{1 \leqslant j \leqslant m_i} c_j \varphi_{ij} \perp P_{\lambda_i}^* H$. Thus, for any $v \in H$,

$$0 = \left\langle \sum_{1 \leqslant j \leqslant m_i} c_j \varphi_{ij}, P_{\lambda_i}^* v \right\rangle = \left\langle P_{\lambda_i} \left(\sum_{1 \leqslant j \leqslant m_i} c_j \varphi_{ij} \right), v \right\rangle = \left\langle \sum_{1 \leqslant j \leqslant m_i} c_j \varphi_{ij}, v \right\rangle,$$

from which we conclude that $c_j = 0$, $1 \leqslant j \leqslant m_i$.

For any $u = \sum_{1 \leqslant j \leqslant m_i} u_{ij} \varphi_{ij} \in P_{\lambda_i} H$, we have $\langle u, \psi_{il} \rangle = \sum_{1 \leqslant j \leqslant m_i} u_{ij} \left\langle \varphi_{ij}, \psi_{il} \right\rangle$, $1 \leqslant l \leqslant m_i$, or

$$\langle u, \boldsymbol{\psi}_i \rangle = \Pi_{\lambda_i} \boldsymbol{u}_i, \quad \text{that is,} \quad \boldsymbol{u}_i = \Pi_{\lambda_i}^{-1} \langle u, \boldsymbol{\psi}_i \rangle. \tag{1.10}$$

Here we have set

$$\langle u, \boldsymbol{\psi}_i \rangle = \Big(\langle u, \psi_{il} \rangle ; \, l \downarrow 1, \ldots, m_i \Big), \quad \text{and} \quad \boldsymbol{u}_i = \Big(u_{ij}; \, j \downarrow 1, \ldots, m_i \Big).$$

For any $v = \sum_{1 \leqslant j \leqslant m_i} v_{ij} \psi_{ij} \in P_i^* H$, we calculate as

$$\left\langle \varphi_{ij}, L^* v \right\rangle = \left\langle \varphi_{ij}, \sum_{1 \leqslant l \leqslant m_i} (\tilde{\Lambda}_i \boldsymbol{v}_i)_l \psi_{il} \right\rangle = \sum_{1 \leqslant l \leqslant m_i} \overline{(\tilde{\Lambda}_i \boldsymbol{v}_i)_l} \left\langle \varphi_{ij}, \psi_{il} \right\rangle, \quad j \downarrow 1, \ldots, m_i,$$

$$\langle \boldsymbol{\varphi}_i, L^* v \rangle = \Pi_{\lambda_i}^{\mathrm{T}} \overline{(\tilde{\Lambda}_i \boldsymbol{v}_i)} \quad \text{or} \quad \langle L^* v, \boldsymbol{\varphi}_i \rangle = \langle v, L \boldsymbol{\varphi}_i \rangle = \Pi_{\lambda_i}^* (\tilde{\Lambda}_i \boldsymbol{v}_i).$$

By noting that $L\boldsymbol{\varphi}_i = \Lambda_i^{\mathrm{T}} \boldsymbol{\varphi}_i$ and $\langle v, \boldsymbol{\varphi}_i \rangle = \Pi_{\lambda_i}^* \boldsymbol{v}_i$, the above last term leads to

$$\Pi_{\lambda_i}^* \tilde{\Lambda}_i \boldsymbol{v}_i = \langle v, L\boldsymbol{\varphi}_i \rangle = \overline{\Lambda_i^{\mathrm{T}}} \langle v, \boldsymbol{\varphi}_i \rangle = \Lambda_i^* \Pi_{\lambda_i}^* \boldsymbol{v}_i,$$
$$\text{or} \quad \tilde{\Lambda}_i \boldsymbol{v}_i = (\Pi_{\lambda_i}^*)^{-1} \Lambda_i^* \Pi_{\lambda_i}^* \boldsymbol{v}_i.$$

We have shown that

$$\tilde{\Lambda}_i = (\Pi_{\lambda_i}^*)^{-1} \Lambda_i^* \Pi_{\lambda_i}^* = \left(\Pi_{\lambda_i} \Lambda_i \Pi_{\lambda_i}^{-1} \right)^*. \tag{1.11}$$

These matrix relationships will be used in the following sections.

3.2 Decomposition of the System

Our stabilization problem is to construct the operators W and G (or w_k and g_k, $1 \leqslant k \leqslant N$) satisfying a decay estimate

$$\left\| e^{-t(L-GW)} \right\| \leqslant \text{const}\, e^{-\mu t}, \quad t \geqslant 0. \tag{2.1}$$

By setting $u_1 = Pu$ and $u_2 = Qu$, Q being the projector $1-P$, the feedback control system (1.4) is decomposed into a system of two differential equations in the product space $H_1 \times H_2$:

$$\begin{cases} \dfrac{du_1}{dt} + L_1 u_1 = PGWu = \displaystyle\sum_{k=1}^{N} \langle u, w_k \rangle Pg_k, \\ \dfrac{du_2}{dt} + L_2 u_2 = QGWu = \displaystyle\sum_{k=1}^{N} \langle u, w_k \rangle Qg_k. \end{cases} \tag{2.2$_1$}$$

Thus, (2.2_1) with state (u_1, u_2) foms a strongly coupled system. Let $W_1 = W|_{PH}$ be the restriction of W onto PH, and let $W_2 = W|_{QH}$. Then (2.2_1) is rewritten as

$$\begin{cases} \dfrac{du_1}{dt} + (L_1 - PGW_1)u_1 = PGW_2 u_2, \\ \dfrac{du_2}{dt} + L_2 u_2 = QGW_1 u_1 + QGW_2 u_2. \end{cases} \tag{2.2$_2$}$$

Let us consider first the simplest case where $QG = 0$, or $Qg_k = 0$, $1 \leqslant k \leqslant N$, so that the actuators g_k are constructed in the subspace H_1. Then, $u_2(t) = e^{-tL_2} Qu_0$, and $u_2(t) \to 0$ as $t \to \infty$ by (1.6). In the equation for u_1, PGW_2u_2 behaves just like a stable perturbation, or

$$u_1(t) = e^{-t(L_1 - PGW_1)} Pu_0 + \int_0^t e^{-(t-s)(L_1 - PGW_1)} PGW_2 u_2(s)\, ds.$$

Let $\mu_1 > \mu$ be an arbitrary number. Let us recall Theorem 2.1 and Proposition 2.2, Chapter 1: If (W_1, L_1) is an observable pair, that is, if

$$\ker \begin{pmatrix} W_1 & W_1 L_1 & \dots & W_1 L_1^{n-1} \end{pmatrix}^{\mathrm{T}} = \{0\}, \tag{2.3}$$

there is an operator PG $(= G)$, or $g_k \in PH$, $1 \leqslant k \leqslant N$, such that

$$\left\| e^{-t(L_1 - PGW_1)} \right\| \leqslant \text{const}\, e^{-\mu_1 t}, \quad t \geqslant 0. \tag{2.4}$$

It is then clear that $\|u_1(t)\| \leqslant \text{const}\, e^{-\mu t} \|u_0\|$, $t \geqslant 0$, and thus $\|u(t)\| \leqslant \text{const}\, e^{-\mu t} \|u_0\|$, $t \geqslant 0$. This is nothing but the desired estimate (2.1).

According to the basis $\{\varphi_{ij}\}$ for H_1, every vector $u \in PH$ is uniquely expressed as $u = \sum_{i,j} u_{ij} \varphi_{ij}$, and the operator W_1 is rewritten as

$$Wu = \hat{W}\hat{u} = \begin{pmatrix} w_{ij}^k; & \begin{matrix} (i,j) & \to & (1,1), \dots, (\nu, m_\nu) \\ k & \downarrow & 1, \dots, N \end{matrix} \end{pmatrix} \hat{u}, \quad w_{ij}^k = \langle \varphi_{ij}, w_k \rangle,$$

where $\hat{u} = \begin{pmatrix} u_{11}\ u_{12}\ \dots\ u_{\nu m_\nu} \end{pmatrix}^{\mathrm{T}}$. Then the restriction $W_i = W|_{P_{\lambda_i} H}$ of W onto $P_{\lambda_i} H$ is clearly

$$W_i u = \hat{W}_i \hat{u} = \begin{pmatrix} w_{ij}^k; & \begin{matrix} j & \to & 1, \dots, m_i \\ k & \downarrow & 1, \dots, N \end{matrix} \end{pmatrix} \hat{u}, \quad u \in P_{\lambda_i} H.$$

As we have seen in Proposition 3.1, Chapter 1, the observability condition (2.3) is equivalent to

$$\begin{aligned} &\ker\left(W_i \quad W_iL_i \quad \ldots \quad W_iL_i^{m_i-1}\right)^{\mathrm{T}} = \{0\}, \quad 1 \leqslant i \leqslant \nu, \quad \text{or} \\ &\operatorname{rank}\left(\hat{W}_i \quad \hat{W}_i\Lambda_i \quad \ldots \quad \hat{W}_i\Lambda^{m_i-1}\right)^{\mathrm{T}} = m_i, \quad 1 \leqslant i \leqslant \nu. \end{aligned} \tag{2.5}$$

We summerize the above result as the first stabilization result:

Theorem 2.1. *Suppose that the actuators g_k, $1 \leqslant k \leqslant N$, are constructed in the finite dimensional subspace $H_1 = PH$. Let $\mu_1 > \mu$ be an arbitrary number. If the weights w_k, $1 \leqslant k \leqslant N$, satisfy the observability condition* (2.5), *then we can find suitable g_k satisfying the decay estimate* (2.1).

Remark: If the parameter μ_1 in (2.4) is chosen such that $\mu_1 > \mu_* = \min_{\lambda\in\sigma_2} \operatorname{Re}\lambda$, a somewhat improved decay

$$\left\|e^{-t(L-GW)}\right\| \leqslant \text{const}\,(1+t^m)e^{-\mu_* t}, \quad t \geqslant 0$$

is obtained (see (1.6_1)), where $m \geqslant 0$ is an integer.

There is an algebraic counterpart of Theorem 2.1. In (2.2_2), we suppose that $W_2 = 0$, that is, $Q^*w_k = 0$, or $P^*w_k = w_k$, $1 \leqslant k \leqslant N$. Then, (2.2_2) is rewritten as

$$\begin{cases} \dfrac{du_1}{dt} + (L_1 - PGW_1)u_1 = 0, \\ \dfrac{du_2}{dt} + L_2u_2 = QGW_1u_1. \end{cases} \tag{2.2$_3$}$$

The equation of u_1 in this case is an autonomous one, and QGW_1u_1 is regarded as a perturbation to the equation of u_2. Note that $(L_1 - PGW_1)^* = {L_1}^* - {W_1}^*(PG)^*$, where ${W_1}^* \in \mathscr{L}(\mathbb{C}^N; P^*H)$; $(PG)^* \in \mathscr{L}(P^*H; \mathbb{C}^N)$; and

$$\begin{aligned} &{W_1}^*f = \sum_{i=1}^{N} f_k P^* w_k, \quad f = \left(f_1 \ \ldots \ f_N\right)^{\mathrm{T}} \in \mathbb{C}^N, \\ &(PG)^*u = \left(\langle u, Pg_1\rangle \quad \ldots \quad \langle u, Pg_N\rangle\right)^{\mathrm{T}} \quad \text{for } u \in P^*H. \end{aligned}$$

Thus P^*w_k and Pg_k turn out, respectively, to be actuators and observation weights. Note that $e^{-t(L_1-PGW_1)^*} = \left(e^{-t(L_1-PGW_1)}\right)^*$, so that the decay property of $e^{-t(L_1-PGW_1)^*}$ is unchanged. Let

$$\hat{G}_i = \left(\left\langle \psi_{ij}, g_k\right\rangle; \begin{array}{ccc} j & \rightarrow & 1\ldots, m_i \\ k & \downarrow & 1, \ldots, N \end{array}\right), \tag{2.6}$$

which corresponds to $\hat{W}_i$. The observability condition with weights Pg_k, $1 \leqslant k \leqslant N$, is then

$$\operatorname{rank}\left(\hat{G}_i\tilde{\Lambda}_i^k; k \downarrow 0, \ldots, m_i - 1\right) = m_i, \quad 1 \leqslant i \leqslant \nu. \tag{2.7}$$

Corollary 2.2. *Suppose that the observation weights w_k, $1 \leqslant k \leqslant N$, are constructed in the finite dimensional subspace P^*H. Let $\mu_1 > \mu$ be an arbitrary number. If the actuators g_k, $1 \leqslant k \leqslant N$, satisfy the condition* (2.7)*, then we can find suitable w_k satisfying the decay estimate* (2.1).

Let us show that the condition (2.7) is nothing but the controllability condition on the actuators g_k, $1 \leqslant k \leqslant N$. In view of the relationship (1.11), the matrix in (2.7) is rewritten as (by setting $\Pi_i = \Pi_{\lambda_i}$ for simplicity)

$$\begin{aligned}
&\overline{\left(\hat{G}_i \, \overline{\left(\Pi_i \Lambda_i^k \Pi_i^{-1}\right)^{\mathrm{T}}}; k \downarrow 0, \dots, m_i - 1\right)} \\
&= \left(\overline{\hat{G}_i} \left(\Pi_i \Lambda_i^k \Pi_i^{-1}\right)^{\mathrm{T}}; k \downarrow 0, \dots, m_i - 1\right) \\
&= \overline{\left(\left(\left(\Pi_i \Lambda_i^k \Pi_i^{-1}\right) \overline{\hat{G}_i}^{\mathrm{T}}\right)^{\mathrm{T}}; k \downarrow 0, \dots, m_i - 1\right)} \\
&= \overline{\left(\left(\Pi_i \Lambda_i^k \Pi_i^{-1}\right) \overline{\hat{G}_i}^{\mathrm{T}}; k \to 0, \dots, m_i - 1\right)}^{\mathrm{T}}.
\end{aligned}$$

Then we see that

$$\begin{aligned}
&\operatorname{rank} \overline{\left(\left(\Pi_i \Lambda_i^k \Pi_i^{-1}\right) \overline{\hat{G}_i}^{\mathrm{T}}; k \to 0, \dots, m_i - 1\right)}^{\mathrm{T}} \\
&= \operatorname{rank} \left(\left(\Pi_i \Lambda_i^k \Pi_i^{-1}\right) \hat{G}_i^*; k \to 0, \dots, m_i - 1\right) \\
&= \operatorname{rank} \Pi_i \left(\Lambda_i^k \Pi_i^{-1} \hat{G}_i^*; k \to 0, \dots, m_i - 1\right) \\
&= \operatorname{rank} \left(\Lambda_i^k \Pi_i^{-1} \hat{G}_i^*; k \to 0, \dots, m_i - 1\right).
\end{aligned}$$

Setting $P_{\lambda_i} g_k = \sum_{1 \leqslant j \leqslant m_i} g_{ij}^k \varphi_{ij}$ for each i, let $\tilde{G}_i$ be an $m_i \times m_i$ matrix defined as

$$\tilde{G}_i = \left(g_{ij}^k; \begin{array}{ccc} j & \downarrow & 1, \dots, m_i \\ k & \to & 1, \dots, N \end{array} \right). \tag{2.8}$$

Then, by the relation (1.10)

$$\tilde{G}_i = \Pi_i^{-1} \left(\langle g_k, \psi_{ij} \rangle; \begin{array}{ccc} j & \downarrow & 1, \dots, m_i \\ k & \to & 1, \dots, N \end{array} \right).$$

Thus,

$$\tilde{G}_i^* = \hat{G}_i \left(\Pi_i^{-1}\right)^*, \quad \text{or } \tilde{G}_i = \Pi_i^{-1} \hat{G}_i^*.$$

The condition (2.7) is finally rewritten as the controllability condition on the actuators g_k:

$$\begin{aligned}
&\operatorname{rank} \left(\Lambda_i^k \tilde{G}_i; k \to 0, \dots, m_i - 1\right) = m_i, \quad 1 \leqslant i \leqslant \nu, \\
&\qquad \text{where } P_{\lambda_i} g_k = \sum_{1 \leqslant j \leqslant m_i} g_{ij}^k \varphi_{ij}.
\end{aligned} \tag{2.9}$$

Example: Let Ω be a bounded domain in $\mathbb{R}^m$ with the boundary Γ which consists of a finite number of smooth components of $(m-1)$-dimension. Our control system has state $u(t,\cdot)$, and is described by the differential equation in $H = L^2(\Omega)$:

$$\begin{cases} \dfrac{\partial u}{\partial t} + \mathscr{L}u = \sum_{k=1}^{N} \langle u, w_k \rangle g_k & \text{in } \mathbb{R}^1_+ \times \Omega, \\ \tau u = 0 \quad \text{on } \mathbb{R}^1_+ \times \Gamma, \\ u(0,\cdot) = u_0(\cdot) \quad \text{in } \Omega. \end{cases} \tag{2.10}$$

Here, $(\mathscr{L}, \tau)$ denotes a pair of differential operators defined by

$$\begin{aligned} \mathscr{L}u &= -\sum_{i,j=1}^{m} \frac{\partial}{\partial x_i}\left(a_{ij}(x)\frac{\partial u}{\partial x_j}\right) + c(x)u, \\ \tau u &= \alpha(\xi)u + (1-\alpha(\xi))\frac{\partial u}{\partial \nu}, \end{aligned} \tag{2.11}$$

where $a_{ij}(x) = a_{ji}(x)$ for $1 \leqslant i, j \leqslant m$, $x \in \overline{\Omega}$; for some positive δ

$$\sum_{i,j=1}^{m} a_{ij}(x)\xi_i\xi_j \geqslant \delta|\xi|^2, \quad \forall \xi = (\xi_1, \ldots, \xi_m) \in \mathbb{R}^m, \quad \forall x \in \overline{\Omega};$$

and

$$0 \leqslant \alpha(\xi) \leqslant 1, \qquad \frac{\partial u}{\partial \nu} = \sum_{i,j=1}^{m} a_{ij}(\xi)\nu_i(\xi)\left.\frac{\partial u}{\partial x_j}\right|_{\Gamma},$$

$\boldsymbol{\nu}(\xi) = (\nu_1(\xi), \ldots, \nu_m(\xi))$ being the unit outer normal at each point $\xi \in \Gamma$. The last term, $\left.\frac{\partial u}{\partial x_j}\right|_{\Gamma}$ means the trace $\gamma\left(\frac{\partial u}{\partial x_j}\right)$ of $\frac{\partial u}{\partial x_j}$ on Γ. Necessary regularities of the coefficients a_{ij}, c, and α are tacitly assumed. Let $\hat{L}$ be the closable operator defined by

$$\begin{aligned} \hat{L}u &= \mathscr{L}u, \quad u \in \mathscr{D}(\hat{L}), \\ \mathscr{D}(\hat{L}) &= \left\{u \in C^2(\Omega) \cap C^1(\overline{\Omega});\ \mathscr{L}u \in L^2(\Omega),\ \tau u = 0\right\}. \end{aligned}$$

The existence of the closure of $\hat{L}$ in $H = L^2(\Omega)$ is ensured in Subsection 3.3, Chapter 2. The closure is denoted as L. The domain $\mathscr{D}(L)$ consists of all $u \in H$ with the following properties: (i) There is a sequence $\{u_n\} \subset \mathscr{D}(\hat{L})$ such that $u_n \to u$ in H, and (ii) $\hat{L}u_n$ converges in H as $n \to \infty$. It is well known that L is a self-adjoint operator, and has a compact resolvent $(\lambda - L)^{-1}$. Thus, there is a set of eigenpairs $\{\lambda_i, \varphi_{ij}\}$ such that

(i) $\sigma(L) = \{\lambda_1, \lambda_2, \ldots, \lambda_i, \ldots\}, \quad \lambda_1 < \lambda_2 < \cdots < \lambda_i < \cdots \to \infty$;

(ii) $(\lambda_i - L)\varphi_{ij} = 0, \quad i \geqslant 1,\ 1 \leqslant j \leqslant m_i\,(< \infty)$; and

(iii) the set $\{\varphi_{ij}\}$ forms an orthonormal basis for H.

There arises no generalized eigenspace in this case, and the projector P is self-adjoint. The feedback control system (2.10) is rewritten as (1.4). Let $\lambda_{\nu+1} > 0$. The observability condition (2.5) on w_k, $1 \leqslant k \leqslant N$, turns out to be

$$\operatorname{rank} \hat{W}_i = m_i, \quad 1 \leqslant i \leqslant \nu,$$

so that the number N must be greater than or equal to $\max_{1\leqslant i\leqslant \nu} m_i$ as a necessary condition. With this condition, there is a set of $g_k \in PH = PL^2(\Omega)$, $1 \leqslant k \leqslant N$ such that

$$\left\| e^{-t(L-GW)} \right\| \leqslant \text{const}\, e^{-\mu_* t}, \quad t \geqslant 0, \quad \text{where } \mu_* = \lambda_{\nu+1}.$$

3.3 Remark on the Choice of the Decay Rate

There is a variety of choice of the parameter μ_1 $(> \mu)$ satisfying the decay estimate (2.4): $\left\| e^{-t(L_1-PGW_1)} \right\| \leqslant \text{const}\, e^{-\mu_1 t}$, $t \geqslant 0$, which leads to (2.1). A rough choice of μ_1 would not give us a satisfactory result. We should pay attention to the choice of μ_1. To see this, let us consider again the feedback control system (2.2_1) in the case where $QG = 0$, or $Pg_k = g_k$, $1 \leqslant k \leqslant N$. For simplicity, we assume that the spectrum $\sigma(L)$ consists of *simple* eigenvalues, so that a single observation $W = \langle \cdot, w \rangle$ is enough. The control system is then rewritten as

$$\frac{du_1}{dt} + (L_1 - \langle \cdot, w \rangle g)u_1 = \langle u_2, w \rangle g, \qquad \frac{du_2}{dt} + L_2 u_2 = 0. \tag{3.1}$$

Let $\sigma_1 = \{\lambda_i\}_{1\leqslant i\leqslant n}$. The observability condition (2.5) is then $w_i = \langle \varphi_i, w \rangle \neq 0$, $1 \leqslant i \leqslant n$. Given a set S of n complex numbers, the condition (2.5) ensures a unique $g = Pg \in H_1$ such that (see Proposition 4.1, Chapter 1)

$$\sigma\big(L_1 - \langle \cdot, P^* w \rangle g\big) = S.$$

More precisely, let $S = \{\mu_i\}_{1\leqslant i\leqslant l}$, where $\mu_i \neq \mu_j$ for $i \neq j$, and each μ_i has multiplicity $m_i \geqslant 1$ with $\sum_{1\leqslant i\leqslant l} m_i = n$. The operator $L_1 - \langle \cdot, P^* w \rangle g$ is, according to the basis $\{\varphi_i\}_{1\leqslant i\leqslant n}$ of H_1, regarded as an $n \times n$ matrix $\Lambda_f = \Lambda - \hat{g}\hat{w}^{\mathrm{T}}$, where $\hat{w} = \begin{pmatrix} w_1 & \dots & w_n \end{pmatrix}^{\mathrm{T}}$, $\hat{g} = \begin{pmatrix} g_1 & \dots & g_n \end{pmatrix}^{\mathrm{T}}$, and $g = \sum_{1\leqslant i\leqslant n} g_i \varphi_i$. As long as $\sigma(L_2) \cap \sigma(\Lambda_f) = \varnothing$, the spectral structure of (3.1) is simple. In fact, the coefficient operator in (3.1) is equivalent to

$$\begin{pmatrix} \Lambda_f & D \\ 0 & L_2 \end{pmatrix}, \quad \text{where } D = -\langle \cdot, Q^* w \rangle \hat{g}, \tag{3.2}$$

Consider Sylvester's equation on $\mathscr{D}(L_2)$

$$\Lambda_f X - X L_2 = D. \tag{3.3}$$

There is a unique solution $X \in \mathscr{L}(H_2;\, \mathbb{C}^n)$, which is expressed as

$$X = \frac{1}{2\pi \mathrm{i}} \int_C (\lambda - \Lambda_f)^{-1} D (\lambda - L_2)^{-1}\, d\lambda,$$

where C denotes a Jordan contour encircling $\sigma(\Lambda_f)$ in its onside with $\sigma(L_2)$ outside C. The operator $T = \begin{pmatrix} 1 & X \\ 0 & 1 \end{pmatrix} \in \mathscr{L}(\mathbb{C}^n \times H_2)$ is boundedly invertible. Then we immediately find that

$$T \begin{pmatrix} \Lambda_f & D \\ 0 & L_2 \end{pmatrix} T^{-1} = \begin{pmatrix} \Lambda_f & 0 \\ 0 & L_2 \end{pmatrix}. \tag{3.4}$$

The spectrum of the control system (3.1) is thus $\sigma(\Lambda_f) \oplus \sigma(L_2)$. The idea is found in [6] (all operators in [6] are, however, limited to bounded ones).

Now what would occur in the case where $\Sigma = \sigma(L_2) \cap \sigma(\Lambda_f) \neq \varnothing$? Let

$$\mu_* = \min_{\mu \in S} \operatorname{Re} \mu, \quad \text{and} \quad \lambda_* = \min_{\lambda \in \sigma_2} \operatorname{Re} \lambda.$$

The parameter λ_* is the one which we cannot manage. If μ_* is chosen such that $\mu_* > \lambda_*$, then the decay of solutions to (3.1) is, of course, $e^{-\lambda_* t}$. For a $\mu \in S = \sigma(\Lambda_f)$, let $W_\mu^{(k)}$, $k = 1, 2, \ldots$, be the generalized eigenspace of Λ_f, i.e., $W_\mu^{(k)} = \{p \in \mathbb{C}^n;\ (\mu - \Lambda_f)^k p = 0\}$. The following result owes to Proposition 4.2, Chapter 1 a lot, and suggests that the decay becomes worse if $\mu_* = \lambda_*$.

Theorem 3.1. (i) *Suppose that $\Sigma = \sigma(L_2) \cap S \neq \varnothing$. Then the set Σ is contained in the spectrum of the operator $L - \langle \cdot, w \rangle g$.*

(ii) *Let $\mu = \lambda_i \in \Sigma$, $i \geqslant n+1$, be an eigenvalue of $L - \langle \cdot, w \rangle g$ such that $\langle \varphi_i, w \rangle \neq 0$. Then μ admits a generalized eigenspace. More precisely, let $m \geqslant 1$ be the smallest integer such that $W_\mu^{(m)} = W_\mu^{(m+1)}$. Setting $\tilde{W}_\mu^{(k)} = \{u;\ (\mu - L + \langle \cdot, w \rangle g)^k x = 0\}$, $k \geqslant 1$, we then have*

$$\dim \tilde{W}_\mu^{(k)} = \begin{cases} k, & 1 \leqslant k \leqslant m+1, \\ m+1, & m+1 < k. \end{cases} \tag{3.5}$$

Thus the ascent of $\mu - L + \langle \cdot, w \rangle g$ is equal to $m+1$.

Remark: Let $\lambda_* = \operatorname{Re} \lambda_{n+1}$ and a $\mu \in \Sigma$ with $\operatorname{Re} \mu = \mu_*$ is chosen so that it coincides with λ_{n+1}, then the decay of (3.1) is dominated by $t^m e^{-\lambda_* t}$ through (3.5). Theorem 3.1 thus suggests a possibility that some choice of the set $S = \sigma(\Lambda_f)$ would cause a worse decay than expected.

Proof. (i) The operator equation (3.3) generally does not admit a solution.

Let P_Σ be the projector corresponding to the set Σ. Instead, we consider another operator equation,

$$\Lambda_f X - XL_2 = D(1-P_\Sigma). \tag{3.6}$$

Let X be an operator defined by

$$X = \frac{1}{2\pi \mathrm{i}} \int_C (\lambda - \Lambda_f)^{-1} D(\lambda - L_2)^{-1} d\lambda,$$

where C is a Jordan contour such that $\sigma(\Lambda_f)$ is inside C with $\sigma(L_2)\setminus\Sigma$ outside C. It is easily seen that the X uniquely solves (3.6) on $\mathscr{D}(L_2)\cap(1-P_\Sigma)H$. Setting then $T = \begin{pmatrix} 1 & X \\ 0 & 1 \end{pmatrix}$, we easily find that

$$T\begin{pmatrix} \Lambda_f & D \\ 0 & L_2 \end{pmatrix} T^{-1} = \begin{pmatrix} \Lambda_f & DP_\Sigma \\ 0 & L_2 \end{pmatrix} = L_f. \tag{3.7}$$

The operator L_f is algebraically similar to $L - \langle \cdot, w \rangle g$. With no loss of generality, we may assume that an element μ of Σ is μ_1 $(=\lambda_i)$, and $m = m_1$. By Proposition 4.2 of Chapter 2, the vector $p_{11} = (\Lambda - \mu_1)^{-1}\hat{g}$ is a basis for $W_{\mu_1}^{(1)}$. It is then clear that $(\mu_1 - L_f)\begin{pmatrix} p_{11} \\ 0 \end{pmatrix} = \begin{pmatrix} 0 \\ 0 \end{pmatrix}$, and thus $\mu = \mu_1$ belongs to $\sigma(L_f) = \sigma(L - \langle \cdot, w \rangle g)$.

(ii) Let $(u\ \ v)^{\mathrm{T}}$ be an eigenvector in $\tilde{W}_{\mu_1}^{(1)}$:

$$(\mu_1 - L_f)\begin{pmatrix} u \\ v \end{pmatrix} = \begin{pmatrix} \mu_1 - \Lambda_f & -DP_\Sigma \\ 0 & \lambda_i - L_2 \end{pmatrix}\begin{pmatrix} u \\ v \end{pmatrix} = \begin{pmatrix} 0 \\ 0 \end{pmatrix}.$$

Since λ_i $(=\mu_1)$ is a simple eigenvalue of L, we see that $v = c\varphi_i$, c being a constant. Thus, $(\mu_1 - \Lambda_f)u + cw_i\hat{g} = 0$, where $w_i = \langle \varphi_i, w \rangle \neq 0$. Since $\Lambda_f = \Lambda - \hat{g}\,\hat{w}^{\mathrm{T}}$, we calculate as

$$(\mu_1 - \Lambda)u + \hat{g}\,\hat{w}^{\mathrm{T}}u + cw_i\hat{g} = 0, \quad \text{or}$$
$$u = (\hat{w}^{\mathrm{T}}u + cw_i)(\Lambda - \mu_1)^{-1}\hat{g} \in W_{\mu_1}^{(1)}.$$

But, this means that $cw_i\hat{g} = (\mu_1 - \Lambda_f)u = 0$. Since $w_i \neq 0$, we see that $c = 0$, or

$$\tilde{W}_{\mu_1}^{(1)} = \left\{ (u\ \ 0)^{\mathrm{T}};\ u \in W_{\mu_1}^{(1)} \right\}, \quad \dim \tilde{W}_{\mu_1}^{(1)} = 1.$$

Supposing inductively that

$$\tilde{W}_{\mu_1}^{(k)} = \left\{ (u\ \ 0)^{\mathrm{T}};\ u \in W_{\mu_1}^{(k)} \right\}, \quad \dim \tilde{W}_{\mu_1}^{(k)} = k, \quad 1 \leqslant k \leqslant m_1 - 1,$$

let us characterize $\tilde{W}_{\mu_1}^{(m_1)}$. For a $(u\ \ v)^{\mathrm{T}} \in \tilde{W}_{\mu_1}^{(m_1)}$, we have

$$\begin{cases} \displaystyle\sum_{k=0}^{m_1-1} \left\langle P_\Sigma(\mu_1 - L_2)^{m_1-1-k}v, w \right\rangle (\mu_1 - \Lambda_f)^k \hat{g} + (\mu_1 - \Lambda_f)^{m_1} u = 0, \\ (\mu_1 - L_2)^{m_1} v = 0. \end{cases} \tag{3.8$_1$}$$

Thus, $v = c\varphi_i$, and by the first relation,

$$(\mu_1 - \Lambda_f)^{m_1-1}\big(cw_i\hat{g} + (\mu_1 - \Lambda_f)u\big) = 0, \quad \text{or}$$
$$cw_i\hat{g} + (\mu_1 - \Lambda_f)u \in W_{\mu_1}^{(m_1-1)}.$$

Recalling that $\{p_{1j}\}_{1\leqslant j\leqslant m_1-1}$ forms a basis for $W_{\mu_1}^{(m_1-1)}$ (see Proposition 4.2 of Chapter 2), we calculate as

$$(\mu_1 - \Lambda)u + \hat{g}\hat{w}^{\mathrm{T}}u + cw_i\hat{g} = \sum_{1\leqslant j\leqslant m_1-1} c_{1j}p_{1j},$$
$$u = (\Lambda - \mu_1)^{-1}\left(\big(\hat{w}^{\mathrm{T}}u + cw_i\big)\hat{g} - \sum_{1\leqslant j\leqslant m_1-1} c_{1j}p_{1j}\right)$$
$$= \big(\hat{w}^{\mathrm{T}}u + cw_i\big)p_{11} - \sum_{1\leqslant j\leqslant m_1-1} c_{1j}p_{1(j+1)} \in W_{\mu_1}^{(m_1)}.$$

Thus, the vector $cw_i\hat{g} = \sum_{1\leqslant j\leqslant m_1-1} c_{1j}p_{1j} - (\mu_1 - \Lambda_f)u$ belongs to $W_{\mu_1}^{(m_1-1)}$. We show that $c = 0$. If not, $\hat{g} = \big(g_1\ g_2\ \ldots\ g_n\big)^{\mathrm{T}}$ belongs to $W_{\mu_1}^{(m_1-1)}$, and must be expressed as

$$\hat{g} = \sum_{1\leqslant j\leqslant m_1-1} \gamma_j p_{1j} = \sum_{1\leqslant j\leqslant m_1-1} \gamma_j(\Lambda - \mu_1)^{-j}\hat{g}.$$

Recall that $g_i \neq 0$, $1 \leqslant i \leqslant n$: This follows from the fact that $\sigma(\Lambda)\cap S = \varnothing$. Thus the above relation means that

$$\begin{pmatrix} \dfrac{1}{\lambda_1-\mu_1} & \dfrac{1}{(\lambda_1-\mu_1)^2} & \cdots & \dfrac{1}{(\lambda_1-\mu_1)^{m_1-1}} \\ \dfrac{1}{\lambda_2-\mu_1} & \dfrac{1}{(\lambda_2-\mu_1)^2} & \cdots & \dfrac{1}{(\lambda_2-\mu_1)^{m_1-1}} \\ \vdots & \vdots & \cdots & \vdots \\ \dfrac{1}{\lambda_n-\mu_1} & \dfrac{1}{(\lambda_n-\mu_1)^2} & \cdots & \dfrac{1}{(\lambda_n-\mu_1)^{m_1-1}} \end{pmatrix} \begin{pmatrix} \gamma_1 \\ \gamma_2 \\ \vdots \\ \gamma_{m_1-1} \end{pmatrix} = \begin{pmatrix} 1 \\ 1 \\ \vdots \\ 1 \end{pmatrix}.$$

Just as in the relation (4.11), Chapter 1, this equation forms an overdetermined system, and has *no* $\{\gamma_j\}_{1\leqslant j\leqslant m_1-1}$ satisfying the above relation. This is a contradiction. We have shown that $c = 0$, and thus,

$$\tilde{W}_{\mu_1}^{(m_1)} = \left\{(u\ \ 0)^{\mathrm{T}};\ u \in W_{\mu_1}^{(m_1)}\right\}, \quad \dim \tilde{W}_{\mu_1}^{(m_1)} = m_1.$$

Next let $(u\ \ v)^{\mathrm{T}}$ be a vector in $\tilde{W}_{\mu_1}^{(m_1+1)}$. Then, we similarly have

$$\begin{cases} \displaystyle\sum_{k=0}^{m_1} \left\langle P_\Sigma(\mu_1 - L_2)^{m_1-k}v, w\right\rangle (\mu_1 - \Lambda_f)^k\hat{g} + (\mu_1 - \Lambda_f)^{m_1+1}u = 0, \\ (\mu_1 - L_2)^{m_1+1}v = 0. \end{cases} \tag{3.8$_2$}$$

Thus, $v = c\varphi_i$, and by the first relation, $cw_i\hat{g} + (\mu_1 - \Lambda_f)u$ belongs to $W_{\mu_1}^{(m_1)}$, and is equal to $\sum_{1\leqslant k\leqslant m_1} c_{1j}p_{1k} = \boldsymbol{p}_1\boldsymbol{c}_1$, where $\boldsymbol{p}_1 = \begin{pmatrix} p_{11} & \dots & p_{1m_1}\end{pmatrix}$, and $\boldsymbol{c}_1 = \begin{pmatrix} c_{11} & \dots & c_{1m_1}\end{pmatrix}^{\mathrm{T}} \in \mathbb{C}^{m_1}$. Since the system $\{p_{jk}\}_{\substack{1\leqslant j\leqslant l,\\ 1\leqslant k\leqslant m_j}} = \{\boldsymbol{p}_j\}_{1\leqslant j\leqslant l}$ forms a basis for $\mathbb{C}^n$, we write as

$$u = \sum_{j,k} u_{jk}p_{jk} = \sum_i \boldsymbol{p}_j\boldsymbol{u}_j, \quad \hat{g} = \sum_{j,k} g_{jk}p_{jk} = \sum_j \boldsymbol{p}_j\boldsymbol{g}_j,$$

where, of course, $\boldsymbol{u}_j = \begin{pmatrix} u_{j1} & \dots & u_{jm_j}\end{pmatrix}^{\mathrm{T}}$ and $\boldsymbol{g}_j = \begin{pmatrix} g_{j1} & \dots & g_{jm_j}\end{pmatrix}^{\mathrm{T}}$. Then, according to Proposition 4.2, Chapter 1, we calculate as

$$cw_i \sum_{1\leqslant j\leqslant l} \boldsymbol{p}_j\boldsymbol{g}_j + \sum_{1\leqslant j\leqslant l} \boldsymbol{p}_j(\mu_1 - M_j)\boldsymbol{u}_j = \boldsymbol{p}_1\boldsymbol{c}_1,$$

$$\boldsymbol{p}_1(cw_i\boldsymbol{g}_1 - N_1\boldsymbol{u}_1 - \boldsymbol{c}_1) + \sum_{2\leqslant j\leqslant l} \boldsymbol{p}_j\Big(cw_i\boldsymbol{g}_j + (\mu_1 - M_j)\boldsymbol{u}_j\Big) = \boldsymbol{0},$$

which means that $cw_i\boldsymbol{g}_1 - N_1\boldsymbol{u}_1 - \boldsymbol{c}_1 = \boldsymbol{0}$, and $cw_i\boldsymbol{g}_j + (\mu_1 - M_j)\boldsymbol{u}_j = \boldsymbol{0}$, $2 \leqslant j \leqslant l$. Thus we have, for arbitrary parameters c and $\boldsymbol{u}_1 \in \mathbb{C}^{m_1}$,

$$\begin{pmatrix} u \\ v \end{pmatrix} = \begin{pmatrix} \boldsymbol{p}_1\boldsymbol{u}_1 - cw_i\sum_{2\leqslant j\leqslant l}\boldsymbol{p}_j(\mu_1 - M_j)^{-1}\boldsymbol{g}_j \\ c\varphi_i \end{pmatrix} = c\begin{pmatrix} u^* \\ \varphi_i \end{pmatrix} + \begin{pmatrix} \boldsymbol{p}_1\boldsymbol{u}_1 \\ 0 \end{pmatrix},$$

where $u^* = -w_i\sum_{2\leqslant j\leqslant l}\boldsymbol{p}_j(\mu_1 - M_j)^{-1}\boldsymbol{g}_j$. The second term $(\boldsymbol{p}_1\boldsymbol{u}_1 \;\; 0)^{\mathrm{T}}$ belongs to $\tilde{W}_{\mu_1}^{(m_1)} \subset \tilde{W}_{\mu_1}^{(m_1+1)}$. It is easily examined that the first term $(u^* \;\; \varphi_i)^{\mathrm{T}}$ actually belongs to $\tilde{W}_{\mu_1}^{(m_1+1)}$. We have shown that

$$\tilde{W}_{\mu_1}^{(m_1+1)} = \left\{ c\begin{pmatrix} u^* \\ \varphi_i \end{pmatrix} + \begin{pmatrix} u \\ 0 \end{pmatrix};\, c \in \mathbb{C},\ u \in W_{\mu_1}^{(m_1)} \right\}, \quad \dim \tilde{W}_{\mu_1}^{(m_1+1)} = m_1 + 1.$$

Finally, let us examine $\tilde{W}_{\mu_1}^{(m_1+2)}$. For a vector $(u \;\; v)^{\mathrm{T}}$ in $\tilde{W}_{\mu_1}^{(m_1+2)}$, we similarly obtain

$$\begin{cases} \displaystyle\sum_{k=0}^{m_1+1} \Big\langle P_\Sigma(\mu_1 - L_2)^{m_1+1-k}v, w\Big\rangle (\mu_1 - \Lambda_f)^k\hat{g} + (\mu_1 - \Lambda_f)^{m_1+2}u = 0, \\ (\mu_1 - L_2)^{m_1+2}v = 0. \end{cases} \tag{3.8$_3$}$$

Thus, $v = c\varphi_i$, and by the first relation, $(\mu_1 - \Lambda_f)^{m_1+1}(cw_i\hat{g} + (\mu_1 - \Lambda_f)u) = 0$, which implies that $(\mu_1 - \Lambda_f)^{m_1}(cw_i\hat{g} + (\mu_1 - \Lambda_f)u) = 0$, since $W_{\mu_1}^{(m_1)} = W_{\mu_1}^{(m_1+1)}$. Thus we immediately find that $(u \;\; v)^{\mathrm{T}}$ belongs to $\tilde{W}_{\mu_1}^{(m_1+1)}$. The proof of Theorem 3.1 is thereby complete. □

3.4 Stability Enhancement

In the preceding sections, we have assumed the control scheme that $Pg_k = g_k$ in Theorem 2.1, or $P^* w_k = w_k$ in Corollary 2.2 to achieve stabilization of the control system (1.4). While the scheme plays an essential role as an *pseudo*-substructure in stabilization problems with boundary observation/boundary feedback schemes, these assumptions are not plausible in practice: For example, the asumption: $Pg_k = g_k$ means that the actuators must be constructed as linear combinations of a finite number of generalizaed eigenvectors of the operator L. However, construction of g_k are usually accompanied by *spillovers* Qg_k, which are serious and non negligible factors affecting the stability property. We study in this section stability improvement or stabilization of the control system (1.4) in the essential presence of spillovers both of w_k and g_k. More precisely, when the actuators g_k satisfy the finite-dimensional controllability conditions, we construct the sensors w_k such that the stability property is improved and enhanced to some extent. In our study both finite- and infinite- dimensional structures are important factors. Especially the evolution of the semigroup in the finite-dimensional substructure plays a central role.

Our control system in this section is somewhat different from (1.4). Let $\gamma > 0$ be a small gain parameter which will be prescribed later. Setting $f = -\gamma W u$ in (1.1), we obtain a feedback control system:

$$\frac{du}{dt} + Lu = -\gamma \sum_{k=1}^{N} \langle u, w_k \rangle g_k, \quad t > 0, \quad u(0) = u_0. \tag{4.1}$$

The precise assumptions on the spectrum is that $\sigma(L)$ consists of two disjoint closed sets σ_1 and σ_2: $\sigma(L) = \sigma_1 \cup \sigma_2$, and $\sigma_1 \cap \sigma_2 = \varnothing$. Here,

(i) σ_1 consists of the eigenvalues λ_i of L, $1 \leqslant i \leqslant \nu$, on the vertical line: $\mathrm{Re}\,\lambda = \omega$;

(ii) for each λ_i, there is a set of eigenvectors φ_{ij} of L, $1 \leqslant j \leqslant m_i\,(< \infty)$, which forms a basis for the subspace $P_{\lambda_i} H$, where P_{λ_i} denotes the projector $\frac{1}{2\pi i}\int_{C_i} (\lambda - L)^{-1}\, d\lambda$, C_i being a small contour encircling λ_i; and

(iii) $\min_{\lambda \in \sigma_2} \mathrm{Re}\,\lambda > \omega$.

When there is no control action, the semigroup of the unperturbed equation satisfies the estimate

$$\left\| e^{-tL} \right\| \leqslant \mathrm{const}\, e^{-\omega t}, \quad t \geqslant 0. \tag{4.2}$$

Henceforth c with or without subscript will denote a various positive constant. We show that the power ω is improved a little for the perturbed equation (4.1) in the essential presence of the spillovers of w_k and g_k.

Some readers might be afraid that the control system (1.4) would not reflect the boundary observation/boundary feedback scheme. Let us show that a class of problems in this scheme is reduced to (4.1) with slight technical modifications. Let Ω be a bounded domain in $\mathbb{R}^m$ with the boundary Γ which consists of a finite number of smooth components of $(m-1)$-dimension. Our boundary control system with state $u(t,\cdot)$ is described by the differential equation:

$$\begin{cases} \dfrac{\partial u}{\partial t} + \mathscr{L}u = 0 \quad \text{in } \mathbb{R}^1_+ \times \Omega, \\ \tau u = -\gamma \displaystyle\sum_{k=1}^{N} \langle u, w_k \rangle_\Gamma g_k \quad \text{on } \mathbb{R}^1_+ \times \Gamma, \\ u(0,\cdot) = u_0(\cdot) \quad \text{in } \Omega. \end{cases} \tag{4.3}$$

Here, $(\mathscr{L}, \tau)$ denotes a pair of differential operators defined by

$$\mathscr{L}u = -\sum_{i,j=1}^{m} \frac{\partial}{\partial x_i}\left(a_{ij}(x) \frac{\partial u}{\partial x_j} \right) + \sum_{i=1}^{m} b_i(x) \frac{\partial u}{\partial x_i} + c(x)u,$$
$$\tau u = \frac{\partial u}{\partial \nu} + \sigma(\xi) u, \tag{4.4}$$

and $a_{ij}(x) = a_{ji}(x)$ for $1 \leqslant i,\, j \leqslant m,\ x \in \overline{\Omega}$; for some positive δ

$$\sum_{i,j=1}^{m} a_{ij}(x) \xi_i \xi_j \geqslant \delta |\xi|^2, \quad \forall \xi = (\xi_1, \ldots, \xi_m) \in \mathbb{R}^m, \quad \forall x \in \overline{\Omega}$$

and $\partial u/\partial \nu = \sum_{i,j=1}^{m} a_{ij}(\xi) \nu_i(\xi)\, \partial u/\partial x_j \big|_\Gamma$, where $(\nu_1(\xi), \ldots, \nu_m(\xi))$ denotes the unit outer normal at $\xi \in \Gamma$. Necessary conditions on the coefficients a_{ij}, b_i, c, and σ are tacitly assumed. Set $H = L^2(\Omega)$. The inner products in $L^2(\Omega)$ and in $L^2(\Gamma)$ are denoted as $\langle \cdot, \cdot \rangle$ and $\langle \cdot, \cdot \rangle_\Gamma$ respectively. In standard manner, set $Lu = \mathscr{L}u$ for $u \in \mathscr{D}(L)$, where $\mathscr{D}(L) = \{ u \in H^2(\Omega);\ \tau u = 0 \text{ on } \Gamma \}$. Choose a constant $c > 0$ large enough so that $-c \in \mathscr{D}(L)$. It is well known that (see (3.32), Chapter 2)

$$\mathscr{D}(L_c^\theta) = H^{2\theta}(\Omega), \qquad 0 \leqslant \theta < \frac{3}{4}, \qquad L_c = L + c.$$

Set

$$x(t) = L_c^{-1/4-\varepsilon} u(t), \quad \text{where } 0 < \varepsilon < \frac{1}{4}.$$

Then $x(t)$, $t > 0$, belongs to $\mathscr{D}(L)$, and satisfies the differential equation in $H = L^2(\Omega)$:

$$\frac{dx}{dt} + Lx = -\gamma \sum_{k=1}^{N} \left\langle L_c^{1/4+\varepsilon} x, w_k \right\rangle_\Gamma L_c^{3/4-\varepsilon} h_k, \quad x(0) = x_0 = L_c^{-1/4-\varepsilon} u_0, \tag{4.5}$$

where $h_k \in H^2(\Omega)$, $1 \leqslant k \leqslant N$, denote unique solutions to the boundary value problems:

$$(c+\mathscr{L})h_k = 0 \quad \text{in } \Omega, \qquad \tau h_k = g_k \quad \text{on } \Gamma.$$

Let L_f be an operator defined as

$$L_f = L + \gamma \sum_{k=1}^{N} \left\langle L_c^{1/4+\varepsilon} \cdot, w_k \right\rangle_\Gamma L_c^{3/4-\varepsilon} h_k, \quad \mathscr{D}(L_f) = \mathscr{D}(L).$$

By choosing a larger constant $c > 0$ if necessary, both L_c and $L_f + c$ are *m*-accretive. Thus we see that (see [25])

$$\mathscr{D}\big((L_f+c)^\omega\big) = \mathscr{D}\big(L_c^\omega\big), \quad 0 \leqslant \omega \leqslant 1.$$

Solutions $u(t,\cdot)$ to (4.3) are then expressed as

$$u(t,\cdot) = L_c^{\alpha/2}(L_f+c)^{-\alpha/2} \cdot e^{-tL_f} \cdot (L_f+c)^{\alpha/2} L_c^{-\alpha/2} u_0, \quad t \geqslant 0, \tag{4.6}$$

where $\alpha = \frac{1}{2} + 2\varepsilon$, and both $L_c^{\alpha/2}(L_f+c)^{-\alpha/2}$ and $(L_f+c)^{\alpha/2}L_c^{-\alpha/2}$ are bounded. Thus the problem is reduced to the problem of the analytic semigroup e^{-tL_f}. In the operator L_f, unboundedness arising from the boundary observation is merely of technical nature.

Let us go back to the control system (4.1). Let P denote the projector associated with the eigenvalues $\lambda_1, \dots, \lambda_\nu$: $P = \sum_{i=1}^{\nu} P_{\lambda_i}$, and set $L_1 = L|_{PH}$ and $L_2 = L|_{QH}$ with $\mathscr{D}(L_2) = \mathscr{D}(L) \cap QH$, where $Q = 1 - P$. We mainly consider (4.1), and then extend the result—via technical modifications—to the case of the boundary control system (4.3). By setting $u_1 = Pu$ and $u_2 = Qu$, (4.1) is decomposed into a system of two differential equations:

$$\begin{cases} \dfrac{du_1}{dt} + L_1 u_1 = -\gamma \displaystyle\sum_{k=1}^{N} \langle u_1, P^* w_k \rangle P g_k - \gamma \sum_{k=1}^{N} \langle u_2, Q^* w_k \rangle P g_k, \\ \dfrac{du_2}{dt} + L_2 u_2 = -\gamma \displaystyle\sum_{k=1}^{N} \langle u_1, P^* w_k \rangle Q g_k - \gamma \sum_{k=1}^{N} \langle u_2, Q^* w_k \rangle Q g_k. \end{cases} \tag{4.7$_1$}$$

We rewrite (4.7$_1$) in appropriate form. Writing $u_1 = \sum_{i,j} u_{ij} \varphi_{ij}$ and $P g_k = \sum_{i,j} g_{ij}^k \varphi_{ij}$, $1 \leqslant k \leqslant N$, according to the basis $\{\varphi_{ij}\}$ for PH, let

$$\begin{aligned} \hat{u} &= \Big(u_{ij};\ (i,j) \downarrow (1,1), \dots, (\nu, m_\nu)\Big), \\ \hat{g}_k &= \Big(g_{ij}^k;\ (i,j) \downarrow (1,1), \dots, (\nu, m_\nu)\Big), \quad \text{and} \\ \Lambda &= \operatorname{diag}\big(\Lambda_1\ \Lambda_2\ \dots\ \Lambda_\nu\big), \quad \Lambda_i = \operatorname{diag}\big(\underbrace{\lambda_i\ \lambda_i\ \dots\ \lambda_i}_{m_i\ \lambda_i}\big). \end{aligned}$$

In the second equation let L_{f2} be the operator in QH defined by

$$L_{f2} = L_2 + \gamma \sum_{k=1}^{N} \langle \cdot, Q^* w_k \rangle Q g_k, \quad \mathscr{D}(L_{f2}) = \mathscr{D}(L_2). \tag{4.8}$$

Then (4.7_1) is rewritten as a system of differential equations in $\mathbb{C}^n \times QH$ $(n = m_1 + \cdots + m_\nu)$:

$$\begin{cases} \dfrac{d\hat{u}}{dt} + (\Lambda + \gamma \hat{G}\hat{W})\hat{u} = -\gamma \displaystyle\sum_{k=1}^{N} \langle u_2, Q^* w_k \rangle \hat{g}_k, & \hat{u}(0) = \hat{u}_0, \\ \dfrac{du_2}{dt} + L_{f2} u_2 = -\gamma (Q g_1 \dots Q g_N) \hat{W} \hat{u}, & u_2(0) = Q u_0, \end{cases} \tag{4.7$_2$}$$

where $\hat{G}$ and $\hat{W}$ denote $n \times N$ and $N \times n$ matrices defined as

$$\hat{G} = \left(\hat{g}_1\, \hat{g}_2\, \dots\, \hat{g}_N\right) = \left(\langle \varphi_{ij}, w_k \rangle ; \begin{array}{ccc} (i,j) & \downarrow & (1,1), \dots, (\nu, m_\nu) \\ k & \rightarrow & 1, \dots, N \end{array}\right),$$

$$\text{and} \quad \hat{W} = \left(\langle \varphi_{ij}, w_k \rangle ; \begin{array}{ccc} (i,j) & \rightarrow & (1,1), \dots, (\nu, m_\nu) \\ k & \downarrow & 1, \dots, N \end{array}\right),$$

respectively. By setting

$$\hat{G}_i = \left(g_{ij}^k ; \begin{array}{ccc} j & \downarrow & 1, \dots, m_i \\ k & \rightarrow & 1, \dots, N \end{array}\right), \quad \text{and}$$

$$\hat{W}_i = \left(\langle \varphi_{ij}, w_k \rangle ; \begin{array}{ccc} j & \rightarrow & 1, \dots, m_i \\ k & \downarrow & 1, \dots, N \end{array}\right),$$

we have the expression:

$$\hat{G} = \left(\hat{G}_i;\, i \downarrow 1, \dots, \nu\right), \quad \text{and } \hat{W} = \left(\hat{W}_i;\, i \rightarrow 1, \dots, \nu\right).$$

Changing the order of λ_i if necessary, we may assume with no loss of generality that

$$m_1 \geqslant m_2 \geqslant \cdots \geqslant m_\nu. \tag{4.9}$$

Our first result is stated as follows:

Theorem 4.1. *Set $N = m_1$. In (4.7_1) or (4.7_2), choose g_k and w_k such that*

$$\begin{aligned} &\hat{G}_i = \left(G_{i1}\ \ 0\right), \quad G_{i1};\ m_i \times m_i, \\ &\operatorname{rank} \hat{G}_i = m_i, \quad \text{and } \hat{W}_i = \begin{pmatrix} G_{i1}^{-1} \\ 0 \end{pmatrix}, \quad 1 \leqslant i \leqslant \nu. \end{aligned} \tag{4.10}$$

Then, as long as $\gamma > 0$ is small enough, there exist a constant $c > 0$ which is independent of γ and an $O(\gamma^2)$, such that

$$\left\| \exp\left(-t\left(L + \gamma \textstyle\sum_{1 \leqslant k \leqslant N} \langle \cdot, w_k \rangle g_k\right)\right) \right\| \leqslant c e^{-(\omega + \gamma + O(\gamma^2))t}, \quad t \geqslant 0. \tag{4.11}$$

Remark: In (4.10), the rank conditions on g_k are the controllability conditions, since each eigenvalue λ_i admits *no* generalized eigenspace. The essential difference between Theorem 4.1 and those in the preceding sections lies in the construction of g_k and w_k: The only requisite in our assertion is that w_k satisfy the finite-dimensional conditions: $\hat{W}_i = \begin{pmatrix} G_{i1}^{-1} \\ 0 \end{pmatrix}$. The resultant spillovers Qg_k and $Q^* w_k$ are the quantities which we *cannot* manipulate in general.

Proof. Eqn. (4.7$_2$) is rewritten as a system of integral equations which is described by

$$\begin{cases} \hat{u}(t) = e^{-t(\Lambda+\gamma\hat{G}\hat{W})}\hat{u}(0) - \gamma \int_0^t e^{-(t-s)(\Lambda+\gamma\hat{G}\hat{W})} \sum_{k=1}^{N} \langle u_2(s), Q^* w_k \rangle \, \hat{g}_k \, ds, \\ u_2(t) = e^{-tL_{f2}} u_2(0) - \gamma \int_0^t e^{-(t-s)L_{f2}} (Qg_1 \dots Qg_N) \hat{W} \hat{u}(s) \, ds. \end{cases} \tag{4.12}$$

Combining these equations, we will derive an integral inequality for $|\hat{u}(t)|$.

Choose an arbitrary β such that $\omega < \beta < \min_{\lambda \in \sigma_2} \operatorname{Re} \lambda$. Then note that

$$\left\| e^{-tL_2} \right\| \leqslant M_1 e^{-\beta t}, \quad t \geqslant 0. \tag{4.13}$$

It is immediately seen via the standard perturbation argument that

$$\left\| e^{-tL_{f2}} \right\| \leqslant M_1 e^{-(\beta - M_1 c_1 \gamma)t}, \quad t \geqslant 0, \quad c_1 = \sum_{k=1}^{N} \|Q^* w_k\| \, \|Qg_k\|. \tag{4.14}$$

The eigenvalues of $\Lambda + \gamma\hat{G}\hat{W}$ are nonlinear functions of γ. According to the choice of w_k, we have the following proposition which forms the key to the theorem. The proof is to be given later.

Proposition 4.2. *There exist a constant $M > 0$ and an $O(\gamma^2)$ such that*

$$\left\| e^{-t(\Lambda+\gamma\hat{G}\hat{W})} \right\|_{\mathscr{L}(\mathbb{C}^n)} \leqslant M e^{-(\omega+\gamma+O(\gamma^2))t}, \quad t \geqslant 0, \tag{4.15}$$

where the constant M is independent of γ.

Set $\alpha(\gamma) = \gamma + O(\gamma^2)$. When γ is small, we may assume that

$$\omega + \alpha(\gamma) < \beta - M_1 c_1 \gamma.$$

Based on the estimates (4.14), (4.15), and the integral equation (4.12), we derive an inequality

$$\begin{aligned} |\hat{u}(t)| \leqslant{} & M e^{-(\omega+\alpha(\gamma))t} |\hat{u}(0)| + \frac{M M_1 c_3 \gamma}{\beta - \omega - M_1 c_1 \gamma - \alpha(\gamma)} e^{-(\omega+\alpha(\gamma))t} \|u_2(0)\| \\ & + M M_1 c_2 c_3 \gamma^2 \int_0^t K(t-\sigma) |\hat{u}(\sigma)| \, d\sigma, \end{aligned} \tag{4.16}$$

where

$$c_2 = \sum_{k=1}^{N} \|P^* w_k\| \, \|Q g_k\|, \quad c_3 = \sum_{k=1}^{N} \|Q^* w_k\| \, \|P g_k\|,$$

$$\text{and} \quad K(t) = \int_0^t e^{-(\omega+\alpha(\gamma))(t-\tau)} e^{-(\beta - M_1 c_1 \gamma)\tau} \, d\tau$$
$$< \frac{e^{-(\omega+\alpha(\gamma))t}}{\beta - \omega - M_1 c_1 \gamma - \alpha(\gamma)}, \quad t \geqslant 0.$$

Thus the estimate (4.16) is rewritten as

$$|\hat{u}(t)| \leqslant M_2 e^{-(\omega+\alpha(\gamma))t} \|u_0\| + \frac{M M_1 c_2 c_3}{\beta - \omega - M_1 c_1 - \alpha(\gamma)} \gamma^2 \int_0^t e^{-(\omega+\alpha(\gamma))(t-\sigma)} |\hat{u}(\sigma)| \, d\sigma,$$

where $M_2\,(> M)$ denotes a constant which is close to M and independent of γ. Applying Gronwall's inequality [22] to the above, we see that

$$|\hat{u}(t)| \leqslant M_2 \|u_0\| \exp\left(- \left(\omega + \gamma + O(\gamma^2) - \frac{M M_1 c_2 c_3}{\beta - \omega - M_1 c_1 \gamma - \alpha(\gamma)} \gamma^2 \right) t \right), \quad t \geqslant 0. \tag{4.17}$$

By going back to the equation for u_2 in (4.12), this leads to a similar estimate for $\|u_2(t)\|$. Thus we have proven the desired estimate (4.11).

Proof of Proposition 4.2. We calculate $e^{-t(\Lambda + \gamma \hat{G}\hat{W})}$ according to the formula:

$$e^{-t(\Lambda+\gamma\hat{G}\hat{W})} = \frac{1}{2\pi i} \int_C e^{-t\lambda} (\lambda - \Lambda - \gamma \hat{G}\hat{W})^{-1} \, d\lambda, \tag{4.18}$$

where C denotes a counterclockwise contour encircling $\sigma(\Lambda + \gamma\hat{G}\hat{W})$. We need to calculate and estimate the residue of the integrand at each singularity. Set

$$\begin{pmatrix} \lambda - \lambda_1 - \gamma & -\gamma & \dots & -\gamma \\ -\gamma & \lambda - \lambda_2 - \gamma & \dots & -\gamma \\ \vdots & \vdots & \ddots & \vdots \\ -\gamma & -\gamma & \dots & \lambda - \lambda_i - \gamma \end{pmatrix}^{-1} = \left(a^i_{jk}; \begin{matrix} j & \downarrow & 1, \dots, i \\ k & \rightarrow & 1, \dots, i \end{matrix} \right) \tag{4.19}$$

for each i, $1 \leqslant i \leqslant \nu$. Extending a^k_{ij} as

$$a^k_{ij} = 0, \qquad \text{if } i > k \quad \text{or } j > k,$$

define $N \times N (= m_1 \times m_1)$ diagonal matrices A_{ij} as

$$A_{ij} = \begin{pmatrix} a_{ij}^{\nu} I_{m_\nu} & O & \dots & O \\ O & a_{ij}^{\nu-1} I_{m_{\nu-1}-m_\nu} & \dots & O \\ \vdots & \vdots & \ddots & \vdots \\ O & O & \dots & a_{ij}^{1} I_{m_1-m_2} \end{pmatrix}, \quad 1 \leqslant i,\ j \leqslant \nu. \tag{4.20}$$

When $\nu = 3$, for example, the structure of A_{ij} is illustrated as follows (by setting $p = m_3,\ q = m_2 - m_3$, and $r = m_1 - m_2$):

$$A_{11} = \begin{pmatrix} a_{11}^3 I_p & O & O \\ O & a_{11}^2 I_q & O \\ O & O & a_{11}^1 I_r \end{pmatrix}, \quad A_{12} = \begin{pmatrix} a_{12}^3 I_p & O & O \\ O & a_{12}^2 I_q & O \\ O & O & O \end{pmatrix}, \quad A_{13} = \begin{pmatrix} a_{13}^3 I_p & O & O \\ O & O & O \\ O & O & O \end{pmatrix},$$

$$A_{21} = \begin{pmatrix} a_{21}^3 I_p & O & O \\ O & a_{21}^2 I_q & O \\ O & O & O \end{pmatrix}, \quad A_{22} = \begin{pmatrix} a_{22}^3 I_p & O & O \\ O & a_{22}^2 I_q & O \\ O & O & O \end{pmatrix}, \quad A_{23} = \begin{pmatrix} a_{23}^3 I_p & O & O \\ O & O & O \\ O & O & O \end{pmatrix},$$

$$A_{31} = \begin{pmatrix} a_{31}^3 I_p & O & O \\ O & O & O \\ O & O & O \end{pmatrix}, \quad A_{32} = \begin{pmatrix} a_{32}^3 I_p & O & O \\ O & O & O \\ O & O & O \end{pmatrix}, \quad A_{33} = \begin{pmatrix} a_{33}^3 I_p & O & O \\ O & O & O \\ O & O & O \end{pmatrix}.$$

Based on these preparations, we show

Lemma 4.3. *The resolvent of* $\Lambda + \gamma \hat{G}\hat{W}$ *is expressed as*

$$(\lambda - \Lambda - \gamma \hat{G}\hat{W})^{-1} = \begin{pmatrix} \hat{G}_1 A_{11} \hat{W}_1 & \hat{G}_1 A_{12} \hat{W}_2 & \dots & \hat{G}_1 A_{1\nu} \hat{W}_\nu \\ \hat{G}_2 A_{21} \hat{W}_1 & \hat{G}_2 A_{22} \hat{W}_2 & \dots & \hat{G}_2 A_{2\nu} \hat{W}_\nu \\ \vdots & \vdots & \ddots & \vdots \\ \hat{G}_\nu A_{\nu 1} \hat{W}_1 & \hat{G}_\nu A_{\nu 2} \hat{W}_2 & \dots & \hat{G}_\nu A_{\nu\nu} \hat{W}_\nu \end{pmatrix}. \tag{4.21}$$

Proof. We only have to show that, for each i and j,

$$\begin{aligned} K_{ij} &= -\gamma \hat{G}_i A_{1j} \hat{W}_j - \gamma \hat{G}_i \hat{W}_2 \hat{G}_2 A_{2j} \hat{W}_j - \dots + (\lambda - \lambda_i - \gamma) \hat{G}_i A_{ij} \hat{W}_j \\ &\quad - \dots - \gamma \hat{G}_i \hat{W}_\nu \hat{G}_\nu A_{\nu j} \hat{W}_j \\ &= \begin{cases} I_{m_i}, & i = j, \\ O_{m_i}, & i \neq j. \end{cases} \end{aligned} \tag{4.22}$$

In view of the definition of a_{ij}^k, it is clear that $\hat{W}_l \hat{G}_l A_{lj} = A_{lj}$. Then,

$$K_{ij} = \hat{G}_i \kappa_{ij} \hat{W}_j, \quad \text{where}$$

$$\kappa_{ij} = -\gamma A_{1j} - \gamma A_{2j} - \dots + (\lambda - \lambda_i - \gamma) A_{ij} - \dots - \gamma A_{\nu j}.$$

By recalling the definition of the a_{ij}^k again, each diagonal block of the $N \times N$ matrix κ_{ij} is calculated, when $i > j$ for example, as

$$\begin{aligned}
&\text{the } (1,1)\text{-block:} \quad -\gamma a_{1j}^{\nu} - \gamma a_{2j}^{\nu} - \cdots + (\lambda - \lambda_i - \gamma) a_{ij}^{\nu} - \cdots - \gamma a_{\nu j}^{\nu} = 0,\\
&\text{the } (2,2)\text{-block:} \quad -\gamma a_{1j}^{\nu-1} - \gamma a_{2j}^{\nu-1} - \cdots + (\lambda - \lambda_i - \gamma) a_{ij}^{\nu-1} - \cdots \\
&\qquad\qquad\qquad\qquad - \gamma a_{(\nu-1)j}^{\nu-1} = 0,\\
&\cdots \quad \cdots\\
&\text{the } (\nu+1-i, \nu+1-i)\text{-block:} \quad -\gamma a_{1j}^{i} - \gamma a_{2j}^{i} - \cdots + (\lambda - \lambda_i - \gamma) a_{ij}^{i} = 0.
\end{aligned}$$

The other cases: $i < j$ and $i = j$ are similarly calculated. Consequently we see that

$$\kappa_{ij} = \begin{cases} \begin{pmatrix} O_{m_i \times N} \\ ** \end{pmatrix}, & i > j, \\ O_N, & i < j, \\ \begin{pmatrix} I_{m_i} & O \\ O & O_{N-m_i} \end{pmatrix}, & i = j. \end{cases}$$

By our choice of $\hat{G}_i$ and $\hat{W}_j$, relation (4.22) is now clear. □

Lemma 4.3 shows that each element of $(\lambda - \Lambda - \gamma HW)^{-1}$ is a rational function of λ with the denominator which is one of the following $d_1, \ldots, d_\nu$:

$$d_i = \begin{vmatrix} \lambda - \lambda_1 - \gamma & -\gamma & \cdots & -\gamma \\ -\gamma & \lambda - \lambda_2 - \gamma & \cdots & -\gamma \\ \vdots & \vdots & \ddots & \vdots \\ -\gamma & -\gamma & \cdots & \lambda - \lambda_i - \gamma \end{vmatrix}, \quad 1 \leqslant i \leqslant \nu. \tag{4.23}$$

Thus singularities of each element of $(\lambda - \Lambda - \gamma HW)^{-1}$ are simple poles as long as $\gamma > 0$ is small enough. Let $\lambda_{i1}(\gamma), \ldots, \lambda_{ii}(\gamma)$ be the distinct solutions to the algebraic equation: $d_i = 0$, where each $\lambda_{ij}(\gamma)$ is close to λ_j in a neighborhood of $\gamma = 0$. In order to know the behavior of the $\lambda_{ij}(\gamma)$, differentiate the both sides of $d_i = 0$ with respect to γ and set $\gamma = 0$. Then we see that

$$\frac{d}{d\gamma} \lambda_{ij}(\gamma) \bigg|_{\gamma=0} = 1, \qquad 1 \leqslant j \leqslant i \leqslant \nu.$$

Thus,

$$\lambda_{ij}(\gamma) = \lambda_j + \gamma + O(\gamma^2), \qquad 1 \leqslant j \leqslant i \leqslant \nu.$$

Calculating the residue of $e^{-t\lambda} a_{ij}^k$ at each possible pole in the integral (4.18), we obtain the estimate (4.15). This finishes the proof of Proposition 4.2. □

The proof of Theorem 4.1 is thereby complete. □

In the case where $m_1 = m_2 = \cdots = m_\nu$, each matrix A_{ij} in (4.20) is reduced to $a_{ij}^\nu I_N$. Thus, $(\lambda - \Lambda - \gamma\hat{G}\hat{W})^{-1}$ is simply expressed as

$$
\begin{aligned}
&(\lambda - \Lambda - \gamma\hat{G}\hat{W})^{-1} \\
&= \begin{pmatrix} I_N & 0 & \cdots & 0 \\ 0 & \hat{G}_2\hat{G}_1^{-1} & \cdots & 0 \\ \vdots & \vdots & \ddots & \vdots \\ 0 & 0 & \cdots & \hat{G}_\nu\hat{G}_1^{-1} \end{pmatrix} \begin{pmatrix} a_{11}^\nu I_N & a_{12}^\nu I_N & \cdots & a_{1\nu}^\nu I_N \\ a_{21}^\nu I_N & a_{22}^\nu I_N & \cdots & a_{2\nu}^\nu I_N \\ \vdots & \vdots & \ddots & \vdots \\ a_{\nu 1}^\nu I_N & a_{\nu 2}^\nu I_N & \cdots & a_{\nu\nu}^\nu I_N \end{pmatrix} \\
&\quad \times \begin{pmatrix} I_N & 0 & \cdots & 0 \\ 0 & \hat{G}_1\hat{G}_2^{-1} & \cdots & 0 \\ \vdots & \vdots & \ddots & \vdots \\ 0 & 0 & \cdots & \hat{G}_1\hat{G}_\nu^{-1} \end{pmatrix}.
\end{aligned}
$$

Corollary 4.4. *Consider the simplest case where $m_1 = \cdots = m_n$. Set $N = m_1$, and suppose that*

$$
\operatorname{rank} \hat{G}_i = m_i, \quad 1 \leqslant i \leqslant \nu. \tag{4.24}
$$

Choose w_k so that $\hat{W}_i = \hat{G}_i^{-1}$, $1 \leqslant i \leqslant \nu$. Then, as long as $\gamma > 0$ is small enough, there exists an $O(\gamma^2)$ such that the estimate (4.11) holds.

Application to a class of boundary control systems:

Let us consider the boundary feedback control system (4.3) in the simplest case where the pair $(\mathscr{L}, \tau)$ is given as

$$
\mathscr{L} = -\Delta, \quad \tau = \frac{\partial}{\partial \nu}. \tag{4.25}
$$

The operator $L = -\Delta|_{\ker\tau}$ is self-adjoint, and the spectrum $\sigma(L)$ lies in $[0, \infty)$. The smallest eigenvalue λ_1 is 0; its multiplicity m_1 is 1; and the corresponding eigenfunction φ_{11} is a non-zero constant, so that the unperturbed system, $u_t + Lu = 0$ is stable, but not exponentially stable. By choosing $N = 1$ in (4.3), the equation for u is now written as

$$
\begin{cases}
\dfrac{\partial u}{\partial t} = \Delta u \quad \text{in } \mathbb{R}_+^1 \times \Omega, \\[2mm]
\dfrac{\partial u}{\partial \nu} = -\gamma\langle u, w\rangle_\Gamma g \quad \text{on } \mathbb{R}_+^1 \times \Gamma, \\[2mm]
u(0,\cdot) = u_0(\cdot) \quad \text{in } \Omega.
\end{cases} \tag{4.3$'$}
$$

Our aim is to raise the stability property of the system (4.3′) a little. For a constant $c > 0$, set $L_c = L + c$ as before, and let $h \in H^2(\Omega)$ be a unique solution to the boundary value problem

$$
(c - \Delta)h = 0 \quad \text{in } \Omega, \qquad \frac{\partial h}{\partial \nu} = g \quad \text{on } \Gamma.
$$

Setting $x(t) = L_c^{-1/4-\varepsilon} u(t,\cdot)$, we derive an equation for $x(t) \in \mathscr{D}(L)$, $t > 0$:

$$\frac{dx}{dt} + Lx = -\gamma \left\langle L_c^{1/4+\varepsilon} x, w \right\rangle_\Gamma L_c^{3/4-\varepsilon} h, \quad x(0) = x_0 = L_c^{-1/4-\varepsilon} u_0, \tag{4.5'}$$

For simplicity, let φ_{11} be normalized in $H = L^2(\Omega)$, i.e., $\|\varphi_{11}\| = 1$, and let $Ph = h_{11}\varphi_{11}$. It is clear that $h_{11} = \langle h, \varphi_{11}\rangle$. Then, $PL_c^{3/4-\varepsilon} h = c^{3/4-\varepsilon} h_{11}\varphi_{11}$ in (4.7_1). The condition (4.10) is reduced to

$$c^{3/4-\varepsilon} h_{11} \neq 0, \quad \left\langle L_c^{1/4+\varepsilon} \varphi_{11}, w \right\rangle_\Gamma = c^{1/4+\varepsilon} \langle \varphi_{11}, w\rangle_\Gamma = \frac{1}{c^{3/4-\varepsilon} h_{11}}.$$

By Green's formula, we note that $ch_{11} = c\langle h, \varphi_{11}\rangle = \langle g, \varphi_{11}\rangle_\Gamma$. Thus the above condition means that

$$\langle g, \varphi_{11}\rangle_\Gamma \langle \varphi_{11}, w\rangle_\Gamma = 1. \tag{4.10'}$$

By Theorem 4.1 or Corollary 4.4, the condition (4.10′) ensures the decay estimate (4.11) for every solution to (4.3′), where $\omega = 0$.

3.5 Some Generalization

We extend in this section Theorem 4.1 to some extent. The assumption on g_k will be somewhat weakened (see (5.2) below). Symbols and basic assumptions are the same as in Section 4: For example, relation (4.9) on the multiplicities is assumed. For positive integers i and j with $2 \leqslant i < j \leqslant \nu$ and $\lambda \in \mathbb{C}$, set

$$\begin{aligned} \Xi_{(i,j)}(\lambda) &= (\lambda - \lambda_{i+1})\cdots(\lambda - \lambda_j) + \cdots + (\lambda - \lambda_i)\cdots(\lambda - \lambda_{j-1}) \\ &= \prod_{k=i}^{j} (\lambda - \lambda_k) \cdot \left(\frac{1}{\lambda - \lambda_i} + \cdots + \frac{1}{\lambda - \lambda_j} \right). \end{aligned} \tag{5.1}$$

Here it is assumed that $\nu \geqslant 3$. Then we have

Theorem 5.1. *Take $N = m_1$, and assume that*

$$\hat{G}_i = (G_{i1}\, G_{i2}), \quad G_{i1};\ m_i \times m_i,$$
$$\det G_{i1} \neq 0 \ (\textit{and thus } \operatorname{rank} \hat{G}_i = m_i), \quad \textit{and } \hat{W}_i = \begin{pmatrix} G_{i1}^{-1} \\ 0 \end{pmatrix}, \quad 1 \leqslant i \leqslant \nu. \tag{5.2}$$

Assume finally that

$$\Xi_{(i,j)}(\lambda_h) \neq 0, \quad 1 \leqslant h < i < j \leqslant \nu, \quad 1 \leqslant i \leqslant \nu. \tag{5.3}$$

Then the assertion of Theorem 4.1 *holds.*

Remark 1: When $\nu = 3$, for example, the algebraic condition (5.3) means that

$$\lambda_1 \neq \frac{\lambda_2 + \lambda_3}{2},$$

and when $\nu = 4$, that

$$\lambda_1 \neq \frac{\lambda_2+\lambda_3}{2}, \quad \lambda_1 \neq \frac{\lambda_3+\lambda_4}{2}, \quad \lambda_2 \neq \frac{\lambda_3+\lambda_4}{2}, \quad \text{and}$$
$$\sum_{2\leqslant i<j\leqslant 4} A_i(\lambda_1)A_j(\lambda_1) = \sum_{2\leqslant i<j\leqslant 4} (\lambda_1-\lambda_i)(\lambda_1-\lambda_j) \neq 0,$$

where $A_i(\lambda) = \lambda - \lambda_i$.

Remark 2: Assumption (5.3) is posed for a technical reason, and seems not essential in our theorem. In fact, when (5.3) is not satisfied, it is shown that an estimate a litte weaker than (4.15)

$$\left\| e^{-t(\Lambda+\gamma\hat{G}\hat{W})} \right\|_{\mathscr{L}(\mathbb{C}^n)} \leqslant Me^{-(\omega+\gamma/2+O(\gamma^2))t}, \quad t \geqslant 0, \tag{5.4}$$

holds for $\nu = 3, 4$ instead. It is the author's conjectute that Theorem 5.1 would generally hold for $\nu \geqslant 3$ without the additional assumption (5.3) and with the estimate (4.15) replaced by (5.4).

Proof. Since the key idea is to obtain the estimate (4.15) for the semigroup $e^{-t(\Lambda+\gamma\hat{G}\hat{W})}$, we concentrate hereafter on the behavior of the resolvent $(\lambda-\Lambda-\gamma\hat{G}\hat{W})^{-1}$ in a neighborhood of each singularity. The rest of the proof is carried out in the same manner as in Theorem 4.1.

The presence of the terms G_{i2} rather complicates the structure of the inverse $(\lambda-\Lambda-\gamma\hat{G}\hat{W})^{-1}$: It seems difficult in this case to obtain an expression similar to (4.21) in Lemma 4.3. We propose as an alternative means a factorization of $(\lambda-\Lambda-\gamma\hat{G}\hat{W})^{-1}$ into the product of two upper-triangular matrices (see (5.8) below). For this, let us introduce $N\times N$ matrices $\langle k\rangle$, which are defined as

$$\langle k\rangle = \gamma A_1\cdots A_{k-1}\hat{W}_k\hat{G}_k + d_kI_N, \quad k=2,\ldots,\nu,$$

where $A_i = A_i(\lambda) = \lambda - \lambda_i$ are defined just above, and d_k the determinants in (4.23). Set

$$\Psi(\lambda,\gamma) =$$
$$\begin{pmatrix} I & 0 & 0 & 0 & \ldots & 0 \\ \gamma\hat{G}_2\hat{G}_1^{-1} & d_1I & 0 & 0 & \ldots & 0 \\ \gamma\hat{G}_3\frac{\langle 2\rangle}{d_1}\hat{G}_1^{-1} & \gamma A_1\hat{G}_3\hat{W}_2 & d_2I & 0 & \ldots & 0 \\ \gamma\hat{G}_4\frac{\langle 3\rangle\langle 2\rangle}{d_2d_1}\hat{G}_1^{-1} & \gamma A_1\hat{G}_4\frac{\langle 3\rangle}{d_2}\hat{W}_2 & \gamma A_1A_2\hat{G}_4\hat{W}_3 & d_3I & \ldots & 0 \\ \vdots & \vdots & \vdots & \vdots & \ddots & \vdots \\ \gamma\hat{G}_\nu\prod_{k=\nu-1}^{2}\frac{\langle k\rangle}{d_{k-1}}\hat{G}_1^{-1} & \gamma A_1\hat{G}_\nu\prod_{k=\nu-1}^{3}\frac{\langle k\rangle}{d_{k-1}}\hat{W}_2 & \gamma A_1A_2\hat{G}_\nu\prod_{k=\nu-1}^{4}\frac{\langle k\rangle}{d_{k-1}}\hat{W}_3 & \ldots & \ldots & d_{\nu-1}I \end{pmatrix}. \tag{5.5}$$

Then we have

Lemma 5.2.

$$\Psi(\lambda,\gamma)(\lambda-\Lambda-\gamma\hat{G}\hat{W})$$
$$=\begin{pmatrix} d_1 I & -\gamma\hat{G}_1\hat{W}_2 & -\gamma\hat{G}_1\hat{W}_3 & -\gamma\hat{G}_1\hat{W}_4 & \dots & -\gamma\hat{G}_1\hat{W}_\nu \\ 0 & d_2 I & -\gamma A_1\hat{G}_2\hat{W}_3 & -\gamma A_1\hat{G}_2\hat{W}_4 & \dots & -\gamma A_1\hat{G}_2\hat{W}_\nu \\ 0 & 0 & d_3 I & -\gamma A_1A_2\hat{G}_3\hat{W}_4 & \dots & -\gamma A_1A_2\hat{G}_3\hat{W}_\nu \\ \vdots & \vdots & \vdots & \vdots & \ddots & \vdots \\ 0 & 0 & 0 & 0 & \dots & d_\nu I \end{pmatrix}. \tag{5.6}$$

The inverse of the last matrix is denoted as $\Phi(\lambda,\gamma)$. Then, $\Phi(\lambda,\gamma)$ is expressed as

$$\Phi(\lambda,\gamma)$$
$$=\begin{pmatrix} \frac{1}{d_1}I & \frac{\gamma}{d_1d_2}\hat{G}_1\hat{W}_2 & \frac{\gamma A_2}{d_2d_3}\hat{G}_1\hat{W}_3 & \frac{\gamma A_2A_3}{d_3d_4}\hat{G}_1\hat{W}_4 & \dots & \frac{\gamma\prod_{k=1}^{\nu-1}A_k}{A_1d_{\nu-1}d_\nu}\hat{G}_1\hat{W}_\nu \\ 0 & \frac{1}{d_2}I & \frac{\gamma A_1}{d_2d_3}\hat{G}_2\hat{W}_3 & \frac{\gamma A_1A_3}{d_3d_4}\hat{G}_2\hat{W}_4 & \dots & \frac{\gamma\prod_{k=1}^{\nu-1}A_k}{A_2d_{\nu-1}d_\nu}\hat{G}_2\hat{W}_\nu \\ 0 & 0 & \frac{1}{d_3}I & \frac{\gamma A_1A_2}{d_3d_4}\hat{G}_3\hat{W}_4 & \dots & \frac{\gamma\prod_{k=1}^{\nu-1}A_k}{A_3d_{\nu-1}d_\nu}\hat{G}_3\hat{W}_\nu \\ 0 & 0 & 0 & \frac{1}{d_4}I & \dots & \frac{\gamma\prod_{k=1}^{\nu-1}A_k}{A_4d_{\nu-1}d_\nu}\hat{G}_4\hat{W}_\nu \\ \vdots & \vdots & \vdots & \vdots & \ddots & \vdots \\ 0 & 0 & 0 & 0 & \dots & \frac{1}{d_\nu}I \end{pmatrix}. \tag{5.7}$$

Proof. The relation (5.6) means that

$$\text{the } i\text{th row of } \Psi(\lambda,\gamma)\times \text{ the } j\text{th column of } (\lambda-\Lambda-\gamma\hat{G}\hat{W})$$
$$=\begin{cases} 0, & i>j, \\ d_i, & i=j, \\ -\gamma(A_1\cdots A_{i-1})\hat{G}_i\hat{W}_j, & i<j. \end{cases}$$

It is elementary but tedious to show the above relation. Let us begin with the case of $i > j$, where we have to show that

$$
\begin{aligned}
&-\gamma^2 \hat{G}_i \frac{\langle i-1\rangle \cdots \langle 2\rangle}{d_{i-2}\cdots d_1} \hat{W}_j - \gamma^2 A_1 \hat{G}_i \frac{\langle i-1\rangle \cdots \langle 3\rangle}{d_{i-2}\cdots d_2} \hat{W}_2 \hat{G}_2 \hat{W}_j \\
&-\gamma^2 A_1 A_2 \hat{G}_i \frac{\langle i-1\rangle \cdots \langle 4\rangle}{d_{i-2}\cdots d_3} \hat{W}_3 \hat{G}_3 \hat{W}_j \\
&-\cdots - \gamma^2 (A_1 \cdots A_{j-2}) \hat{G}_i \frac{\langle i-1\rangle \cdots \langle j\rangle}{d_{i-2}\cdots d_{j-1}} \hat{W}_{j-1} \hat{G}_{j-1} \hat{W}_j \\
&+\gamma (A_1 \cdots A_{j-1}) \hat{G}_i \frac{\langle i-1\rangle \cdots \langle j+1\rangle}{d_{i-2}\cdots d_j} (A_j - \gamma) \hat{W}_j \hat{G}_j \hat{W}_j \\
&-\gamma^2 (A_1 \cdots A_j) \hat{G}_i \frac{\langle i-1\rangle \cdots \langle j+2\rangle}{d_{i-2}\cdots d_{j+1}} \hat{W}_{j+1} \hat{G}_{j+1} \hat{W}_j - \cdots - \gamma d_{i-1} \hat{G}_i \hat{W}_j = 0.
\end{aligned}
$$

The key through the proof is the relation:

$$
\hat{W}_l \hat{G}_l \hat{W}_j = \hat{W}_j, \quad 1 \leqslant l \leqslant j.
$$

Deleting the common factor $\gamma \hat{G}_i$ for simplicity, we calculate as

$$
\begin{aligned}
&\text{the first term} + \text{the second term} \\
&= -\gamma \frac{\langle i-1\rangle \cdots \langle 3\rangle}{d_{i-2}\cdots d_2} \left(\frac{1}{d_1} (d_2 + \gamma A_1 \hat{W}_2 \hat{G}_2) + A_1 \right) \hat{W}_j \\
&= -\gamma \frac{\langle i-1\rangle \cdots \langle 3\rangle}{d_{i-2}\cdots d_2} \left(\frac{1}{d_1} (d_2 + \gamma A_1) + A_1 \right) \hat{W}_j \\
&= -\gamma \frac{\langle i-1\rangle \cdots \langle 3\rangle}{d_{i-2}\cdots d_2} \left(\frac{1}{d_1} d_1 A_2 + A_1 \right) \hat{W}_j \\
&= -\gamma \frac{\langle i-1\rangle \cdots \langle 3\rangle}{d_{i-2}\cdots d_2} \Xi_{(1,2)} \hat{W}_j .
\end{aligned}
$$

Inductively we find that

$$
\sum_{k=1}^{j-1} (\text{the } k\text{th term}) = -\gamma \frac{\langle i-1\rangle \cdots \langle j\rangle}{d_{i-2}\cdots d_{j-1}} \Xi_{(1,j-1)} \hat{W}_j .
$$

Thus, $\sum_{k=1}^{j}$ (the kth term) becomes

$$
\begin{aligned}
&\frac{\langle i-1\rangle\cdots\langle j+1\rangle}{d_{i-2}\cdots d_j}\left(-\gamma\Xi_{(1,j-1)}\frac{\langle j\rangle}{d_{j-1}}+(A_1\cdots A_{j-1})(A_j-\gamma)\right)\hat{W}_j\\
&=\frac{\langle i-1\rangle\cdots\langle j+1\rangle}{d_{i-2}\cdots d_j}\\
&\quad\times\left(-\frac{\gamma\Xi_{(1,j-1)}}{d_{j-1}}\left(d_j+\gamma(A_1\cdots A_{j-1})\hat{W}_j\hat{G}_j\right)+(A_1\cdots A_{j-1})(A_j-\gamma)\right)\hat{W}_j\\
&=\frac{\langle i-1\rangle\cdots\langle j+1\rangle}{d_{i-2}\cdots d_j}\\
&\quad\times\left(-\frac{\gamma\Xi_{(1,j-1)}}{d_{j-1}}\left(d_j+\gamma(A_1\cdots A_{j-1})\right)+(A_1\cdots A_{j-1})(A_j-\gamma)\right)\hat{W}_j\\
&=\frac{\langle i-1\rangle\cdots\langle j+1\rangle}{d_{i-2}\cdots d_j}\left(-\frac{\gamma\Xi_{(1,j-1)}}{d_{j-1}}d_{j-1}A_j+(A_1\cdots A_{j-1})(A_j-\gamma)\right)\hat{W}_j\\
&=\frac{\langle i-1\rangle\cdots\langle j+1\rangle}{d_{i-2}\cdots d_{j+1}}\hat{W}_j.
\end{aligned}
$$

Continuing further the above calculation, we see that

$$
\begin{aligned}
\sum_{k=1}^{i-3}(\text{the }k\text{th term})&=\frac{\langle i-1\rangle\langle i-2\rangle}{d_{i-2}}\hat{W}_j;\\
\sum_{k=1}^{i-2}(\text{the }k\text{th term})&=\frac{\langle i-1\rangle}{d_{i-2}}\left(\langle i-2\rangle-\gamma(A_1\cdots A_{i-3})\hat{W}_{i-2}\hat{G}_{i-2}\right)\hat{W}_j\\
&=\frac{\langle i-1\rangle}{d_{i-2}}d_{i-2}\hat{W}_j=\langle i-1\rangle\hat{W}_j;
\end{aligned}
$$

and finally

$$
\begin{aligned}
\sum_{k=1}^{i}(\text{the }k\text{th term})&=\langle i-1\rangle\hat{W}_j-\gamma(A_1\cdots A_{i-2})\hat{W}_{i-1}\hat{G}_{i-1}\hat{W}_j-d_{i-1}\hat{W}_j\\
&=\left(\langle i-1\rangle-\gamma(A_1\cdots A_{i-2})\hat{W}_{i-1}\hat{G}_{i-1}-d_{i-1}\right)\hat{W}_j\\
&=(d_{i-1}-d_{i-1})W_j=0.
\end{aligned}
$$

Next let us consider the case of $i=j$. Similarly we calculate inductively as (counting the factor $\gamma\hat{G}_i$ at this time)

$$
\sum_{k=1}^{i-2}(\text{the }k\text{th term})=-\gamma^2\hat{G}_i\frac{\langle i-1\rangle}{d_{i-2}}\Xi_{(1,i-2)}\hat{W}_i.
$$

Then, by recalling that $\hat{W}_{i-1}\hat{G}_{i-1}\hat{W}_i = \hat{W}_i$,

$$\begin{aligned}
\sum_{k=1}^{i-1} (\text{the } k\text{th term}) &= -\gamma^2 \hat{G}_i \frac{\Xi_{(1,i-2)}}{d_{i-2}} \langle i-1\rangle \hat{W}_i - \gamma^2 (A_1 \cdots A_{i-2}) \hat{G}_i \hat{W}_{i-1} \hat{G}_{i-1} \hat{W}_i \\
&= -\gamma^2 \hat{G}_i \left(\frac{\Xi_{(1,i-2)}}{d_{i-2}} (d_{i-1} + \gamma(A_1 \cdots A_{i-2})) + (A_1 \cdots A_{i-2}) \right) \hat{W}_i \\
&= -\gamma^2 \hat{G}_i \left(\frac{\Xi_{(1,i-2)}}{d_{i-2}} d_{i-2} A_{i-1} + (A_1 \cdots A_{i-2}) \right) \hat{W}_i \\
&= -\gamma^2 \Xi_{(1,i-1)} \hat{G}_i \hat{W}_i = -\gamma^2 \Xi_{(1,i-1)},
\end{aligned}$$

and finally we have

$$\begin{aligned}
\sum_{k=1}^{i} (\text{the } k\text{th term}) &= -\gamma^2 \Xi_{(1,i-1)} + d_{i-1}(A_i - \gamma) \\
&= -\gamma^2 \Xi_{(1,i-1)} + \left((A_1 \cdots A_{i-1}) - \gamma \Xi_{(1,i-1)} \right) (A_i - \gamma) \\
&= (A_1 \cdots A_i) - \gamma (A_1 \cdots A_{i-1}) - \gamma A_i \Xi_{(1,i-1)} \\
&= d_i.
\end{aligned}$$

Finally let us consider the case of $i < j$. We inductively calculate as

$$\sum_{k=1}^{i-2} (\text{the } k\text{th term}) = -\gamma^2 \hat{G}_i \frac{\Xi_{(1,i-2)}}{d_{i-2}} \langle i-1\rangle \hat{W}_j.$$

Then,

$$\begin{aligned}
&\sum_{k=1}^{i-1} (\text{the } k\text{th term}) \\
&= -\gamma^2 \hat{G}_i \frac{\Xi_{(1,i-2)}}{d_{i-2}} \langle i-1\rangle \hat{W}_j - \gamma^2 (A_1 \cdots A_{i-2}) \hat{G}_i \hat{W}_{i-1} \hat{G}_{i-1} \hat{W}_j \\
&= -\gamma^2 \hat{G}_i \left(\frac{\Xi_{(1,i-2)}}{d_{i-2}} (d_{i-1} + \gamma(A_1 \cdots A_{i-2})) + (A_1 \cdots A_{i-2}) \right) \hat{W}_j \\
&= -\gamma^2 \hat{G}_i \left(\frac{\Xi_{(1,i-2)}}{d_{i-2}} d_{i-2} A_{i-1} + (A_1 \cdots A_{i-2}) \right) \hat{W}_j \\
&= -\gamma^2 \Xi_{(1,i-1)} \hat{G}_i \hat{W}_j,
\end{aligned}$$

and finally we have

$$\begin{aligned}
\sum_{k=1}^{i} (\text{the } k\text{th term}) &= -\gamma^2 \Xi_{(1,i-1)} \hat{G}_i \hat{W}_j - \gamma d_{i-1} \hat{G}_i \hat{W}_j \\
&= -\gamma (A_1 \cdots A_{i-1}) \hat{G}_i \hat{W}_j.
\end{aligned}$$

The calculation of $\Phi(\lambda,\gamma)$ is similarly carried out. Thus it is omitted. □

In view of Lemma 5.2, we obtain the following factorization of $(\lambda - \Lambda - \gamma\hat{G}\hat{W})^{-1}$:

$$(\lambda - \Lambda - \gamma\hat{G}\hat{W})^{-1} = \Phi(\lambda,\gamma)\Psi(\lambda,\gamma). \tag{5.8}$$

In calculating the contour integral in (4.18), we evaluate the behavior of each element of $(\lambda - \Lambda - \gamma\hat{G}\hat{W})^{-1}$ at every singularity. Let us recall the definition of d_i in (4.23). The following lemma holds regardless of the assumption (5.3):

Lemma 5.3. *When $\gamma \neq 0$ is small enough, we have the implication:*

$$d_i = 0 \quad \Rightarrow \quad d_j \neq 0, \quad i \neq j. \tag{5.9}$$

Thus, the singularities of each element of the matrices $\Phi(\lambda,\gamma)$ and $\Psi(\lambda,\gamma)$ consist of simple poles.

Proof. By definition,

$$d_i = A_1 \cdots A_i - \gamma \Xi_{(1,i)}.$$

As long as $\gamma \neq 0$ is small enough, the solutions λ to the algebraic equation $d_i = 0$ are close to, but not equal to any of $\lambda_1, \ldots, \lambda_i$. This follows from the fact: $\lambda'(0) = 1$. Thus when $d_i = 0$, we see that $A_j = \lambda(\gamma) - \lambda_j \neq 0$, $1 \leqslant j \leqslant \nu$. Note that

$$\begin{aligned} d_i &= d_1 A_2 A_3 \cdots A_i - \gamma A_1 \Xi_{(2,i)} \\ &= d_2 A_3 \cdots A_i - \gamma A_1 A_2 \Xi_{(3,i)} \\ &= \cdots \cdots \\ &= d_{i-1} A_i - \gamma A_1 A_2 \cdots A_{i-1}, \end{aligned} \tag{5.10}$$

a part of which has been employed in the previous lemma. Let $\lambda(\gamma)$ be one of the solutions to the equation: $d_i = 0$. For any pair of integers $1 \leqslant p < q \leqslant n$, consider the function $\Xi_{(p,q)}(\lambda(\gamma))$. As an analytic function of γ, we show that, in general,

$$\exists \ell \geqslant 0; \quad \frac{d^\ell}{d\gamma^\ell} \Xi_{(p,q)}(\lambda(0)) \neq 0. \tag{5.11}$$

In fact, if this were not true, we would obtain $\Xi_{(p,q)}(\lambda(\gamma)) \equiv 0$. Set $m = q - p$, and let

$$\Xi_{(p,q)}(\lambda) = (m+1)\lambda^m + a_1\lambda^{m-1} + \cdots + a_{m-2}\lambda^2 + a_{m-1}\lambda + a_m,$$

where the coefficients a_i are the polynomials of $\lambda_p, \ldots, \lambda_q$, e.g., $a_1 = -m(\lambda_p + \cdots + \lambda_q)$. Differentiating the both sides of $\Xi_{(p,q)}(\lambda(\gamma)) \equiv 0$ with respect to γ, we have

$$(m+1)m\lambda^{m-1}\lambda' + (m-1)a_1\lambda^{m-2}\lambda' + \cdots + 2a_{m-2}\lambda\lambda' + a_{m-1}\lambda' \equiv 0.$$

Noting that $\lambda'(0) = 1$, we have, through analytic continuation,

$$(m+1)m\lambda^{m-1}(\gamma) + (m-1)a_1\lambda^{m-2}(\gamma) + \cdots + 2a_{m-2}\lambda(\gamma) + a_{m-1} \equiv 0.$$

Continueing the same procedure repeatedly in the above relation, we finally obtain

$$(m+1)!\lambda + (m-1)!a_1 \equiv 0, \quad \text{or}$$

$$\lambda(\gamma) \equiv \frac{-a_1}{(m+1)m} = \frac{1}{q-p+1}(\lambda_p + \cdots + \lambda_q),$$

which contradicts the property: $\lambda'(0) = 1$. Thus there is an integer ℓ satisfying the relation (5.11). As a result, there is a function $f_{(p,q)}(\gamma)$ which is analytic at $\gamma = 0$ such that

$$\Xi_{(p,q)}(\lambda(\gamma)) = \gamma^{\ell} f_{(p,q)}(\gamma), \quad f_{(p,q)}(0) \neq 0.$$

We go back to (5.10). When $d_i = 0$, then (5.10) implies that

$$d_j = \frac{\gamma A_1 \cdots A_j}{A_{j+1} \cdots A_i} \Xi_{(j+1,i)} = \frac{A_1 \cdots A_j}{A_{j+1} \cdots A_i} \gamma^{\ell+1} f_{(j+1,i)}(\gamma).$$

As long as $\gamma \neq 0$ is small enough, the above expression implies that $d_j \neq 0$ for $1 \leqslant j \leqslant i-1$. Similarly we see that $d_j \neq 0$ for $i+1 \leqslant j \leqslant n$, when $d_i = 0$. □

Based on Lemma 5.3, it is convenient to write down the d_i-d_j table which describes the behavior of d_j, $j \neq i$ when $d_i = 0$. The table is written down at the end of this section. In view of Lemma 5.3, the singularities of $\Phi(\lambda, \gamma)$ arise at points different from those of $\Psi(\lambda, \gamma)$. We first consider the singularities of $\Phi(\lambda, \gamma)$. The mth column has the sigularities at the points where $d_{m-1} = 0$ and $d_m = 0$. We need to calculate the residues of the matrix $e^{-t\lambda}(\lambda - \Lambda - \gamma\hat{G}\hat{W})^{-1}$ when $d_{m-1} = 0$ and $d_m = 0$. When $d_{m-1} = 0$, we see that[1]

$$\begin{aligned} &\left\| d_{m-1}\Phi(\lambda, \gamma)_{(i,m)} \right\| \\ &\leqslant \begin{cases} O(1)\left|\dfrac{\gamma A_1 \cdots A_{m-1}}{A_i d_m}\right| \leqslant \dfrac{O(1)}{|A_i|} = O\left(\dfrac{1}{\gamma}\right), & 1 \leqslant i \leqslant m-1, \\ 0, & m \leqslant i \leqslant \nu. \end{cases} \end{aligned}$$

Similarly, when $d_m = 0$, the element $\Phi(\lambda, \gamma)_{(i,m)}$ times d_m is estimated as follows:

$$\begin{aligned} &\left\| d_m\Phi(\lambda, \gamma)_{(i,m)} \right\| \\ &\leqslant \begin{cases} O(1)\left|\dfrac{\gamma A_1 \cdots A_{m-1}}{A_i d_{m-1}}\right| \leqslant \dfrac{O(1)|A_m|}{|A_i|} = O\left(\dfrac{1}{\gamma}\right)|A_m|, & 1 \leqslant i \leqslant m-1, \\ 1, & i = m, \\ 0, & m+1 \leqslant i \leqslant \nu. \end{cases} \end{aligned}$$

[1] We are calculating the residue: $\lim_{\lambda \to \lambda_{(m-1)j}(\gamma)}(\lambda - \lambda_{(m-1)j}(\gamma))\Phi(\lambda, \gamma)_{(i,m)}$, $1 \leqslant j \leqslant m-1$. The same convension appears just below when $d_m = 0$.

In the calculation of the residues, the corresponding terms are the mth row of $\Psi(\lambda,\gamma)$. By (5.5), they are written down as

$$\Psi(\lambda,\gamma)_{(m,j)} = \begin{cases} \hat{G}_m \gamma \prod_{k=m-1}^{2} \dfrac{\langle k\rangle}{d_{k-1}} \hat{G}_1^{-1}, & j=1, \\ \hat{G}_m \gamma A_1 \cdots A_{j-1} \prod_{k=m-1}^{j+1} \dfrac{\langle k\rangle}{d_{k-1}} \hat{W}_j, & 2 \leqslant j \leqslant m-2, \\ \hat{G}_m \gamma A_1 \cdots A_{m-2} \hat{W}_{m-1}, & j=m-1, \\ d_{m-1}, & j=m, \\ O, & m+1 \leqslant j \leqslant \nu. \end{cases} \tag{5.12}$$

As we have seen in the estimate of $d_{m-1}\Phi(\lambda,\gamma)_{(i,m)}$ and $d_m\Phi(\lambda,\gamma)_{(i,m)}$, $1 \leqslant i \leqslant \nu$, the above elements $\Psi(\lambda,\gamma)_{(m,j)}$ are desired to be at least of order $O(\gamma)$ when $d_{m-1}=0$ and $d_m=0$. To examine this we need to know the behavior of the matrices $\langle k\rangle = \gamma A_1 \cdots A_{k-1}\hat{W}_k\hat{G}_k + d_k I_N$, as the functions of $\gamma \sim 0$, when $d_i = 0$. According to the d_i-d_j table, we complete the d_i-$\langle j\rangle$ table which is also written down at the end of this section. In view of this table, it is immediately seen that, when $d_{m-1}=0$ and $d_m=0$,

$$\left\| \frac{\langle k\rangle}{d_{k-1}} \right\| = O(1)|A_k|, \quad 2 \leqslant k \leqslant m-1.$$

Thus we see that

$$\left\|\Psi(\lambda,\gamma)_{(m,j)}\right\| = O(\gamma), \quad 1 \leqslant j \leqslant \nu, \ j \neq m \quad \text{when } d_{m-1}=0 \text{ and } d_m=0.$$

The only exception is the case where $j=m$. When $d_m=0$, we remark however that, in a neighborhood of $\gamma=0$,

$$\left\|d_m\Phi(\lambda,\gamma)_{(i,m)} \cdot \Psi(\lambda,\gamma)_{(m,m)}\right\| = O\left(\frac{1}{\gamma}\right)|A_m| \cdot |d_{m-1}| = O(1).$$

These are the desired estimates, and

$$\begin{aligned} &\left|\text{the residues of each element of } e^{-t\lambda}(\lambda - \Lambda - \gamma\hat{G}\hat{W})^{-1}\right| \\ &= O(1)e^{-(\omega+\gamma+O(\gamma^2))t}, \quad \text{when } d_{m-1}=0 \text{ and } d_m=0. \end{aligned} \tag{5.13}$$

Let us turn to the singularities of $\Psi(\lambda,\gamma)_{(m,j)}$ in (5.12). The corresponding terms in $\Phi(\lambda,\gamma)$ are the mth column. We have to evaluate

$$\hat{G}_i \frac{\gamma A_1 \cdots A_{m-1}}{A_i d_{m-1} d_m} \hat{W}_m \Psi(\lambda,\gamma)_{(m,j)}, \quad 1 \leqslant i \leqslant m-1, \quad \text{and} \quad \frac{1}{d_m}\Psi(\lambda,\gamma)_{(m,j)},$$

which amounts to the estimate of the residues of

$$\gamma \prod_{k=m-1}^{2} \frac{\langle k\rangle}{d_{k-1}} \cdot \frac{\gamma A_1 \cdots A_{m-1}}{d_{m-1}d_m} \frac{1}{A_i}, \qquad \gamma \prod_{k=m-1}^{2} \frac{\langle k\rangle}{d_{k-1}} \cdot \frac{1}{d_m}, \tag{5.14}$$
$$1 \leqslant i \leqslant m-1, \quad j=1,$$

and

$$\gamma A_1\cdots A_{j-1}\prod_{k=m-1}^{j+1}\frac{\langle k\rangle}{d_{k-1}}\cdot\frac{\gamma A_1\cdots A_{m-1}}{d_{m-1}d_m}\frac{1}{A_i},\qquad \gamma A_1\cdots A_{j-1}\prod_{k=m-1}^{j+1}\frac{\langle k\rangle}{d_{k-1}}\cdot\frac{1}{d_m},$$
$$1\leqslant i\leqslant m-1,\quad 2\leqslant j\leqslant m-2, \tag{5.15}$$

when $d_l=0,\ 1\leqslant l\leqslant m-2$. Let us consider first (5.14). It is plain that, when $d_l=0,\ 1\leqslant l\leqslant m-2$,

$$\begin{cases}\left\|\dfrac{\langle k\rangle}{d_{k-1}}\right\| &= O(1)|A_k|, \qquad 2\leqslant k\leqslant l,\\ \|\langle l+1\rangle\| &= \gamma|A_1\cdots A_l|,\\ \left\|\dfrac{\langle k\rangle}{d_{k-1}}\right\| &= O(1), \qquad l+2\leqslant k\leqslant m-1.\end{cases}$$

Thus, we see that, when $d_l=0$,

$$\begin{aligned}\left\|d_l\left(\gamma\prod_{k=m-1}^{2}\frac{\langle k\rangle}{d_{k-1}}\cdot\frac{\gamma A_1\cdots A_{m-1}}{d_{m-1}d_m}\frac{1}{A_i}\right)\right\| &= O(1)\frac{\gamma|A_2\cdots A_{m-1}|}{|A_i|}\\ &= O(1), \qquad 1\leqslant i\leqslant m-1,\\ \left\|d_l\left(\gamma\prod_{k=m-1}^{2}\frac{\langle k\rangle}{d_{k-1}}\cdot\frac{1}{d_m}\right)\right\| &= O(1)\gamma\end{aligned}$$

Each term of (5.15) satisfies a similar estimate. Thus we see that

$$\begin{aligned}&\left|\text{residues of each element of } e^{-t\lambda}(\lambda-\Lambda-\gamma\hat{G}\hat{W})^{-1}\right|\\ &= O(1)e^{-(\omega+\gamma+O(\gamma^2))t}, \quad \text{when } d_l=0,\ 1\leqslant l\leqslant m-2.\end{aligned} \tag{5.16}$$

Combining this with (5.13), we finally obtain the desired estimate (4.15). The proof of Theorem 5.1 is thereby complete. □

As we mentioned in Remark 2, the assumption (5.3) seems just of a technical nature. Let us show the estimate (4.15) in the case where (5.3) is lost with $\nu=3$. When $\nu=4$, however, its examination is lengthy; takes a lot of pages; and thus omitted: Details are found in [46]. We have three eigenvalues λ_1, λ_2, and λ_3 in question. When (5.3) is lost, we have the relation: $\lambda_1=(\lambda_2+\lambda_3)/2$, which might cause a possible singularity regarding γ in calculation of the residues of $e^{-t(\Lambda+\gamma\hat{G}\hat{W})}$. When $\nu=4$, we have to examine three more relations regarding λ_i (see Remark 1 following Theorem 5.1).

The matrix $(\lambda-\Lambda-\gamma\hat{G}\hat{W})^{-1}$ has nine blocks according to the factorization (5.8). We examine these blocks in each case. Possible singularities of $(\lambda-\Lambda-\gamma\hat{G}\hat{W})^{-1}$, however, arise only in the $(1,1)$-block. The other blocks have no problem, the proof of which is omitted to save spaces. Otherwise we have to write down 9×1 cases altogether ($16 \times 4 = 64$ cases in the case where $\nu = 4$).

On the residues of the block $e^{-t\lambda}(\lambda - \Lambda - \gamma\hat{G}\hat{W})^{-1}|_{(1,1)}$:

According to the expressions (5.5) and (5.7), we see that

$$(\lambda - \Lambda - \gamma\hat{G}\hat{W})^{-1}|_{(1,1)} = \frac{1}{d_1} I + \frac{\gamma^2}{d_1 d_2} \hat{G}_1 \hat{W}_2 \hat{G}_2 \hat{G}_1^{-1} + \frac{\gamma^2 A_2}{d_1 d_2 d_3} \hat{G}_1 \hat{W}_3 \hat{G}_3 \langle 2 \rangle \hat{G}_1^{-1}, \tag{5.17}$$

where $\langle 2 \rangle = \gamma A_1 \hat{W}_2 \hat{G}_2 + d_2 I$. Thus we have to evaluate the residues of the functions:

$$\frac{\gamma^3 A_1 A_2}{d_1 d_2 d_3}, \quad \frac{\gamma^2 A_2}{d_1 d_3}, \quad \text{and} \quad \frac{\gamma^2}{d_1 d_2}$$

times $e^{-t\lambda}$ at each singularity. In the following, we encounter similar calculations repeatedly. Thus we only examine typical cases. The other calculations will be left to the readers. Based on the d_i-d_j table, we need to analyze precisely the properties of d_j, $j \neq i$ when $d_i = 0$.

	d_1	d_2	d_3
The case where $d_1 = 0$ $(\lambda_{11}(\gamma) = \lambda_1 + \gamma)$			
$\lambda_{11}(\gamma)$:	0	$O(\gamma^2)$	$O(\gamma^3)$
The case where $d_2 = 0$ $(\lambda = \lambda_{21}(\gamma) \sim \lambda_1,\ \lambda_{22}(\gamma) \sim \lambda_2)$			
$\lambda_{21}(\gamma)$:	$O(\gamma^2)$	0	$O(\gamma^2)$
$\lambda_{22}(\gamma)$:	$O(1)$	0	$O(\gamma^2)$
The case where $d_3 = 0$ $(\lambda = \lambda_{31}(\gamma) \sim \lambda_1,\ \lambda_{32}(\gamma) \sim \lambda_2,\ \lambda_{33}(\gamma) \sim \lambda_3)$			
$\lambda_{31}(\gamma)$:	$O(\gamma^3)$	$O(\gamma^2)$	0
$\lambda_{32}(\gamma)$:	$O(1)$	$O(\gamma^2)$	0
$\lambda_{33}(\gamma)$:	$O(1)$	$O(1)$	0

The residues of $\dfrac{\gamma^3 A_1 A_2}{d_1 d_2 d_3} e^{-t\lambda}$

As we see below, singularities regarding γ arise when $d_1 = 0$ and $d_3 = 0$. In view of the behaviors of the d_i in the above table, no singularity arises in the residues at $\lambda = \lambda_{21}(\gamma)$, $\lambda_{22}(\gamma)$ (or $d_2 = 0$), $\lambda_{32}(\gamma)$, $\lambda_{33}(\gamma)$. At $\lambda = \lambda_{11}(\gamma) = \lambda_1 + \gamma$, we calculate the residue as

$$\begin{aligned}\operatorname{Res}\left(\frac{\gamma^3 A_1 A_2}{d_1 d_2 d_3} e^{-t\lambda}; \lambda_1 + \gamma\right) &= \left.\frac{\gamma^3 A_1 A_2}{d_2 d_3} e^{-t\lambda}\right|_{\lambda=\lambda_1+\gamma} \\ &= \frac{1}{2\gamma}(\lambda_1 - \lambda_2 + \gamma) e^{-(\lambda_1+\gamma)t}.\end{aligned}$$

At $\lambda = \lambda_{31}(\gamma) \sim \lambda_1$, the residue is

$$\begin{aligned}\operatorname{Res}\left(\frac{\gamma^3 A_1 A_2}{d_1 d_2 d_3} e^{-t\lambda}; \lambda_{31}(\gamma)\right) &= \left.\frac{\gamma^3 A_1 A_2}{d_1 d_2 (\lambda - \lambda_{32})(\lambda - \lambda_{33})} e^{-t\lambda}\right|_{\lambda=\lambda_{31}} \\ &= \frac{1}{2\gamma}(\lambda_1 - \lambda_3 + O(\gamma)) e^{-\lambda_{31}(\gamma)t} \\ &= -\frac{1}{2\gamma}(\lambda_1 - \lambda_2 + O(\gamma)) e^{-\lambda_{31}(\gamma)t}.\end{aligned}$$

Recalling that $\lambda_{31}(\gamma) = \lambda_1 + \gamma + O(\gamma^2)$, we obtain the estimate:

$$\begin{aligned}&\left|\operatorname{Res}\left(\frac{\gamma^3 A_1 A_2}{d_1 d_2 d_3} e^{-t\lambda}; \lambda_1 + \gamma\right) + \operatorname{Res}\left(\frac{\gamma^3 A_1 A_2}{d_1 d_2 d_3} e^{-t\lambda}; \lambda_{31}(\gamma)\right)\right| \\ &\qquad \leqslant \text{const}\, e^{-(\omega+\gamma/2+O(\gamma^2))t}, \quad t \geqslant 0.\end{aligned} \tag{5.18}$$

The residues of $\dfrac{\gamma^2 A_2}{d_1 d_3} e^{-t\lambda}$

At $\lambda = \lambda_{11}(\gamma) = \lambda_1 + \gamma$, the residue is

$$\begin{aligned}\operatorname{Res}\left(\frac{\gamma^2 A_2}{d_1 d_3} e^{-t\lambda}; \lambda_1 + \gamma\right) &= \left.\frac{\gamma^2 A_2}{d_3} e^{-t\lambda}\right|_{\lambda=\lambda_1+\gamma} \\ &= -\frac{1}{2\gamma}(\lambda_1 - \lambda_2 + \gamma) e^{-(\lambda_1+\gamma)t}.\end{aligned}$$

At $\lambda = \lambda_{31}(\gamma) \sim \lambda_1$, the residue is

$$\begin{aligned}\operatorname{Res}\left(\frac{\gamma^2 A_2}{d_1 d_3} e^{-t\lambda}; \lambda_{31}(\gamma)\right) &= \left.\frac{\gamma^2 A_2}{d_1 (\lambda - \lambda_{32})(\lambda - \lambda_{33})} e^{-t\lambda}\right|_{\lambda=\lambda_{31}} \\ &= \frac{1}{2\gamma}(\lambda_1 - \lambda_2 + O(\gamma)) e^{-\lambda_{31}(\gamma)t}.\end{aligned}$$

Similarly we obtain

$$\begin{aligned}&\left|\operatorname{Res}\left(\frac{\gamma^2 A_2}{d_1 d_3} e^{-t\lambda}; \lambda_1 + \gamma\right) + \operatorname{Res}\left(\frac{\gamma^2 A_2}{d_1 d_3} e^{-t\lambda}; \lambda_{31}(\gamma)\right)\right| \\ &\qquad \leqslant \text{const}\, e^{-(\omega+\gamma/2+O(\gamma^2))t}, \quad t \geqslant 0.\end{aligned} \tag{5.19}$$

Combining these estimates together, we obtain in (4.18)

$$\left\| e^{-t(\Lambda+\gamma\hat{G}\hat{W})}\Big|_{(1,1)} \right\|_{\mathscr{L}(\mathbb{C}^n)} = \left\| \frac{1}{2\pi i} \int_C e^{-t\lambda} (\lambda - \Lambda - \gamma\hat{G}\hat{W})^{-1}\Big|_{(1,1)} d\lambda \right\|_{\mathscr{L}(\mathbb{C}^n)}$$
$$\leqslant \text{const}\, e^{-(\omega+\gamma/2+O(\gamma^2))t}, \quad t \geqslant 0. \tag{5.20}$$

Remark: Due to the presence of the matrix $\hat{W}_2\hat{G}_2 \neq I_N$ in $\langle 2 \rangle$ (see (5.17)), simple calculations such as

$$\text{Res}\left(\frac{\gamma^3 A_1 A_2}{d_1 d_2 d_3} e^{-t\lambda}; \lambda_1 + \gamma\right) + \text{Res}\left(\frac{\gamma^2 A_2}{d_1 d_3} e^{-t\lambda}; \lambda_1 + \gamma\right)$$

are not generally allowed.

Table 3.1: The d_i-d_j table

	d_1	d_2	d_3	d_4	d_5	$\cdots$	d_ν
$d_1 = 0$	0	$-\gamma A_1$	$-\gamma A_1 \Xi_{(2,3)}$	$-\gamma A_1 \Xi_{(2,4)}$	$-\gamma A_1 \Xi_{(2,5)}$	$\cdots$	$-\gamma A_1 \Xi_{(2,\nu)}$
$d_2 = 0$	$\frac{\gamma A_1}{A_2}$	0	$-\gamma A_1 A_2$	$-\gamma A_1 A_2 \Xi_{(3,4)}$	$-\gamma A_1 A_2 \Xi_{(3,5)}$	$\cdots$	$-\gamma A_1 A_2 \Xi_{(3,\nu)}$
$d_3 = 0$	$\frac{\gamma A_1}{A_2 A_3} \Xi_{(2,3)}$	$\frac{\gamma A_1 A_2}{A_3}$	0	$-\gamma A_1 A_2 A_3$	$-\gamma A_1 A_2 A_3 \Xi_{(4,5)}$	$\cdots$	$-\gamma A_1 A_2 A_3 \Xi_{(4,\nu)}$
$d_4 = 0$	$\frac{\gamma A_1}{A_2 A_3 A_4} \Xi_{(2,4)}$	$\frac{\gamma A_1 A_2}{A_3 A_4} \Xi_{(3,4)}$	$\frac{\gamma A_1 A_2 A_3}{A_4}$	0	$-\gamma A_1 A_2 A_3 A_4$	$\cdots$	$-\gamma A_1 A_2 A_3 A_4 \Xi_{(5,\nu)}$
$d_5 = 0$	$\frac{\gamma A_1}{A_2 A_3 A_4 A_5} \Xi_{(2,5)}$	$\frac{\gamma A_1 A_2}{A_3 A_4 A_5} \Xi_{(3,5)}$	$\frac{\gamma A_1 A_2 A_3}{A_4 A_5} \Xi_{(4,5)}$	$\frac{\gamma A_1 A_2 A_3 A_4}{A_5}$	0	$\cdots$	$-\gamma A_1 A_2 A_3 A_4 A_5 \Xi_{(6,\nu)}$
$\vdots$	$\vdots$	$\vdots$	$\vdots$	$\vdots$	$\vdots$	$\vdots$	$\vdots$
$d_\nu = 0$	$\frac{\gamma A_1}{A_2 A_3 \cdots A_\nu} \Xi_{(2,\nu)}$	$\frac{\gamma A_1 A_2}{A_3 \cdots A_\nu} \Xi_{(3,\nu)}$	$\frac{\gamma A_1 A_2 A_3}{A_4 \cdots A_\nu} \Xi_{(4,\nu)}$	$\frac{\gamma A_1 A_2 A_3 A_4}{A_5 \cdots A_\nu} \Xi_{(5,\nu)}$	$\frac{\gamma A_1 A_2 A_3 A_4 A_5}{A_6 \cdots A_\nu} \Xi_{(6,\nu)}$	$\cdots$	0

Table 3.2: The d_i-$\langle j \rangle$ table

	$\langle 2 \rangle$	$\langle 3 \rangle$	$\langle 4 \rangle$	$\cdots$	$\langle \nu - 1 \rangle$
$d_1 = 0$	$\gamma A_1(\hat{W}_2\hat{G}_2 - 1)$	$\gamma A_1 A_2 \left(\hat{W}_3\hat{G}_3 - \frac{\Xi_{(2,3)}}{A_2}\right)$	$\gamma A_1 A_2 A_3 \left(\hat{W}_4\hat{G}_4 - \frac{\Xi_{(2,4)}}{A_2 A_3}\right)$	$\cdots$	$\gamma A_1 \cdots A_{\nu-2} \left(\hat{W}_{\nu-1}\hat{G}_{\nu-1} - \frac{\Xi_{(2,\nu-1)}}{A_2 \cdots A_{\nu-2}}\right)$
$d_2 = 0$	$\gamma A_1 \hat{W}_2\hat{G}_2$	$\gamma A_1 A_2(\hat{W}_3\hat{G}_3 - 1)$	$\gamma A_1 A_2 A_3 \left(\hat{W}_4\hat{G}_4 - \frac{\Xi_{(3,4)}}{A_3}\right)$	$\cdots$	$\gamma A_1 \cdots A_{\nu-2} \left(\hat{W}_{\nu-1}\hat{G}_{\nu-1} - \frac{\Xi_{(3,\nu-1)}}{A_3 \cdots A_{\nu-2}}\right)$
$d_3 = 0$	$\gamma A_1 \left(\hat{W}_2\hat{G}_2 + \frac{A_2}{A_3}\right)$	$\gamma A_1 A_2 \hat{W}_3\hat{G}_3$	$\gamma A_1 A_2 A_3(\hat{W}_4\hat{G}_4 - 1)$	$\cdots$	$\gamma A_1 \cdots A_{\nu-2} \left(\hat{W}_{\nu-1}\hat{G}_{\nu-1} - \frac{\Xi_{(4,\nu-1)}}{A_4 \cdots A_{\nu-2}}\right)$
$d_4 = 0$	$\gamma A_1 \left(\hat{W}_2\hat{G}_2 + \frac{A_2 \Xi_{(3,4)}}{A_3 A_4}\right)$	$\gamma A_1 A_2 \left(\hat{W}_3\hat{G}_3 + \frac{A_3}{A_4}\right)$	$\gamma A_1 A_2 A_3 \hat{W}_4\hat{G}_4$	$\cdots$	$\gamma A_1 \cdots A_{\nu-2} \left(\hat{W}_{\nu-1}\hat{G}_{\nu-1} - \frac{\Xi_{(5,\nu-1)}}{A_5 \cdots A_{\nu-2}}\right)$
$d_5 = 0$	$\gamma A_1 \left(\hat{W}_2\hat{G}_2 + \frac{A_2 \Xi_{(3,5)}}{A_3 A_4 A_5}\right)$	$\gamma A_1 A_2 \left(\hat{W}_3\hat{G}_3 + \frac{A_3 \Xi_{(4,5)}}{A_4 A_5}\right)$	$\gamma A_1 A_2 A_3 \left(\hat{W}_4\hat{G}_4 + \frac{A_4}{A_5}\right)$	$\cdots$	$\gamma A_1 \cdots A_{\nu-2} \left(\hat{W}_{\nu-1}\hat{G}_{\nu-1} - \frac{\Xi_{(6,\nu-1)}}{A_6 \cdots A_{\nu-2}}\right)$
$\vdots$	$\vdots$	$\vdots$	$\vdots$	$\vdots$	$\vdots$
$d_\nu = 0$	$\gamma A_1 \left(\hat{W}_2\hat{G}_2 + \frac{A_2 \Xi_{(3,\nu)}}{A_3 A_4 \cdots A_\nu}\right)$	$\gamma A_1 A_2 \left(\hat{W}_3\hat{G}_3 + \frac{A_3 \Xi_{(4,\nu)}}{A_4 \cdots A_\nu}\right)$	$\gamma A_1 A_2 A_3 \left(\hat{W}_4\hat{G}_4 + \frac{A_4 \Xi_{(5,\nu)}}{A_5 \cdots A_\nu}\right)$	$\cdots$	$\gamma A_1 \cdots A_{\nu-2} \left(\hat{W}_{\nu-1}\hat{G}_{\nu-1} + \frac{A_{\nu-1}}{A_\nu}\right)$

Chapter 4

Stabilization of linear systems of infinite dimension: Dynamic feedback

4.1 Introduction

In Chapter 3, stabilization problems are discussed in the framework of the static feedback scheme. While the scheme is simple, an essential assumption is that at least one of the sensorss w_k and the actuatots g_k must be constructed in a finite-dimensional subspace, that is, $w_k \in P^*H$ or $g_k \in PH$, $1 \leqslant k \leqslant N$. The assumption seems not plausible in engineering implementations, especially when considering the scheme of boundary observation/boundary feedback. In the case where both w_k and g_k admit spillovers, a stability improvement is discussed in Theorems 4.1 and 5.1, and also applied to the scheme of boundary observation/boundary feedback. However, the improvement is just a little. As mentioned in the beginning of Chapter 3, other recent stabilization results by static feedback scheme have serious difficulties in engineering implementation: There is little viewpoint on guaranteeing narrower supports of sensors and actuators.

We develop in this chapter stabilization problems for linear parabolic control systems when all control actions are executed on the boundary, i.e., the scheme of boundary observation/boundary feedback. In achieving stabilization, we employ the *dynamic feedback control scheme* containing dynamic compensators in the feedback loop. The concept of a dynamic compensator originates from D. G. Luenberger [33], who studied linear control systems of finite dimension when desirable outputs from sensors are not expected: The designed scheme makes the outputs of the compensator asymptotically approach the desired outputs as $t \to \infty$. In our study, outputs from the boundary are not desirable ones. By assuming a finite number of observability conditions on the sensors w_k and controllability conditions on the actuators g_k, we design a finite dimensional dynamic compensator such that the state of the controlled plant and the compensator decays exponentially with the designated decay rate as $t \to \infty$. While the pair $(\mathscr{L}, \tau)$ of coefficient operators in this chapter is a standard one, it has enough generality in the sense that

(i) *no* Riesz basis is generally expected associated with $(\mathscr{L}, \tau)$;

(ii) Jordan spectral substructures naturally arise; and

(iii) enough information on the fractional powers of the associated elliptic operator is *not* available, due to the complexity of the boundary operator τ.

There are two kinds of compensators: One is an *identity* compensator, and the other a *generalized* one. In this chapter, feedback schemes containing generalized compensators are studied to cope with a considerably general class of boundary control systems. In the case where a Riesz basis exists, however, an identity compensator is also useful, since the Riesz basis makes it possible to approximate the system quantitatively by finite-dimensional linear systems. Stabilization problems for linear boundary systems with Riesz bases will be discused in Chapter 5. The results of this chapter are based on those discussed in [37, 39, 41, 42, 44, 45, 50].

Let us begin with preliminary results which characterizes the boundary control systems. Let Ω be a bounded domain in $\mathbb{R}^m$ with the boundary Γ which consists of a finite number of smooth components of $(m-1)$-dimension, and let $(\mathscr{L}, \tau)$ be the pair of differential operators which appeared in Sections 1 and 3, Chapter 2, that is,

$$\begin{aligned} \mathscr{L}u &= -\sum_{i,j=1}^{m} \frac{\partial}{\partial x_i}\left(a_{ij}(x)\frac{\partial u}{\partial x_j}\right) + \sum_{i=1}^{m} b_i(x)\frac{\partial u}{\partial x_i} + c(x)u, \\ \tau u &= \alpha(\xi)u + (1-\alpha(\xi))\frac{\partial u}{\partial \nu}, \end{aligned} \tag{1.1}$$

where $a_{ij}(x) = a_{ji}(x)$ for $1 \leqslant i, j \leqslant m$, $x \in \overline{\Omega}$; for some positive δ

$$\sum_{i,j=1}^{m} a_{ij}(x)\xi_i\xi_j \geqslant \delta|\xi|^2, \quad \forall \xi = (\xi_1, \dots, \xi_m) \in \mathbb{R}^m, \quad \forall x \in \overline{\Omega};$$

and

$$0 \leqslant \alpha(\xi) \leqslant 1, \qquad \frac{\partial u}{\partial \nu} = \sum_{i,j=1}^{m} a_{ij}(\xi)\nu_i(\xi) \left. \frac{\partial u}{\partial x_j} \right|_{\Gamma},$$

$\boldsymbol{\nu}(\xi) = (\nu_1(\xi), \ldots, \nu_m(\xi))$ being the unit outer normal at each point $\xi \in \Gamma$. As in the preceding chapters, the last term $\left.\frac{\partial u}{\partial x_j}\right|_{\Gamma}$ means the trace $\gamma\left(\frac{\partial u}{\partial x_j}\right)$ of $\frac{\partial u}{\partial x_j}$ on Γ. As for the regularity of the coefficients, it is enough to assume that $a_{ij}(\cdot)$, $b_i(\cdot)$, $c(\cdot)$, and $\alpha(\cdot)$ belong to $C^2(\overline{\Omega})$, $C^2(\overline{\Omega})$, $C^{\omega}(\overline{\Omega})$, and $C^{2+\omega}(\Gamma)$, respectively, where ω, $0 < \omega < 1$ denote respective constants. Throughout the chapter, all arguments are based on the $L^2(\Omega)$-framework. As in Chapter 3, the pair $(\mathscr{L}, \tau)$ equipped with the homogeneous boundary condition, $\tau u = 0$ defines an elliptic operator L: Let $\hat{L}$ be the operator defined by

$$\hat{L}u = \mathscr{L}u, \quad u \in \mathscr{D}(\hat{L}),$$
$$\mathscr{D}(\hat{L}) = \left\{u \in C^2(\Omega) \cap C^1(\overline{\Omega});\ \mathscr{L}u \in L^2(\Omega),\ \tau u = 0\right\}.$$

The closure of $\hat{L}$ in $L^2(\Omega)$, the existence of which is ensured (see Subsection 3.3, Chapter 2), is denoted by L. The domain $\mathscr{D}(L)$ consists of all $u \in L^2(\Omega)$ with the following properties: (i) There is a sequence $\{u_n\} \subset \mathscr{D}(\hat{L})$ such that $u_n \to u$ in $L^2(\Omega)$, and (ii) $\hat{L}u_n$ converges in $L^2(\Omega)$ as $n \to \infty$. It is well known (see Chapter 2) that L has a compact resolvent $(\lambda - L)^{-1}$, and that the spectrum $\sigma(L)$ lies in the complement $(\overline{\Sigma} - b)^c$ of some sector $\overline{\Sigma} - b$, where $\overline{\Sigma} = \{\lambda \in \mathbb{C};\ \theta_0 \leqslant |\arg\lambda| \leqslant \pi\}$, $0 < \theta_0 < \pi/2$, $b \in \mathbb{R}^1$ (actually, the spectrum $\sigma(L)$ lies inside some parabola with the real axis as the axis of symmetry). There is a set of *generalized* eigenpairs $\{\lambda_i,\ \varphi_{ij}\}$ such that

(i) $\sigma(L) = \{\lambda_1, \lambda_2, \ldots, \lambda_i, \ldots\}$, $\quad \mathrm{Re}\,\lambda_1 \leqslant \mathrm{Re}\,\lambda_2 \leqslant \cdots \leqslant \mathrm{Re}\,\lambda_i \leqslant \cdots \to \infty$; and

(ii) $L\varphi_{ij} = \lambda_i\varphi_{ij} + \sum_{k<j} \alpha^i_{jk}\varphi_{ik}, \quad i \geqslant 1,\ 1 \leqslant j \leqslant m_i\,(<\infty)$.

Let $(\mathscr{L}^*, \tau^*)$ be the formal adjoint of $(\mathscr{L}, \tau)$:

$$\mathscr{L}^*\varphi = -\sum_{i,j=1}^{m} \frac{\partial}{\partial x_i}\left(a_{ij}(x)\frac{\partial \varphi}{\partial x_j}\right) - \mathrm{div}(\boldsymbol{b}(x)\varphi) + c(x)\varphi,$$
$$\tau^*\varphi = \alpha(\xi)\varphi + (1-\alpha(\xi))\left(\frac{\partial \varphi}{\partial \nu} + (\boldsymbol{b}(\xi)\cdot\boldsymbol{\nu}(\xi))\varphi\right), \tag{1.1'}$$

where $\boldsymbol{b}(x) = (b_1(x), \ldots, b_m(x))$. The pair $(\mathscr{L}^*, \tau^*)$ similarly defines the operator $\hat{L}^*$. Then the adjoint of L, denoted by L^*, is given as the closure of $\hat{L}^*$ in $L^2(\Omega)$. There is a set of generalized eigenpairs $\{\overline{\lambda_i},\ \psi_{ij}\}$ such that

(i) $\sigma(L^*) = \left\{\overline{\lambda_1}, \overline{\lambda_2}, \ldots, \overline{\lambda_i}, \ldots\right\}$; and

(ii) $L^*\psi_{ij} = \overline{\lambda_i}\psi_{ij} + \sum_{k<j} \beta^i_{jk}\psi_{ik}, \quad i \geqslant 1,\ 1 \leqslant j \leqslant m_i\,(<\infty)$.

In the specific case where $\alpha(\xi) \equiv 1$ or $\alpha(\xi) < 1$ on Γ, we note that the elliptic theory for L is standard, and that deeper results are well known (see, e.g., [1, 20, 35]). For example, the domain $\mathscr{D}(L)$ is simply characterized by the set $\{u \in H^2(\Omega);\ \tau u = 0 \text{ on } \Gamma\}$.

Remark: The set $\{\varphi_{ij};\ i \geqslant 1,\ 1 \leqslant j \leqslant m_i\}$ spans $L^2(\Omega)$ (see [1]). In the general case, however, it is not clear if $\{\varphi_{ij}\}$ would be a Riesz basis.

Note that the estimates

$$\begin{aligned} &\|(\lambda - L)^{-1}\| \leqslant \frac{\text{const}}{1+|\lambda|}, \qquad \text{and} \\ &\|(\lambda - L)^{-1}\|_{\mathscr{L}(L^2(\Omega);H^1(\Omega))} \leqslant \frac{\text{const}}{1+|\lambda|^{1/2}}, \qquad \lambda \in \overline{\Sigma} - b \end{aligned} \tag{1.2}$$

hold, where the norm $\|\cdot\|$ denotes the $L^2(\Omega)$- or the $\mathscr{L}(L^2(\Omega))$-norm. The latter estimate is derived from the relation (see (3.30), Chapter 2)

$$\mathscr{D}(L_c^{\omega/2}) \subset H^{\omega}(\Omega), \quad 0 \leqslant \omega \leqslant 1,$$

where $L_c = L + c$ is m-accretive, $c > 0$ being chosen so that $\sigma(L_c) \subset \mathbb{C}_+$. By the former estimate in (1.2), $-L$ is an infinitesimal generator of an analytic semigroup e^{-tL}, $t > 0$.

Before stating our stabilization problem, it is worthwhile to remark that elliptic operators admitting Jordan spectral substructures are not limited to the above general L: Jordan substructures also appear very naturally, for example, when control systems are composed, as a coupling system, of finite-dimensional ode and parabolic pde. Here we raise two such examples.

Example 1: Let L_1 be a sectorial operator in a Hilbert space H_1 with dense domain, and L_2 an $n \times n$ general square matrix admitting Jordan structures. Replace our L in $L^2(\Omega)$ by another L:

$$L = \begin{pmatrix} L_1 & 0 \\ M & L_2 \end{pmatrix}$$

in the product space $H_1 \times \mathbb{R}^n$. Here M denotes a linear operator connecting the pde and the ode which is subordinate to L_1. As long as $\sigma(L_1) \cap \sigma(L_2) = \varnothing$, the operator L is algebraically similar to $\begin{pmatrix} L_1 & 0 \\ 0 & L_2 \end{pmatrix}$. Thus, $\sigma(L) = \sigma(L_1) \cup \sigma(L_2)$, and the Jordan structures of L_2 is added to L as a substructure. In fact, Sylvester's equation: $XL_1 - L_2X = -M$ admits a unique operator solution given by the Rosenblum formula [6]:

$$X = \frac{1}{2\pi i} \int_C (\lambda - L_2)^{-1} M (\lambda - L_1)^{-1}\, d\lambda,$$

where C denotes a Jordan contour encircling only $\sigma(L_2)$. Setting then $T = \begin{pmatrix} 1 & 0 \\ X & 1 \end{pmatrix}$, we immediately find that

$$TLT^{-1} = \begin{pmatrix} L_1 & 0 \\ 0 & L_2 \end{pmatrix}.$$

The above assumption, $\sigma(L_1) \cap \sigma(L_2) = \varnothing$ is not a necessity. In fact, let us consider another example of such composite systems.

Example 2: Let L_1 be a sectorial operator in a Hilbert space H_1, such that $\lambda_1 \in \sigma(L_1)$ is a *simple* eigenvalue and φ_1 the corresponding eigenvector. Let L be an operator in $H_1 \times \mathbb{R}^2$ defined by

$$L = \begin{pmatrix} L_1 & 0 \\ M & L_2 \end{pmatrix},$$

where

$$L_2 = \begin{pmatrix} \lambda_1 & 1 \\ 0 & \lambda_1 \end{pmatrix}, \quad \text{and } M = \begin{pmatrix} \langle \cdot, a \rangle \\ \langle \cdot, b \rangle \end{pmatrix}, \quad a,\, b \in H_1.$$

Then, $\sigma(L_1) \cap \sigma(L_2) \neq \varnothing$. Assuming that $\langle \varphi_1, b \rangle = 0$, we easily find that λ_1 is an eigenvalue of L, and that the dimension of the eigenspace $W_{\lambda_1}^{(1)}$ is equal to 2. In addition, L admits a generalized eigenspace $W_{\lambda_1}^{(k)} = \{u \in \mathscr{D}(L^k) \times \mathbb{R}^2;\ (\lambda_1 - L)^k u = 0\}$, $k \geqslant 2$. After calculation, we see that $W_{\lambda_1}^{(1)} \subsetneq W_{\lambda_1}^{(2)} = W_{\lambda_1}^{(3)} = \ldots$, and that $\dim W_{\lambda_1}^{(2)} = 3$. Laurent's expansion of the resolvent $(\lambda - L)^{-1}$ at $\lambda = \lambda_1$ is then

$$(\lambda - L)^{-1} = \frac{A_{-2}}{(\lambda - \lambda_1)^2} + \frac{A_{-1}}{\lambda - \lambda_1} + \sum_{k=0}^{\infty} (\lambda - \lambda_1)^i A_i, \quad A_{-2} \neq 0.$$

The operator A_{-1}, of course, means the projector corresponding to λ_1.

Let us go back to our problem. Our aim is to construct a boundary feedback control system which is finally described by the following system of linear differential equations:

$$\begin{cases} \dfrac{\partial u}{\partial t} + \mathscr{L} u = 0 \quad \text{in } \mathbb{R}_+^1 \times \Omega, \\ \tau u = \displaystyle\sum_{k=1}^{M} \langle v, \rho_k \rangle_{\mathbb{R}^\ell}\, g_k \quad \text{on } \mathbb{R}_+^1 \times \Gamma, \\ \dfrac{dv}{dt} + B_1 v = \displaystyle\sum_{k=1}^{N} p_k(u) \xi_k \quad \text{in } \mathbb{R}_+^1, \\ u(0,\cdot) = u_0(\cdot) \quad \text{in } \Omega, \qquad v(0) = v_0. \end{cases} \tag{1.3}$$

Equation (1.3) will be derived from the stabilization problem later in Section 2. In (1.3), the controlled plant Σ_p with state $u = u(t,\cdot)$ is characterized by

(i) the pair of linear differential operators $(\mathscr{L}, \tau)$ in (1.1),
(ii) outputs $p_k(u)$, $1 \leqslant k \leqslant N$, on Γ defined through sensors w_k in $L^2(\Gamma)$ as

$$p_k(u) = \begin{cases} \langle u, w_k \rangle_\Gamma, & \text{in the case where } \alpha(\xi) \not\equiv 1, \quad 1 \leqslant k \leqslant N, \\ \left\langle \dfrac{\partial u}{\partial \nu}, w_k \right\rangle_\Gamma, & \text{in the case where } \alpha(\xi) \equiv 1, \quad 1 \leqslant k \leqslant N, \end{cases} \tag{1.4}$$

and

(iii) actuators g_k, $1 \leqslant k \leqslant M$, on Γ belonging to $C^{2+\omega}(\Gamma)$.

The *compensator* Σ_c with state $v = v(t)$ is described by a differential equation in $\mathbb{R}^\ell$, where B_1 denotes an $\ell \times \ell$ coefficient matrix, ρ_k sensors in $\mathbb{R}^\ell$, and ξ_k actuators in $\mathbb{R}^\ell$. The stabilization problem for (1.3) is stated as follows:

> *"Given the parameters of Σ_p, i.e., $(\mathscr{L}, \tau)$, w_k, and g_k satisfying a finite number of algebraic conditions, determine the parameters of Σ_c, i.e., dimension ℓ, B_1, ρ_k, and ξ_k, such that the state $(u(t,\cdot), v(t))$ decays with a designated decay rate as $t \to \infty$ for every initial state (u_0, v_0)."*

An advantage of the dynamic feedback control scheme studied here over the static scheme is that, as long as both w_k and g_k on Γ satisfy the first two algebraic conditions of (3.5) in Theorem 3.1, there is no geometric restriction on their structures, especially on their locations, i.e., supports on Γ. The relationship between u and v in (1.3) is shown in the following figure. As before, the symbol $(\dots)^{\mathrm{T}}$ in Figure 2 means the transpose of a vector $(\dots)$:

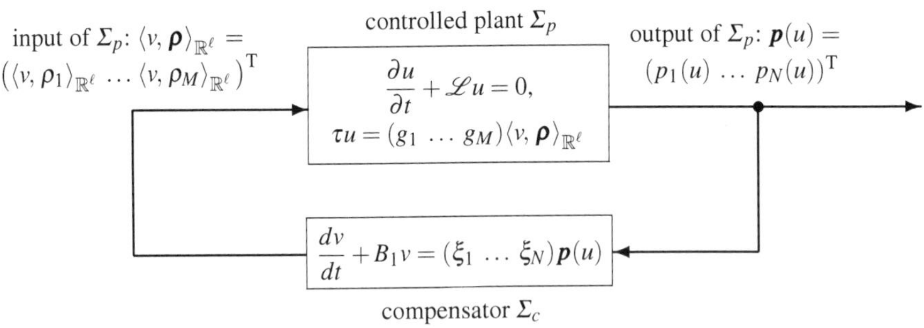

Figure 2

For better understanding of various parameters appearing in (1.3), let us illustrate a simple example in one space dimension. Let Ω be an interval $I = (a, b) \subset \mathbb{R}^1$, and the pair $(\mathscr{L}, \tau)$ characterizing Σ_p be such that

$$\mathscr{L}u = -(a(x)u_x)_x + b(x)u_x + c(x)u,$$
$$\tau u = (u_\nu + \sigma(\xi)u)|_{\xi = a, b},$$

where $a(x) \geqslant \delta$ (> 0) on I, and τ is of the Robin type, that is, the case where $\alpha(\xi) < 1$ in (1.1). Note that there is a Riesz basis associated with L (through the Sturm-Liouville transform [11]). However, we do not use this fact in our approach. In this case, we choose $M = N = 1$, since the spectrum of L consists only of simple eigenvalues. The output $p(u)$ is considered at $x = a$, i.e., $p(u) = u(t,a)$, or $(w(a), w(b)) = (1, 0)$ in (1.4). The input is considered at $x = b$, i.e., $(g(a), g(b)) = (0, 1)$ in (1.3). Thus, eqn. (1.3) is described as

$$\begin{cases} u_t + \mathscr{L}u = 0, \quad u(0,\cdot) = u_0(\cdot), \\ (u_x + \sigma u)\big|_{x=a} = 0, \quad (u_x + \sigma u)\big|_{x=b} = \langle v, \rho \rangle_{\mathbb{R}^\ell}, \\ v_t + B_1 v = u(t,a)\xi, \quad v(0) = v_0. \end{cases} \tag{1.3'}$$

In (1.3′), the dimension ℓ of Σ_c, the $\ell \times \ell$ matrix B_1, and the vectors ρ, $\xi \in \mathbb{R}^\ell$ in Σ_c are the parameters to be designed. Since the first two conditions of (3.5) in Section 3 are satisfied for the above w and g, Theorem 3.1 stated later will enable us to design these parameters, such that the state $(u, v) \in L^2(I) \times \mathbb{R}^\ell$ is stabilized with a designated decay rate. The equation for v and the parameters therein are derived from a differential equation in a separable Hilbert space H. To give an example of ℓ, B_1, and q, we have to solve numerically an *ill-posed problem* derived from an infinite-dimensional Sylvester's equation (see (3.8), and Chapter 8 for a numerical approximation algorithm).

As defined just above, the assumption on the pair $(\mathscr{L}, \tau)$ is fairly general, so that we can *no more* expect a Riesz basis associated with L when the dimension m is greater than or equal to 2. Thus, the operator L studied here would work as a prototype of its abstract versions assuming only that L is a sectorial operator, and that the unstable part of $\sigma(L)$ consists of eigenvalues of finite multiplicities. We remark the following observation: There is an attempt to draw out a class of elliptic operators equipped with Riesz bases (see, for example, [29, 34]). It seems that these operators are limited to a narrow class such as self-adjoint operators plus *relatively small* perturbations. Along a similar line, there are also classical results on when a *little* perturbed system of a Riesz basis would form a Riesz basis [72].

We hope to establish a stabilization scheme and the sharpest criterion effective to a very broad class of controlled plants Σ_p such as highly complicated composite systems of general pde-ode (and/or pde-pde), flexible arms with multiple joints, and others in physical and/or chemical applications. To cope with such systems, the compensator Σ_c is of *generalized type*, and described in an arbitrary separable Hilbert space H. The feedback control system is thus described as a differential equation in the product space $L^2(\Omega) \times H$. With the dynamic feedback control scheme containing this type of compensators, we first establish the stabilization by assuming a finite number of the observability conditions on w_k and a finite number of the controllability conditions on g_k. It is essential in our stabilization process to introduce an operator equation, so called Sylvester's equation, and its operator solution $X \in \mathscr{L}(L^2(\Omega); H)$. Let F be the operator $\mathscr{L}$ equipped with the boundary condition: $\tau u = \sum_{k=1}^{M} \langle u, y_k \rangle g_k$. Based on the controllability conditions on g_k, we find a set of $y_k \in L^2(\Omega)$, $1 \leqslant k \leqslant M$ such that the semigroup e^{-tF}, generated by $-F$, decays exponentially as $t \to \infty$. There are two roles of the operator X: One is to ensure suitable vectors $\rho_k \in H$ such that $X^* \rho_k$ arbitrarily approximate y_k in the $L^2(\Omega)$-topology, and the other to ensure the decay of $\|Xu - v\|_H$ as $t \to \infty$, where v denotes the state of the compensator Σ_c. Here, the vectors ρ_k can stay in a finite dimensional subspace of H, the dimension of which determines the dimension of the compensator Σ_c. Since the set of y_k in F is determined by g_k, the dimension of Σ_c is, in other words, determined only by the actuators g_k.

In achieving our stabilization, an infinite dimensional version of Sylvester's equation and Carleman's theorem [31, 54, 67, 72] known in classical Fourier analysis play a central role: In stabilization problems in Chapter 1, we have realized effectiveness of Sylvester's equation of finite dimension. Carleman's theorem has been recognized as an important means to show completeness of a set of exponential functions of the form $e^{i\lambda_n x}$ on a finite interval, $\{\lambda_n\}$ being a sequence of distinct complex numbers such that $\sum_n |\lambda_n|^{-1} = \infty$ (details on an additional condition on $\{\lambda_n\}$ are found, e.g., [31, 54, 64, 72]). It is now applied to our stabilization studies, and stated as follows. The proof is detailed for the readers' convenience.

Theorem 1.1 (Carleman). *For a given positive number R, let $D = \{\lambda \in \mathbb{C};\ |\lambda| < R,\ \mathrm{Im}\,\lambda > 0\}$ be a bounded domain. Suppose that $f(\lambda)$ is analytic in $\overline{D} = D \cup \partial D$ such that* (i) $f(0) = 1$, *and* (ii) *f has the zeros $r_k e^{i\theta_k}$ of order m_k, $1 \leqslant k \leqslant p$, in D. Then we have the relation:*

$$\begin{aligned}\sum_{k=1}^{p} m_k \left(\frac{1}{r_k} - \frac{r_k}{R^2} \right) \sin\theta_k &= \frac{1}{\pi R} \int_0^{\pi} \log |f(Re^{i\theta})| \sin\theta \, d\theta \\ &\quad + \frac{1}{2\pi} \int_0^R \log |f(-x) f(x)| \left(\frac{1}{x^2} - \frac{1}{R^2} \right) dx \\ &\quad + \frac{1}{2} \,\mathrm{Im}\, f'(0).\end{aligned} \tag{1.5}$$

Proof. The boundary ∂D consists of the semicircle: $|\lambda| = R$, $0 < \arg \lambda < \pi$ and the segment: $|\lambda| \leqslant R$ on the real axis. There is a possibility that $f(\lambda)$ might have a finite number of zeros on ∂D. For simplicity, we may assume that the point a on the semicircle is a zero of order m; b on the segment $[-R, R]$ a zero of order n; and there is no other zero on ∂D. By setting

$$h(\lambda) = \frac{f(\lambda)}{\left(1 - \frac{\lambda}{a}\right)^m \left(1 - \frac{\lambda}{b}\right)^n} = \frac{f(\lambda)}{h_a(\lambda)^m h_b(\lambda)^n}, \tag{1.6}$$

where

$$h_a(\lambda) = 1 - \frac{\lambda}{a}, \quad h_b(\lambda) = 1 - \frac{\lambda}{b},$$

the function $h(\lambda)$ is analytic in $\overline{D}$; $h(\lambda) \neq 0$, $\lambda \in \partial D$; and $h(0) = 1$. The zeros of $h(\lambda)$ in D coincide with the zeros of $f(\lambda)$ of the same order: $\lambda = r_k e^{i\theta_k}$, $1 \leqslant k \leqslant p$, $0 < \theta_k < \pi$ (of order m_k). For a small $\rho > 0$, we modify ∂D a little around the origin 0, so that the modified contour C_ρ contains 0 in its inside. The domain inside C_ρ is denoted as D_ρ (see Figure 3). Let us consider the contour integral,

$$I = \frac{1}{2\pi} \int_{C_\rho} \log h(\lambda) \left(\frac{1}{\lambda^2} - \frac{1}{R^2} \right) d\lambda, \tag{1.7}$$

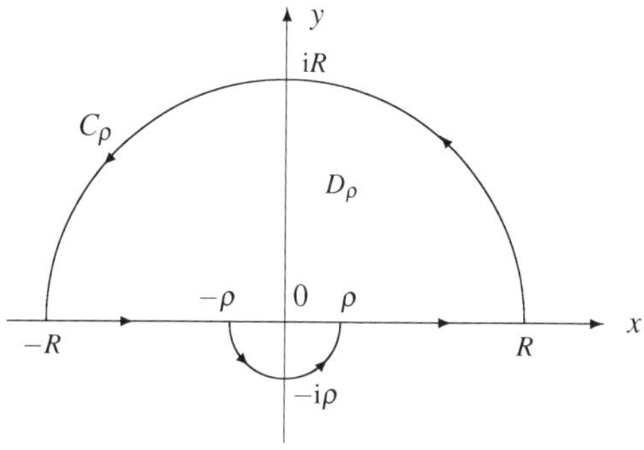

Figure 3

where C_ρ is positively oriented, and the integral begins and ends at $\lambda = R$ with a fixed determination of the logarithm. Let $\arg h(R) = \theta_0$, $0 \leqslant \theta_0 < 2\pi$ at the initial point R for a moment. When λ goes toward $-\rho \sim 0$ along C_ρ, then $h(-\rho)$ is close to $h(0) = 1$. Assume that $h(\lambda)$ turns l_1 times around 0 at that time (Figure 4 below shows an example of the contour of $h(\lambda)$ in the case of $l_1 = 2$): More precisely, it is assumed that the resultant $\arg h(-\rho)$ ($\sim \arg 1$) is close to $2l_1\pi$. Then, by shifting the argument, set $\arg h(R) = \theta_0 - 2l_1\pi$, so that $\log h(R) = \log |h(R)| + i(\theta_0 - 2l_1\pi)$ at the initial point. Since $\operatorname{Im} \log h(\lambda) = \arg h(\lambda)$ is a continuous function of λ, we see that

$$\begin{array}{rll} \lambda: & R \quad \xrightarrow{C_\rho} & -\rho \\ \hline \arg h(\lambda): & \theta_0 - 2l_1\pi \quad \longrightarrow & \arg h(-\rho) \\ & & \sim \arg h(0) = \arg 1 = 0 \end{array}$$

Thus, in a neighborhood of $\lambda = 0$, $\log h(\lambda)$ is expressed as

$$\begin{aligned} \log h(\lambda) &= \log h(0) + \frac{h'(0)}{h(0)} \lambda + \lambda^2 \varphi(\lambda) \\ &= h'(0)\lambda + \lambda^2 \varphi(\lambda), \end{aligned} \tag{1.8}$$

$\varphi(\lambda)$ being analytic at 0.

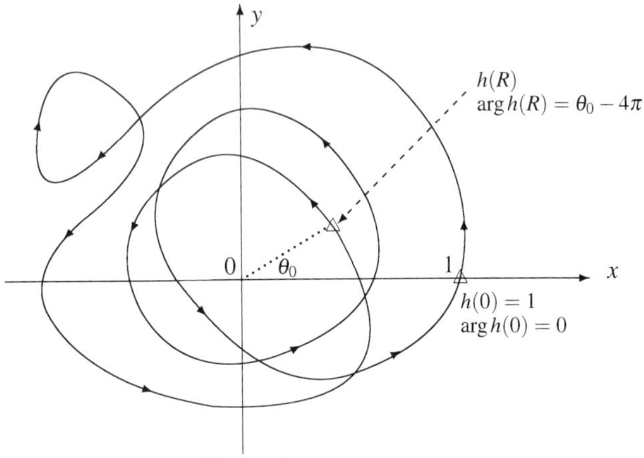

Figure 4

The integral I in (1.7) is calculated in two different manners. By setting $\lambda = Re^{i\theta}$, θ; $0 \to \pi$ on the semicircle of radius R,

$$\begin{aligned} &\frac{1}{2\pi}\int_0^{\pi} \log h(Re^{i\theta})\left(\frac{1}{R^2e^{2i\theta}} - \frac{1}{R^2}\right) Rie^{i\theta}\,d\theta \\ &= \frac{1}{2\pi}\int_0^{\pi} \log h(Re^{i\theta})\,\frac{1}{R}(-2i\sin\theta)i\,d\theta \\ &= \frac{1}{\pi R}\int_0^{\pi} \log h(Re^{i\theta})\,\sin\theta\,d\theta. \end{aligned}$$

The integral on the segments $[-R, -\rho]$ and $[\rho, R]$ becomes

$$\frac{1}{2\pi}\int_{\rho}^{R} \log h(-x)\left(\frac{1}{x^2} - \frac{1}{R^2}\right) dx,$$

and

$$\frac{1}{2\pi}\int_{\rho}^{R} \log h(x)\left(\frac{1}{x^2} - \frac{1}{R^2}\right) dx,$$

respectively. By applying Taylor's expansion (1.8) of $\log h(\lambda)$, the integral on the semicircle of radius ρ around 0 is calculated as follows:

$$\begin{aligned} &\frac{1}{2\pi}\int \left(h'(0)\lambda + \lambda^2\varphi(\lambda)\right)\left(\frac{1}{\lambda^2} - \frac{1}{R^2}\right) d\lambda \\ &= \frac{1}{2\pi}\int_{\pi}^{2\pi} \left(h'(0)\rho e^{i\theta} + \rho^2 e^{2i\theta}\varphi(\rho e^{i\theta})\right)\left(\frac{1}{\rho^2 e^{2i\theta}} - \frac{1}{R^2}\right) \rho i e^{i\theta}\,d\lambda \\ &= \frac{1}{2\pi}\int_{\pi}^{2\pi} \left(h'(0)i - \frac{h'(0)}{R^2}\rho^2 e^{2i\theta} i + \varphi(\rho e^{i\theta})\rho i e^{i\theta} - \frac{\rho^3 e^{3i\theta}}{R^2}\, i\varphi(\rho e^{i\theta})\right) d\theta \\ &= \frac{i}{2}h'(0) + O(\rho). \end{aligned}$$

Taking the real parts of these integrals, we have

$$\begin{aligned}\mathrm{Re}\,\frac{1}{2\pi}\int_{C_\rho}\log h(\lambda)\left(\frac{1}{\lambda^2}-\frac{1}{R^2}\right)d\lambda &= \frac{1}{\pi R}\int_0^{\pi}\log|h(Re^{\mathrm{i}\theta})|\sin\theta\, d\theta\\ &+\frac{1}{2\pi}\int_\rho^R \log|h(-x)h(x)|\left(\frac{1}{x^2}-\frac{1}{R^2}\right)dx\\ &-\frac{1}{2}\,\mathrm{Im}\,h'(0)+O(\rho).\end{aligned}$$

On the other hand, integration by parts implies that

$$\begin{aligned}\frac{1}{2\pi}\int_{C_\rho}\log h(\lambda)\left(\frac{1}{\lambda^2}-\frac{1}{R^2}\right)d\lambda &= \frac{1}{2\pi}\log h(\lambda)\left(-\frac{1}{\lambda}-\frac{\lambda}{R^2}\right)\Big|_R^R\\ &+\frac{1}{2\pi}\int_{C_\rho}\frac{h'(\lambda)}{h(\lambda)}\left(\frac{1}{\lambda}+\frac{\lambda}{R^2}\right)d\lambda.\end{aligned} \tag{1.9}$$

Let the difference between $\arg h(R)$ at the terminal point R and $\arg h(R)$ at the initial point R be $2l_2\pi$ (e.g., $l_2=3$ in the above figure). Then, the first term of the right-hand side of (1.9) is equal to

$$\frac{1}{2\pi}2l_2\pi\mathrm{i}\,\frac{-2}{R}=-\frac{2l_2\mathrm{i}}{R}.$$

In the second term, the integrand is analytic in $\overline{D_\rho}$ except at 0 and the zeros of $h(\lambda)$. It is apparent that the zero $r_ke^{\mathrm{i}\theta_k}$ of $h(\lambda)$ is a pole of $h'(\lambda)/h(\lambda)$ of order 1. By calculating the residues, (1.9) is rewritten as

$$\begin{aligned}&\frac{1}{2\pi}\int_{C_\rho}\log h(\lambda)\left(\frac{1}{\lambda^2}-\frac{1}{R^2}\right)d\lambda\\ &=-\frac{2l_2}{R}\mathrm{i}+\mathrm{i}\left(\sum_{k=1}^{p}m_k\left(\frac{1}{r_ke^{\mathrm{i}\theta_k}}+\frac{r_ke^{\mathrm{i}\theta_k}}{R^2}\right)+h'(0)\right).\end{aligned}$$

Taking the real parts of both sides, we have

$$\begin{aligned}\mathrm{Re}\,\frac{1}{2\pi}\int_{C_\rho}\log h(\lambda)\left(\frac{1}{\lambda^2}-\frac{1}{R^2}\right)d\lambda &= \sum_{k=1}^{p}m_k\left(\frac{1}{r_k}-\frac{r_k}{R^2}\right)\sin\theta_k\\ &-\mathrm{Im}\,h'(0).\end{aligned}$$

Thus,

$$\begin{aligned}\sum_{k=1}^{p}m_k\left(\frac{1}{r_k}-\frac{r_k}{R^2}\right)\sin\theta_k &= \frac{1}{\pi R}\int_0^{\pi}\log|h(Re^{i\theta})|\sin\theta\, d\theta\\ &+\frac{1}{2\pi}\int_\rho^R\log|h(-x)h(x)|\left(\frac{1}{x^2}-\frac{1}{R^2}\right)dx\\ &+\frac{1}{2}\,\mathrm{Im}\,h'(0)+O(\rho).\end{aligned}$$

In view of (1.7), note that $\log|h(-x)h(x)| = \mathrm{Re}\,(\varphi(-x)+\varphi(x))x^2$ in a neighborhood of $x=0$, so that the second term of the above right-hand side is integrable at $x=0$. Letting $\rho \downarrow 0$, we finally obtain the relation

$$\begin{aligned}\sum_{k=1}^{p} m_k\left(\frac{1}{r_k}-\frac{r_k}{R^2}\right)\sin\theta_k &= \frac{1}{\pi R}\int_0^{\pi}\log|h(Re^{i\theta})|\sin\theta\,d\theta\\ &\quad+\frac{1}{2\pi}\int_0^{R}\log|h(-x)h(x)|\left(\frac{1}{x^2}-\frac{1}{R^2}\right)dx \qquad (1.10)\\ &\quad+\frac{1}{2}\,\mathrm{Im}\,h'(0).\end{aligned}$$

Next, consider the denominators $h_a(\lambda)=1-\lambda/a$ and $h_b(\lambda)=1-\lambda/b$ in (1.6). As for $h_a(\lambda)$, let $a=Re^{\mathrm{i}\alpha}$, $0<\alpha<\pi$. For a small $\varepsilon>0$, we modify ∂D a little around the points a and 0, so that the modified contour C_a contains both a and 0 inside (see Figure 5 below). Set

$$\arg h_a(R) = \frac{\pi-\alpha}{2}-2\pi = -\frac{3}{2}\pi-\frac{\alpha}{2}$$

at the initial point R, so that $\arg h_a(\lambda)$ is close to 0 in a neighborhood of $\lambda=0$.

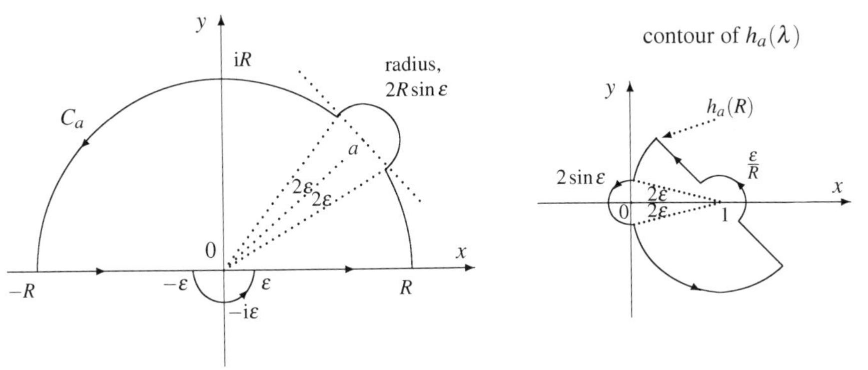

Figure 5

Consider the contour integral

$$I=\frac{1}{2\pi}\int_{C_a}\log h_a(\lambda)\left(\frac{1}{\lambda^2}-\frac{1}{R^2}\right)d\lambda, \qquad (1.11)$$

where the setting of the integral is similar to the one in (1.7). On the semicircle of radius R, we calculate as

$$\begin{aligned}&\frac{1}{2\pi}\int_{[0,\pi]\setminus(\alpha-2\varepsilon,\alpha+2\varepsilon)}\log h_a(Re^{\mathrm{i}\theta})\left(\frac{1}{R^2e^{2\mathrm{i}\theta}}-\frac{1}{R^2}\right)R\mathrm{i}e^{\mathrm{i}\theta}\,d\theta\\ &=\frac{1}{\pi R}\int_{[0,\pi]\setminus(\alpha-2\varepsilon,\alpha+2\varepsilon)}\log h_a(Re^{\mathrm{i}\theta})\,\sin\theta\,d\theta.\end{aligned}$$

The above right-hand side is integrable in a neighborhood of $\theta = \alpha$:

$$\begin{aligned}
\left| \int_{\alpha-2\varepsilon}^{\alpha} \log h_a(Re^{\mathrm{i}\theta}) \sin\theta \, d\theta \right| &\leqslant \int_{\alpha-2\varepsilon}^{\alpha} \left| \log h_a(Re^{\mathrm{i}\theta}) \right| \sin\theta \, d\theta \\
&\leqslant \int_{\alpha-2\varepsilon}^{\alpha} \left(\left| \log |h_a(Re^{\mathrm{i}\theta})| \right| + \left| \arg h_a(Re^{\mathrm{i}\theta}) \right| \right) \sin\theta \, d\theta \\
&\leqslant \int_{\alpha-2\varepsilon}^{\alpha} \left(\left| \log 2\sin\frac{\alpha-\theta}{2} \right| + \left(\frac{3\pi}{2} + \varepsilon \right) \right) \sin\theta \, d\theta \\
&\leqslant \mathrm{const}\, \varepsilon + \int_0^{2\sin\varepsilon} \frac{|\log\varphi|}{\sqrt{1-\varphi^2/4}} \, d\varphi \\
&\to 0, \quad \varepsilon \downarrow 0.
\end{aligned}$$

A similar estimate holds on $(\alpha, \alpha + 2\varepsilon)$. Thus,

$$\begin{aligned}
&\mathrm{Re}\, \frac{1}{2\pi} \int_{[0,\pi]\setminus(\alpha-2\varepsilon,\alpha+2\varepsilon)} \log h_a(Re^{\mathrm{i}\theta}) \left(\frac{1}{R^2 e^{2\mathrm{i}\theta}} - \frac{1}{R^2} \right) R\mathrm{i}e^{\mathrm{i}\theta} d\theta \\
&\quad \to \frac{1}{\pi R} \int_0^{\pi} \log |h_a(Re^{\mathrm{i}\theta})| \sin\theta \, d\theta, \quad \varepsilon \downarrow 0.
\end{aligned}$$

As for the integral around a, set $\lambda - a = 2R\sin\varepsilon\, e^{\mathrm{i}\theta}$, $\theta;\ \alpha - \frac{\pi}{2} - \varepsilon \to \alpha + \frac{\pi}{2} + \varepsilon$. Note that

$$\arg h_a(\lambda) = \theta - \alpha - \pi;\ -\frac{3\pi}{2} - \varepsilon \to -\frac{\pi}{2} + \varepsilon.$$

Then,

$$\begin{aligned}
&\left| \frac{1}{2\pi} \int \log h_a(\lambda) \left(\frac{1}{\lambda^2} - \frac{1}{R^2} \right) d\lambda \right| \\
&= \left| \int_{\alpha-\frac{\pi}{2}-\varepsilon}^{\alpha+\frac{\pi}{2}+\varepsilon} (\log(2\sin\varepsilon) + \mathrm{i}(\theta - \alpha - \pi)) \left(\frac{1}{\lambda^2} - \frac{1}{R^2} \right) 2R\sin\varepsilon\, \mathrm{i}\, e^{\mathrm{i}\theta} d\theta \right| \\
&\leqslant \mathrm{const} \sin\varepsilon \int_{\alpha-\frac{\pi}{2}-\varepsilon}^{\alpha+\frac{\pi}{2}+\varepsilon} \left(|\log(2\sin\varepsilon)| + \frac{3\pi}{2} + \varepsilon \right) d\theta \to 0, \quad \varepsilon \downarrow 0.
\end{aligned}$$

The integrals on the segments $[-R, -\varepsilon]$, and $[\varepsilon, R]$ are

$$\frac{1}{2\pi} \int_{\varepsilon}^{R} \log h_a(-x) \left(\frac{1}{x^2} - \frac{1}{R^2} \right) dx,$$

and

$$\frac{1}{2\pi} \int_{\varepsilon}^{R} \log h_a(x) \left(\frac{1}{x^2} - \frac{1}{R^2} \right) dx,$$

respectively. Since $\log h_a(0) = \log 1 + \mathrm{i}0 = 0$ by our setting of the logarithm, note that $\log h_a(\lambda) = -\frac{1}{a}\lambda + \varphi(\lambda)\lambda^2$ in a neighborhood of $\lambda = 0$. Then, the integral on the semicircle of radius ε is ($\lambda = \varepsilon e^{\mathrm{i}\theta},\ \theta;\ \pi \to 2\pi$)

$$\begin{aligned}
\frac{1}{2\pi}\int \log h_a(\lambda)\left(\frac{1}{\lambda^2} - \frac{1}{R^2}\right) d\lambda &= \frac{1}{2\pi}\int \left(-\frac{1}{a}\lambda + \varphi(\lambda)\lambda^2\right)\left(\frac{1}{\lambda^2} - \frac{1}{R^2}\right) d\lambda \\
&= \frac{1}{2\pi}\int \left(-\frac{1}{a\lambda} + \frac{\lambda}{aR^2} + \varphi(\lambda)\left(1 - \frac{\lambda^2}{R^2}\right)\right) d\lambda \\
&= \frac{1}{2\pi}\int_\pi^{2\pi} \left(-\frac{1}{a\varepsilon e^{\mathrm{i}\theta}} + \cdots\right) \varepsilon \mathrm{i} e^{\mathrm{i}\theta}\, d\theta \\
&= -\frac{\mathrm{i}}{2a} + O(\varepsilon).
\end{aligned}$$

Thus,

$$\begin{aligned}
&\mathrm{Re}\,\frac{1}{2\pi}\int_{C_a} \log h_a(\lambda)\left(\frac{1}{\lambda^2} - \frac{1}{R^2}\right) d\lambda \\
&= \frac{1}{\pi R}\int_{[0,\pi]\setminus(\alpha-2\varepsilon,\alpha+2\varepsilon)} \log|h_a(Re^{\mathrm{i}\theta})| \sin\theta\, d\theta + O(\varepsilon) \\
&\quad + \frac{1}{2\pi}\int_\varepsilon^R \log|h_a(-x)h_a(x)|\left(\frac{1}{x^2} - \frac{1}{R^2}\right) dx \\
&\quad + \frac{1}{2}\,\mathrm{Im}\,\frac{1}{a} + O(\varepsilon).
\end{aligned}$$

On the other hand, integration by parts implies that

$$\begin{aligned}
\frac{1}{2\pi}\int_{C_a} \log h_a(\lambda)\left(\frac{1}{\lambda^2} - \frac{1}{R^2}\right) d\lambda &= \frac{1}{2\pi}\log h_a(\lambda)\left(-\frac{1}{\lambda} - \frac{\lambda}{R^2}\right)\Bigg|_R^R \\
&\quad + \frac{1}{2\pi}\int_{C_a} \frac{h_a'(\lambda)}{h_a(\lambda)}\left(\frac{1}{\lambda} + \frac{\lambda}{R^2}\right) d\lambda \\
&= \frac{-2\mathrm{i}}{R} + \mathrm{i}\left(\frac{2}{R}\cos\alpha - \frac{1}{a}\right),
\end{aligned}$$

the right-hand side of which lies on the imaginary axis. Finally, letting $\varepsilon \downarrow 0$, we obtain the relation

$$\begin{aligned}
0 &= \frac{1}{\pi R}\int_0^\pi \log|h_a(Re^{\mathrm{i}\theta})| \sin\theta\, d\theta \\
&\quad + \frac{1}{2\pi}\int_0^R \log|h_a(-x)h_a(x)|\left(\frac{1}{x^2} - \frac{1}{R^2}\right) dx \qquad (1.12) \\
&\quad - \frac{1}{2}\,\mathrm{Im}\,\frac{1}{a}.
\end{aligned}$$

As for $h_b(\lambda)$, the contour integral

$$I = \frac{1}{2\pi}\int_{C_b} \log h_b(\lambda)\left(\frac{1}{\lambda^2} - \frac{1}{R^2}\right) d\lambda \qquad (1.13)$$

is considered along the modified contour C_b, where both b and 0 are inside C_b. At the initial point R, set $\arg h_b(R) = -\pi$ if $0 < b < R$, and $= -2\pi$ if $-R < b < 0$. Similar calculations show that

$$0 = \frac{1}{\pi R} \int_0^{\pi} \log |h_b(Re^{i\theta})| \sin\theta \, d\theta + \frac{1}{2\pi} \int_0^R \log |h_b(-x)h_b(x)| \left(\frac{1}{x^2} - \frac{1}{R^2} \right) dx. \tag{1.14}$$

In the case where $b = R$ or $-R$, the relation (1.14) also holds through the integral along a suitably modified contour C_R or C_{-R}.

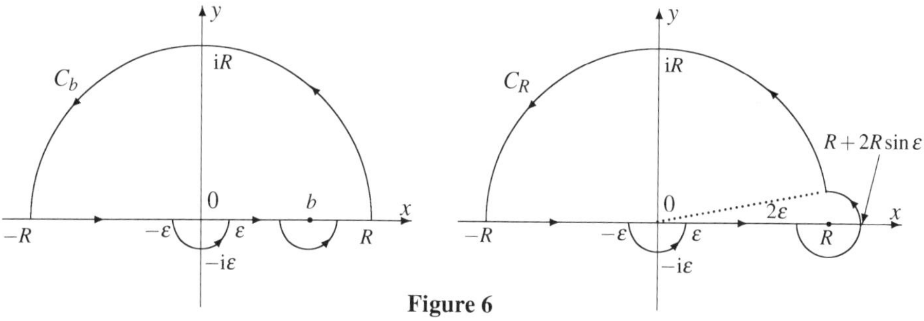

Figure 6

Finally, the desired relation (1.5) for our $f(\lambda)$ is obtained by combining the relation (1.10) with (1.12) and (1.14), and noting that $\log|f| = \log|h| + m\log|h_a| + n\log|h_b|$, and

$$f'(0) = h'(0) + mh'_a(0) + nh'_b(0) = h'(0) - \frac{m}{a} - \frac{n}{b}.$$

□

4.2 Boundary Control Systems

The purpose of this section is to formulate a boundary control system (see (2.7) below) which finally leads to (1.3). The basic space describing the controlled plant Σ_p is $L^2(\Omega)$, and no space with weaker topology appears throughout. The state $u(t,\cdot)$ of the plant Σ_p always stays in $L^2(\Omega)$. In the spectrum $\sigma(L)$, let $\nu \geqslant 1$ be the integer such that

$$\operatorname{Re}\lambda_1 \leqslant 0, \quad \text{and } 0 < \operatorname{Re}\lambda_{\nu+1}, \tag{2.1}$$

so that the semigroup e^{-tL} is not exponentially stable without any control. To obtain the compensator Σ_c in (1.3) as a differential equation in $\mathbb{R}^{\ell}$, we first introduce an infinite dimensional compensator leading to a finite dimensional

Σ_c. Let H be an *arbitrary* separable Hilbert space with inner product $\langle \cdot, \cdot \rangle_H$ and norm $\|\cdot\|_H$. Choose an orthonormal basis for H, and relabel the basis as

$$\left\{ \eta_{ij}^{\pm};\ i \geqslant 1,\ 1 \leqslant j \leqslant n_i \right\},$$

where $n_i < \infty$ for each $i \geqslant 1$. The choice of the integers n_i will be affected by the number N of the sensors $w_k \in L^2(\Gamma)$ or the number M of the actuators $g_k \in C^{2+\omega}(\Gamma)$, but is in our hand. Every vector $v \in H$ is expressed as a Fourier series in terms of $\{\eta_{ij}^{\pm}\}$,

$$v = \sum_{i,j} \left(v_{ij}^{+} \eta_{ij}^{+} + v_{ij}^{-} \eta_{ij}^{-} \right), \quad v_{ij}^{\pm} = \left\langle v, \eta_{ij}^{\pm} \right\rangle_H.$$

Let P_n^H, $n \geqslant 1$, be the projector in H such that

$$P_n^H v = \sum_{i(\leqslant n),\, j} \left(v_{ij}^{+} \eta_{ij}^{+} + v_{ij}^{-} \eta_{ij}^{-} \right) \quad \text{for } v = \sum_{i,j} \left(v_{ij}^{+} \eta_{ij}^{+} + v_{ij}^{-} \eta_{ij}^{-} \right).$$

Let $\{\mu_i\}_{i \geqslant 1}$ be a sequence of increasing positive numbers: $0 < \mu_1 < \mu_2 < \cdots \to \infty$, and define an operator B as

$$Bv = \sum_{i,j} \left(\mu_i \omega^{+} v_{ij}^{+} \eta_{ij}^{+} + \mu_i \omega^{-} v_{ij}^{-} \eta_{ij}^{-} \right) \left(= \sum_{i,j} \mu_i \omega^{\pm} v_{ij}^{\pm} \eta_{ij}^{\pm} \right), \tag{2.2}$$
$$\text{where } \omega^{\pm} = a \pm \mathrm{i}\sqrt{1 - a^2}.$$

Here, a denotes a constant, $0 < a < 1$. It is easily seen that B is a closed operator with dense domain $\mathscr{D}(B) = \left\{ v \in H;\ \sum_{i,j} |\mu_i v_{ij}^{\pm}|^2 < \infty \right\}$. In addition,

(i) $\sigma(B) = \{\mu_i \omega^{\pm};\ i \geqslant 1\}$; and

(ii) $(\mu_i \omega^{\pm} - B)\eta_{ij}^{\pm} = 0, \quad i \geqslant 1,\ 1 \leqslant j \leqslant n_i.$

Thus, $-B$ is the infinitesimal generator of an analytic semigroup e^{-tB}, $t > 0$, which is expressed as

$$e^{-tB} v = \sum_{i,j} e^{-\mu_i \omega^{+} t} v_{ij}^{+} \eta_{ij}^{+} + \sum_{i,j} e^{-\mu_i \omega^{-} t} v_{ij}^{-} \eta_{ij}^{-},$$

and it satisfies the estimate

$$\left\| e^{-tB} \right\|_H \leqslant e^{-a\mu_1 t}, \quad t \geqslant 0. \tag{2.3}$$

It is easily seen that the adjoint operator B^* of B is described as

$$B^* v = \sum_{i,j} \left(\mu_i \omega^{-} v_{ij}^{+} \eta_{ij}^{+} + \mu_i \omega^{+} v_{ij}^{-} \eta_{ij}^{-} \right) \left(= \sum_{i,j} \mu_i \omega^{\mp} v_{ij}^{\pm} \eta_{ij}^{\pm} \right) \tag{2.4}$$

for $v \in \mathscr{D}(B^*) = \mathscr{D}(B)$, and thus $B^*\eta_{ij}^{\pm} = \mu_i \omega^{\mp} \eta_{ij}^{\pm}$.

For our stabilization process, we construct the operator B such that

$$\begin{aligned} &\sigma(L) \cap \sigma(B) = \varnothing; \quad \arg \omega^+ > \theta_0; \quad a\mu_1 > \operatorname{Re} \lambda_{\nu+1}; \\ &\text{and} \quad \mu_i \leqslant \operatorname{const} i^{\gamma}, \quad i \geqslant 1, \quad \text{for } 0 < \exists \gamma < 2. \end{aligned} \tag{2.5}$$

While the structure of $\sigma(L)$ is a fixed one we cannot manage, the above condition (2.5) is fulfilled by adjusting the parameters $\omega^{\pm}$ and μ_1, both of which are in our hand. The relationship between $\sigma(L)$ and $\sigma(B)$ is illustrated in the following Figure 7.

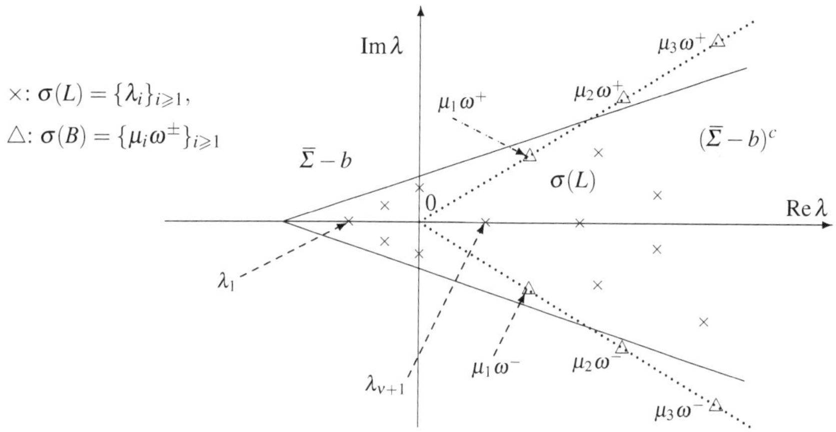

Figure 7 Relationship between $\sigma(L)$ and $\sigma(B)$.

For $g \in C^{2+\omega}(\Gamma)$, let R be a non-unique operator of prolongation such that

$$Rg \in C^{2+\omega}(\overline{\Omega}), \quad Rg|_{\Gamma} = \frac{\partial}{\partial \nu} Rg\bigg|_{\Gamma} = g,$$

where $Rg|_{\Gamma}$, for example, means the trace $\gamma(Rg)$ of Rg on Γ. Then, $\tau Rg = g$ on Γ. Let c be a real number such that $-c \in \rho(L)$. Now we recall the following classical result: If f is in $C^{\omega}(\overline{\Omega})$, then the boundary value problem

$$(c + \mathscr{L})u = f \quad \text{in } \Omega, \qquad \tau u = 0 \quad \text{on } \Gamma$$

admits a unique solution $u \in \mathscr{D}(\hat{L})$ [24, Theorem 19.2]. In other words, $u = L_c^{-1} f$ *is a genuine solution in* $\mathscr{D}(\hat{L})$ *as long as* f *is Hölder continuous*. Thus, the boundary value problem

$$(c + \mathscr{L})u = 0 \quad \text{in } \Omega, \quad \tau u = g \quad \text{on } \Gamma$$

admits a unique solution $u \in C^2(\Omega)\cap C^1(\overline{\Omega})$ (see [24]). The solution is expressed in a non unique manner as $u = Rg - L_c^{-1}(c+\mathscr{L})Rg$. In fact,

$$(c+\mathscr{L})\left(Rg - L_c^{-1}(c+\mathscr{L})Rg\right) = (c+\mathscr{L})Rg - (c+\mathscr{L})Rg = 0 \quad \text{in } \Omega,$$
$$\tau\left(Rg - L_c^{-1}(c+\mathscr{L})Rg\right) = g - 0 = g \quad \text{on } \Gamma.$$

For $\lambda \in \rho(L)$, the function

$$N_\lambda g = Rh - (\lambda - L)^{-1}(\lambda - \mathscr{L})Rg \tag{2.6}$$

is analytic, and coincides with the above genuine solution when $\lambda = -c$ [1]. For our actuators g_k, we thus define $N_{-c}g_k$. Then, $N_{-c}g_k \in C^2(\Omega)\cap C^1(\overline{\Omega})$ and $\mathscr{L}N_{-c}g_k \in L^2(\Omega)$.

With these preparations, we are ready to describe the control system which is finally reduced to (1.3). Let us consider the coupled system of differential equations:

$$\begin{cases} \dfrac{\partial u}{\partial t} + \mathscr{L}u = 0 \quad \text{in } \mathbb{R}^1_+ \times \Omega, \\ \tau u = \displaystyle\sum_{k=1}^{M} \langle v, \rho_k \rangle_H g_k \quad \text{on } \mathbb{R}^1_+ \times \Gamma, \\ \dfrac{dv}{dt} + Bv = \displaystyle\sum_{k=1}^{N} p_k(u)\xi_k + \sum_{k=1}^{M} \langle v, \rho_k \rangle_H \zeta_k \quad \text{in } \mathbb{R}^1_+ \times H, \\ u(0,\cdot) = u_0(\cdot) \in L^2(\Omega), \qquad v(0) = v_0 \in H. \end{cases} \tag{2.7}$$

Here, the differential equation for v denotes a compensator Σ_c in H; ξ_k, $1 \leqslant k \leqslant N$, and ζ_k, $1 \leqslant k \leqslant M$, actuators in H; and a set of linear functionals $\langle v, \rho_k \rangle_H$, $1 \leqslant k \leqslant M$, the output of the compensator, which enters the controlled plant Σ_p as the input through g_k located on Γ. The output $p_k(u)$ of Σ_p is defined in (1.4). In the stabilization procedure, the vectors ρ_k are chosen as linear combinations of a finite number of $\eta_{ij}^{\pm}$. Thus, we assume that ρ_k belong to $\mathscr{D}(B^*)$.

It is our strategy that we always have solutions $u(t,\cdot)$ as well as $\mathscr{L}u(t,\cdot)$ stay in $L^2(\Omega)$. It is essential in our framework that the function $u - \sum_{k=1}^{M} \langle v, \rho_k \rangle_{\mathbb{R}^\ell} N_{-c}g_k$ belongs to $\mathscr{D}(L)$ (see the description just below): This function, being neither decomposed nor transformed into another, is studied as it stands in both regularity and stabilization problems. Our point is as follows: Given the closed operator B satisfying the separation condition (2.5), and an output operator C, we construct a unique operator solution X to the operator equation: $XL - BX = C$ such that the range of X is contained in $\mathscr{D}(B)$ (Proposition 3.2, (ii) in Section 3), which compensates the difficulty arising

[1]More is true. In fact, N_λ belongs to $\mathscr{L}(H^{3/2}(\Gamma); H^2(\Omega))$ in the case where $\alpha(\xi) \equiv 1$ and to $\mathscr{L}(H^{1/2}(\Gamma); H^2(\Omega))$ in the case where $0 \leqslant \alpha(\xi) < 1$ (see, e.g., [32]).

from the operator L and the boundary controls on Γ. In Section 3 we achieve stabilization in this framework. In Section 4, another formulation of boundary control systems is proposed, so that it also leads to achieving stabilization.

First of all, we will show that the problem (2.7) is well posed in $L^2(\Omega)\times H$. Actually we have the following result:

Theorem 2.1. *The problem* (2.7) *is well posed in* $L^2(\Omega)\times H$, *and the solution* $u(t,\cdot)$ *is in* $C^2(\Omega)\cap C^1(\overline{\Omega})$, $t>0$. *The semigroup generated by* (2.7) *is analytic in* $t>0$.

Proof. Assume first that there is a solution $(u(t,\cdot),v(t))$ to (2.7), such that $u(t,\cdot)\in C^2(\Omega)\cap C^1(\overline{\Omega})$ and $v(t)\in\mathscr{D}(B)$, $t>0$. Setting

$$z=u-\sum_{k=1}^{M}\langle v,\rho_k\rangle_H N_{-c}g_k,$$

we obtain the equation for (z,v). It is clear that z belongs to $\mathscr{D}(L)$. Since ρ_k belong to $\mathscr{D}(B^*)$ by our assumption, (z,v) satisfies the equation

$$\begin{aligned}\frac{\partial z}{\partial t}+Lz&=\sum_{k=1}^{M}\langle v,B_c^*\rho_k\rangle_H N_{-c}g_k-\sum_{k=1}^{M}\langle v,\rho_k\rangle_H\left(\beta_k+\sum_{l=1}^{N}p_l(N_{-c}g_k)\alpha_l\right)\\&\quad-\sum_{k=1}^{N}p_k(z)\alpha_k,\qquad \tau z=0,\qquad\qquad (2.8)\\ \frac{dv}{dt}+Bv&=\sum_{k=1}^{M}\langle v,\rho_k\rangle_H\left(\zeta_k+\sum_{l=1}^{N}p_l(N_{-c}g_k)\xi_l\right)+\sum_{k=1}^{N}p_k(z)\xi_k,\end{aligned}$$

where the functions α_k and $\beta_k\in C^2(\Omega)\cap C^1(\overline{\Omega})$ are given, respectively, by

$$\alpha_k=\sum_{l=1}^{M}\langle \xi_k,\rho_l\rangle_H N_{-c}g_l,\quad\text{and}\quad \beta_k=\sum_{l=1}^{M}\langle \zeta_k,\rho_l\rangle_H N_{-c}g_l.$$

By setting $A=\begin{pmatrix}L&0\\0&B\end{pmatrix}$, the equation corresponding to (2.8) is simply written as

$$\frac{d}{dt}\begin{pmatrix}z\\v\end{pmatrix}+A\begin{pmatrix}z\\v\end{pmatrix}=D\begin{pmatrix}z\\v\end{pmatrix}=\begin{pmatrix}F(z,v)\\G(z,v)\end{pmatrix},\qquad \begin{pmatrix}z(0)\\v(0)\end{pmatrix}=\begin{pmatrix}z_0\\v_0\end{pmatrix},\qquad (2.9)$$

where the meaning of $\partial z/\partial t$ is changed to the differentiation of z in $L^2(\Omega)$: dz/dt, and the meaning of the operators D, F, and G will be self-explanatory. There is a sector $\overline{\Sigma}=\{\lambda\in\mathbb{C};\ \theta_0\leqslant|\arg\lambda|\leqslant\pi\}$, $0<\theta_0<\pi/2$, such that $\overline{\Sigma}-a$ for some $a\in\mathbb{R}^1$ is contained in $\rho(A)$ and that

$$\|(\lambda-A)^{-1}\|_{\mathscr{L}(L^2(\Omega)\times H)}\leqslant\frac{\text{const}}{1+|\lambda|},\qquad \lambda\in\overline{\Sigma}-a.$$

If λ is in $\rho(A)$ and the norm of $D(\lambda - A)^{-1}$ is less than 1, we have

$$(\lambda - A + D)^{-1} = (\lambda - A)^{-1}(1 + D(\lambda - A)^{-1})^{-1}.$$

Let us evaluate $D(\lambda - A)^{-1}\begin{pmatrix} z \\ v \end{pmatrix}$. It contains the terms

$$\left\langle (\lambda - B)^{-1}v, B_c^*\rho_k \right\rangle_H, \quad \left\langle (\lambda - B)^{-1}v, \rho_k \right\rangle_H, \quad \text{and} \quad p_k((\lambda - L)^{-1}z).$$

All these terms are easy to handle, since $(\lambda - B)^{-1}v$ goes to 0 as $\lambda \in \overline{\Sigma} - a \to \infty$. Owing to the estimate (1.2), the last term is, in the case where $\alpha(\xi) \not\equiv 1$, estimated as

$$\begin{aligned} \left| p_k((\lambda - L)^{-1}z) \right| &\leqslant \text{const} \, \|(\lambda - L)^{-1}z\|_{H^s(\Omega)} \\ &\leqslant \text{const} \, \|L_c^{s/2}(\lambda - L)^{-1}z\| \leqslant \frac{\text{const}}{1 + |\lambda|^{1-s/2}} \|z\|, \quad s > \frac{1}{2}, \end{aligned}$$

and similarly, in the case where $\alpha(\xi) \equiv 1$,

$$\begin{aligned} \left| p_k((\lambda - L)^{-1}z) \right| &\leqslant \text{const} \, \|(\lambda - L)^{-1}z\|_{H^{3/2+2\varepsilon}(\Omega)} \\ &\leqslant \text{const} \, \|L_c^{3/4+\varepsilon}(\lambda - L)^{-1}z\| \leqslant \frac{\text{const}}{1 + |\lambda|^{1/4-\varepsilon}} \|z\| \end{aligned}$$

for $\lambda \in \overline{\Sigma} - a$ and $0 < \varepsilon < 1/4$. In the latter estimate we have used the standard results: m-accretiveness of L_c and a generalization of Heinz's inequality imply that $\mathscr{D}(L_c^{\omega})$ is contained in $H^{2\omega}(\Omega)$, $0 \leqslant \omega \leqslant 1$ (see [25]). This shows that $\|D(\lambda - A)^{-1}\|_{\mathscr{L}(L^2(\Omega)\times H)}$ goes to 0 as $\lambda \in \overline{\Sigma} - a \to \infty$. We have proven that there is a sector $\overline{\Sigma} - b$ with some $b \in \mathbb{R}^1$ such that

$$\|(\lambda - A + D)^{-1}\|_{\mathscr{L}(L^2(\Omega)\times H)} \leqslant \frac{\text{const}}{1 + |\lambda|}, \quad \lambda \in \overline{\Sigma} - b.$$

Thus eqn. (2.9) determines an analytic semigroup $e^{-t(A-D)}$, $t > 0$, generated by $-A + D$.

Let $\begin{pmatrix} z \\ v \end{pmatrix} = e^{-t(A-D)}\begin{pmatrix} z_0 \\ v_0 \end{pmatrix}$ be a solution to (2.9). Since $Ae^{-t(A-D)}$ is analytic in $t > 0$, both $\begin{pmatrix} z(t+\varepsilon) \\ v(t+\varepsilon) \end{pmatrix}$ and $\begin{pmatrix} Lz(t+\varepsilon) \\ Bv(t+\varepsilon) \end{pmatrix}$ are analytic in $t \geqslant 0$ in the space $L^2(\Omega) \times H$ for any $\varepsilon > 0$. Let us consider the initial boundary value problem for z^{ε}:

$$\begin{aligned} &\frac{\partial z^{\varepsilon}}{\partial t} + \mathscr{L}z^{\varepsilon} = F(z(t+\varepsilon), v(t+\varepsilon)) \quad \text{in } \mathbb{R}^1_+ \times \Omega, \\ &\tau z^{\varepsilon} = 0 \quad \text{on } \mathbb{R}^1_+ \times \Gamma, \\ &z^{\varepsilon}(0,x) = z(\varepsilon,x) \quad \text{in } \Omega. \end{aligned} \tag{2.10}$$

It is clear that $F(z(t+\varepsilon), v(t+\varepsilon))$ is Lipschitz continuous in $[0,\infty)\times\overline{\Omega}$. In fact, we have the inequality:

$$\begin{aligned}|p_k(z(t)) - p_k(z(s))| &\leqslant \text{const } \|z(t)-z(s)\|_{H^1(\Omega)}\\ &\leqslant \text{const } \|L_c(z(t)-z(s))\|\\ &\leqslant \text{const } |t-s|, \qquad t,\, s \geqslant \varepsilon.\end{aligned}$$

Thus the problem admits a unique genuine solution $z^{\varepsilon}(t,x)$ such that $\mathscr{L}z^{\varepsilon}(t,x)$ is bounded in $(t_1,t_2)\times\Omega$ for $0<\forall t_1<\forall t_2$ and so is $\partial z^{\varepsilon}/\partial t$ (see [24]). This means that dz^{ε}/dt also exists in the topology of $L^2(\Omega)$ and it is equal to $\partial z^{\varepsilon}/\partial t$. Thus, taking the difference between $z^{\varepsilon}(t)$ and $z(t+\varepsilon)$, we see that

$$\frac{d}{dt}(z^{\varepsilon}(t)-z(t+\varepsilon)) + L(z^{\varepsilon}(t)-z(t+\varepsilon)) = 0, \quad z^{\varepsilon}(0)-z(0+\varepsilon)=0,$$

in other words,

$$z^{\varepsilon}(t)-z(t+\varepsilon) = e^{-tL}0 = 0, \quad t\geqslant 0.$$

Thus $z(t+\varepsilon)$ satisfies the equation

$$\begin{aligned}&\frac{\partial z(t+\varepsilon,x)}{\partial t} + \mathscr{L}z(t+\varepsilon,x) = F(z(t+\varepsilon,x), v(t+\varepsilon)) \quad \text{in } \mathbb{R}^1_+\times\Omega,\\ &\tau z(t+\varepsilon,\xi) = 0 \quad \text{on } \mathbb{R}^1_+\times\Gamma,\\ &z(0+\varepsilon,x) = z(\varepsilon,x) \quad \text{in } \Omega.\end{aligned}$$

Since $\varepsilon>0$ is arbitrary, $z(t,x)$ satisfies the first equation in (2.8). Thus $(u(t,x),v(t))$ satisfies the system (2.7) by setting $u = z+\sum_{k=1}^{M}\langle v, \rho_k\rangle_H N_{-c}g_k$, and the solution is unique. □

4.3 Stabilization

Assuming that the semigroup e^{-tL}, $t>0$ is not exponentially stable (see (2.1)), let us achieve stabilization of the boundary control system (2.7) with a prescribed decay rate $-r$, $r>0$. Since both of the sensors w_k and the actuators g_k are located on the boundary Γ at this time, we need to interpret the observability condition (2.5) and the controllability condition (2.9) in Chapter 3 in terms of the generalized eigenfunctions φ_{ij} and ψ_{ij} on Γ. According to the type of the boundary condition, let us define matrices $\hat{W}_i$ and $\hat{G}_i$ as

$$\begin{aligned}\hat{W}_i &= \left(p_k(\varphi_{ij});\ \begin{matrix} j & \rightarrow & 1,\dots,m_i\\ k & \downarrow & 1,\dots,N\end{matrix}\right), \quad \text{and}\\ \hat{G}_i &= \left(\langle g_k, \sigma\psi_{ij}\rangle_\Gamma;\ \begin{matrix} j & \downarrow & 1,\dots,m_i\\ k & \rightarrow & 1,\dots,M\end{matrix}\right),\end{aligned} \tag{3.1}$$

respectively, where $p_k(\cdot)$ are defined in (1.4), and σ a boundary operator defined as

$$\sigma\psi_{ij} = (1-\boldsymbol{b}\cdot\boldsymbol{\nu})\psi_{ij} - \frac{\partial\psi_{ij}}{\partial\nu}. \tag{3.2}$$

In (2.7), let us construct the actuators $\xi_k \in H$ in the form,

$$\xi_k = \sum_{i=1}^{\infty}\sum_{j=1}^{n_i}\left(\xi_{ij}^k\eta_{ij}^+ + \overline{\xi_{ij}^k}\eta_{ij}^-\right), \quad 1\leqslant k\leqslant N,$$

such that

$$\begin{cases} \sum_{i,j}\left|\xi_{ij}^k\mu_i^{1/4+\varepsilon}\right|^2 < \infty, & \text{in the case where } \alpha(\xi)\not\equiv 1, \\ \sum_{i,j}\left|\xi_{ij}^k\mu_i^{3/4+\varepsilon}\right|^2 < \infty, & \text{in the case where } \alpha(\xi)\equiv 1 \end{cases} \tag{3.3}$$

for small $\varepsilon > 0$, where the upper bar means the complex conjugate of complex numbers. Define $n_i \times N$ matrices Ξ_i as

$$\Xi_i = \left(\xi_{ij}^k; \begin{array}{ccc} j & \downarrow & 1,\dots,n_i \\ k & \rightarrow & 1,\dots,N \end{array}\right), \quad i\geqslant 1. \tag{3.4}$$

Our stabilization result is then stated as follows:

Theorem 3.1. *We assume the basic conditions* (2.1) *on L and* (2.5) *on B. Let r be an arbitrary number such that* $0 < r < \operatorname{Re}\lambda_{\nu+1}$. *Suppose that* w_k, g_k, *and* ξ_k *satisfy the rank conditions*

$$\begin{aligned} &\operatorname{rank}\left(\hat{W}_i \quad \hat{W}_i\Lambda_i \quad \dots \quad \hat{W}_i\Lambda_i^{m_i-1}\right)^{\mathrm{T}} = m_i, \quad 1\leqslant i\leqslant\nu, \\ &\operatorname{rank}\left(\hat{G}_i \quad \tilde{\Lambda}_i^*\hat{G}_i \quad \dots \quad (\tilde{\Lambda}_i^*)^{m_i-1}\hat{G}_i\right) = m_i, \quad 1\leqslant i\leqslant\nu, \quad \textit{and} \\ &\operatorname{rank}\Xi_i = N, \quad i\geqslant 1, \end{aligned} \tag{3.5}$$

respectively. Then we find a suitable integer $n\geqslant 1$; $\rho_k\in P_n^H H$, $1\leqslant k\leqslant M$; *and* ζ_k, $1\leqslant k\leqslant M$, *such that every solution* $(u(t,\cdot), v(t))$ *to* (2.7) *satisfies the decay estimate*

$$\|u(t,\cdot)\| + \|v(t)\|_H \leqslant \text{const}\, e^{-rt}\left(\|u_0\| + \|v_0\|_H\right), \quad t\geqslant 0 \tag{3.6}$$

for every solution $(u(t,\cdot),v(t))$ *to* (2.7).

(ii) *Eqn.* (1.3) *is derived from* (2.7) *by setting* $\ell = \dim P_n^H H$, *and it is well posed in* $L^2(\Omega)\times\mathbb{R}^\ell$, *where the solution* $u(t,\cdot)$ *is in* $C^2(\Omega)\cap C^1(\overline{\Omega})$, $t>0$. *Every solution* (u,v) *to* (1.3) *satisfies the decay estimate*

$$\|u(t,\cdot)\| + |v(t)|_\ell \leqslant \text{const}\, e^{-rt}\left(\|u_0\| + |v_0|_\ell\right), \quad t\geqslant 0. \tag{3.7}$$

Remark: (i) The best possible number N of the sensors is obtainable in the case where $\alpha^i_{(j+1)j} \neq 0$, $1 \leqslant j < m_i$, $1 \leqslant i \leqslant \nu$ (see Section 1 for α^i_{jk}). Actually, if $p_1(\varphi_{i1}) \neq 0$, $1 \leqslant i \leqslant \nu$, in this case, the above observability condition in (3.5) is fulfilled with $N = 1$. In the case where Λ_i are the diagonal matrices, however, the observability condition means that rank $\hat{W}_i = m_i$, $1 \leqslant i \leqslant \nu$, which requires N equal to or greater than $\max(m_1, \ldots, m_\nu)$. A similar discussion is possible on the number M of the actuators g_k.

(ii) Can we choose $r = \mathrm{Re}\,\lambda_{\nu+1}$? The answer to this question is unclear at present, since it depends heavily on the geometric property (3.15) of the operator X discussed below.

Proof of Theorem 3.1:

The proof is divided into five steps; *Step* I through V for clearer statement: In *Step* I, existence of a unique solution X to the operator equation (3.8) below and its geometric property are discussed. In *Step* II, stabilization of a boundary feedback control system is discussed, where the outputs are considered in Ω. In *Step* III, stabilization for eqn. (2.7) is achieved by determining necessary parameters, where a geometric property of X, playing a central role in the theorem, is stated as Proposition 3.3 without proof. It is also shown in this step that the vectors ρ_k can be chosen in a suitable finite-dimensional subspace $P^H_n H$. In *Step* IV, the control system (2.7) is reduced to another one in the form of (1.3), where the compensator is described as a finite-dimensional differential equation, and the decay of solutions is unchanged. Finally in *Step* V, proofs of miscellaneous results in the preceding steps are given.

Step I (*Operator equation*): Let us first consider the operator equation (Sylvester's equation) of infinite dimension:

$$XL - BX = C \quad \text{on } \mathscr{D}(L), \quad \text{where} \quad C = -\sum_{k=1}^{N} p_k(\cdot)\xi_k. \tag{3.8}$$

Here, $\mathscr{D}(C) = \cup_{s>1/2} H^s(\Omega)$ in the case where $\alpha(\xi) \not\equiv 1$, and $\mathscr{D}(C) = \cup_{s>3/2} H^s(\Omega)$ in the case where $\alpha(\xi) \equiv 1$. A finite-dimensional version has appeared in (1.5) of Chapter 1. The following existence result is based on the separation condition (2.5) between $\sigma(L)$ and $\sigma(B)$:

Proposition 3.2. (i) *The operator equation* (3.8) *admits a unique operator solution* $X \in \mathscr{L}(L^2(\Omega); H)$. *The solution* X *is expressed as*

$$Xu = \sum_{i,j}\sum_{k=1}^{N} f_k(\mu_i\omega^+; u)\xi^k_{ij}\eta^+_{ij} + \sum_{i,j}\sum_{k=1}^{N} f_k(\mu_i\omega^-; u)\overline{\xi^k_{ij}}\eta^-_{ij},$$
$$u \in L^2(\Omega), \quad \textit{where } f_k(\lambda; u) = p_k\left((\lambda - L)^{-1}u\right), \quad 1 \leqslant k \leqslant N. \tag{3.9}$$

(ii) *The ranges of* X *and its adjoint* X^* *are contained, respectively, in* $\mathscr{D}(B)$ *and* $\mathscr{D}(L_c^{*\omega})$, $0 \leqslant \omega < 3/4$.

Proposition 3.3. *Under the first and the third assumptions in* (3.5), *we have the inclusion relation:*

$$P_\nu^* L^2(\Omega) \subset \overline{X^* H}, \tag{3.10}$$

where the overline on the right-hand side means the closure in $L^2(\Omega)$, *and the left-hand side is a finite-dimensional subspace spanned by* ψ_{ij}, $1 \leqslant i \leqslant \nu$, $1 \leqslant j \leqslant m_i$.

By Theorem 2.1, eqn. (2.7) admits a unique genuine solution $(u(t,\cdot), v(t)) \in L^2(\Omega) \times H$ such that $u(t,\cdot)$ belongs to $C^2(\Omega) \cap C^1(\overline{\Omega})$, $t > 0$. We rewrite the equation for u as

$$\frac{du}{dt} + L_c \left(u - \sum_{k=1}^{M} \langle v, \rho_k \rangle_H N_{-c} g_k \right) = cu.$$

In view of (ii) of Proposition 3.2, apply the operator X to the both sides. Then, we have

$$\frac{d}{dt} Xu + (B_c X + C) \left(u - \sum_{k=1}^{M} \langle v, \rho_k \rangle_H N_{-c} g_k \right) = cXu, \quad \text{or}$$

$$\frac{d}{dt} Xu + (B_c X + C) u = \sum_{k=1}^{M} \langle v, \rho_k \rangle_H (B_c X + C) N_{-c} g_k + cXu.$$

Note that, if u were in $\mathscr{D}(L)$, we could have $(B_c X + C)u = XL_c u$. But, this is not true in our problem. At this stage we define the vectors ζ_k as

$$\zeta_k = (B_c X + C) N_{-c} g_k, \quad 1 \leqslant k \leqslant M. \tag{3.11}$$

Then we see that

$$\frac{d}{dt}(Xu - v) + B(Xu - v) = 0, \quad \text{or}$$
$$Xu(t,\cdot) - v(t) = e^{-tB}(Xu_0 - v_0), \quad t \geqslant 0.$$

Due to the decay property of e^{-tB} (see (2.3)), the above right-hand side goes to 0 exponentially as $t \to \infty$:

$$\|Xu(t,\cdot) - v(t)\| \leqslant e^{-a\mu_1 t} \|Xu_0 - v_0\|, \quad t \geqslant 0. \tag{3.12}$$

Step II (*Operator* F). In view of the decay (3.12), we rewrite the equation for u again in the form:

$$\begin{aligned} &\frac{du}{dt} + \mathscr{L} u = 0, \quad u(0,\cdot) = u_0, \\ &\tau u - \sum_{k=1}^{M} \langle u, X^* \rho_k \rangle g_k = \sum_{k=1}^{M} \left\langle e^{-tB}(v_0 - Xu_0), \rho_k \right\rangle_H g_k. \end{aligned} \tag{3.13}$$

Given a set of functions y_k, $1 \leqslant k \leqslant M$, we define the operator $\hat{F}$ as

$$\begin{aligned} \hat{F}u &= \mathscr{L}u, \quad u \in \mathscr{D}(\hat{F}), \\ \mathscr{D}(\hat{F}) &= \left\{ u \in C^2(\Omega) \cap C^1(\overline{\Omega});\ \mathscr{L}u \in L^2(\Omega),\ \tau_f u = 0 \ \text{on}\, \Gamma \right\}, \\ \text{where } \tau_f u &= \tau u - \sum_{k=1}^{M} \langle u, y_k \rangle g_k, \quad y_k \in L^2(\Omega). \end{aligned} \tag{3.14}$$

The boundary condition for $\hat{F}$ is thus characterized in feedback form. A specific feature of the operator $\hat{F}$ is stated as follows. The proof is to be given later.

Proposition 3.4. (i) *The operator $\hat{F}$ admits the closure F in $L^2(\Omega)$. The closure F is densely defined, and generates an analytic semigroup e^{-tF}, $t > 0$. If in addition y_k, $1 \leqslant k \leqslant M$, belongs to $\mathscr{D}(L_c^{*\beta})$, $\beta > 0$, then $e^{-tF}u_0$ is a genuine solution to the equation:*

$$\frac{\partial u}{\partial t} + \hat{F}u = 0, \quad u(0,\cdot) = u_0$$

for each $u_0 \in L^2(\Omega)$.

(ii) *Suppose that the actuators g_k satisfy the second condition of* (3.5), *i.e., the controllability condition. Then there exists a set of $y_k \in P_\nu^* L^2(\Omega)$, $1 \leqslant k \leqslant M$ which ensures the decay estimate,*

$$\left\| e^{-tF} \right\| \leqslant \text{const}\, e^{-r_1 t}, \quad t \geqslant 0, \quad r < r_1 < \operatorname{Re} \lambda_{\nu+1}. \tag{3.15}$$

Remark: If all the eigenvalues of L on the vertical line: $\operatorname{Re} \lambda = \operatorname{Re} \lambda_{\nu+1}$ are poles of $(\lambda - L)^{-1}$ of order 1, the above r_1 is replaced by $\operatorname{Re} \lambda_{\nu+1}$.

We add a small perturbation to y_k in Proposition 3.4, (ii). The perturbed functions, denoted by $\tilde{y}_k$, define another elliptic operator, say, $\tilde{F}$. For later convenience, however, it is still denoted by the same symbol F without confusion. The following result looks merely like a standard perturbation result in the case where the coefficient $\alpha(\xi)$ of the boundary operator τ satisfies the condition $\alpha(\xi) \equiv 1$ or $0 \leqslant \alpha(\xi) < 1$. We need, however, a more careful consideration in our general case.

Proposition 3.5. *If $\sum_{k=1}^{M} \|\tilde{y}_k - y_k\|$ is small enough, we have the estimate*

$$\left\| e^{-tF} \right\| \leqslant \text{const}\, e^{-rt}, \quad t \geqslant 0. \tag{3.16}$$

Step III (*Stabilization*). Let $y_k \in P_\nu^* L^2(\Omega)$, $1 \leqslant k \leqslant M$, be the functions stated in Proposition 3.4, (ii). Proposition 3.3 guarantees suitable sequences of functions $X^* \rho_k$ which are arbitrarily close to y_k. In addition, the set $\{\eta_{ij}^{\pm}\}$ forms an orthonormal basis for H. Thus we can choose suitable ρ_k, which are expressed as linear combinations of a finite number of $\eta_{ij}^{\pm}$, say, $1 \leqslant i \leqslant n$, such that the decay (3.16) is ensured for the semigroup e^{-tF}, F being the closure of $\hat{F}$ with y_k replaced by $X^*\rho_k$.

Given a function $g \in C^{2+\omega}(\Gamma)$, let us consider the boundary value problem

$$(c+\mathscr{L})u = 0 \quad \text{in } \Omega, \quad \tau_f u = g \quad \text{on } \Gamma. \tag{3.17}$$

Then we have

Lemma 3.6. *Choose a $c > 0$ large enough so that $-c$ is in $\rho(L)$. Then the boundary value problem (3.17) admits a unique solution $u \in C^2(\Omega) \cap C^1(\overline{\Omega})$. The solution is denoted by $u = N^f_{-c}g$.*

We go back to eqn. (3.13). Choose a $c > 0$ in Lemma 3.6, and set

$$p(t) = u(t,\cdot) - \sum_{k=1}^{M} f_k(t) N^f_{-c} g_k, \quad f_k(t) = \left\langle e^{-tB}(v_0 - Xu_0), \rho_k \right\rangle_H.$$

The function $p(t)$, $t > 0$, belongs to $\mathscr{D}(\hat{F})$ and satisfies the equation

$$\frac{dp}{dt} + Fp = \sum_{k=1}^{M} \Big(c f_k(t) - \frac{d}{dt} f_k(t) \Big) N^f_{-c} g_k, \quad p(0) = u_0 - \sum_{k=1}^{M} f_k(0) N^f_{-c} g_k.$$

Since the vectors ρ_k belong to $\mathscr{D}(B^*)$, both $f_k(t)$ and $df_k(t)/dt = -\left\langle e^{-tB}(v_0 - Xu_0), B^*\rho_k \right\rangle$ go to 0 exponentially as $t \to \infty$. According to Proposition 3.5, we see that

$$\|p(t)\| \leqslant \text{const}\, e^{-rt} \left(\|u_0\| + \|v_0\|_H \right), \quad t \geqslant 0.$$

This immediately leads to the decay estimate (3.6), and the stabilization of eqn. (2.7) has been achieved. Combining this result with Theorem 2.1, we also obtain the estimate (see (2.9))

$$\|(\lambda - A + D)^{-1}\|_{\mathscr{L}(L^2(\Omega)\times H)} \leqslant \frac{\text{const}}{1+|\lambda|}, \quad \lambda \in (\overline{\Sigma} - a) \cup \{\lambda;\, \operatorname{Re}\lambda \leqslant r\}. \tag{3.18}$$

Step IV (*Reduction to a finite-dimensional compensator*). We reduce eqn. (2.7) to (1.3) in this step. Recall that the vectors ρ_k are chosen in the subspace P^H_n (see *Step* III). In (2.7), set $v_1(t) = P^H_n v(t)$. Applying P^H_n to the both sides of the equation for v, we obtain the coupled system of differential equations

$$\begin{cases} \dfrac{du}{dt} + \mathscr{L}u = 0 \quad \text{in } \mathbb{R}^1_+ \times \Omega, \\ \tau u = \displaystyle\sum_{k=1}^{M} \langle v_1, \rho_k \rangle_H g_k \quad \text{on } \mathbb{R}^1_+ \times \Gamma, \\ \dfrac{dv_1}{dt} + B_1 v_1 = \displaystyle\sum_{k=1}^{N} p_k(u) P^H_n \xi_k + \sum_{k=1}^{M} \langle v_1, \rho_k \rangle_H P^H_n \zeta_k \quad \text{in } \mathbb{R}^1_+ \times P^H_n H, \\ u(0,\cdot) = u_0(\cdot) \in L^2(\Omega), \qquad v_1(0) = P^H_n v_0 \in P^H_n H. \end{cases} \tag{3.19}$$

Here B_1 denotes the restriction of B onto the invariant subspace $P_n^H H$. In exactly the same manner as in Theorem 2.1, it is shown that eqn. (3.19) is well posed in $L^2(\Omega) \times P_n^H H$, and the solution $u(t,\cdot)$ is in $C^2(\Omega) \cap C^1(\overline{\Omega})$, $t > 0$. The semigroup generated by (3.19) is analytic in $t > 0$. In other words, every solution $(u(t,\cdot), v_1(t))$ to (3.19) is derived from the corresponding solution to (2.7), and satisfies the decay estimate (3.6). The equation for v_1 in (3.19) means the finite-dimensional compensator with $\ell = \dim P_n^H H$, where the terms on v_1 on the right-hand side are absorbed into B_1 in the expression of (1.3).

Step V (*Proofs of the preceding results*). Let us turn to the proofs of the preceding propositions and lemmas.

Proof of Proposition 3.3. To prove (3.10), we show the implication:

$$Xu = 0 \quad \Rightarrow \quad P_\mathrm{v} u = 0.$$

Setting $Xu = 0$ in (3.9), we see that

$$\sum_{k=1}^{N} f_k(\mu_i \omega^+; u)\, \xi_{ij}^k = \sum_{k=1}^{N} f_k(\mu_i \omega^-; u)\, \overline{\xi_{ij}^k} = 0, \quad i \geqslant 1, \;\; 1 \leqslant j \leqslant n_i.$$

Since rank $\Xi_i = N$, $i \geqslant 1$, by (3.5), this implies that

$$f_k(\mu_i \omega^\pm; u) = p_k\left((\mu_i \omega^\pm - L)^{-1} u\right) = 0, \quad 1 \leqslant k \leqslant N, \; i \geqslant 1.$$

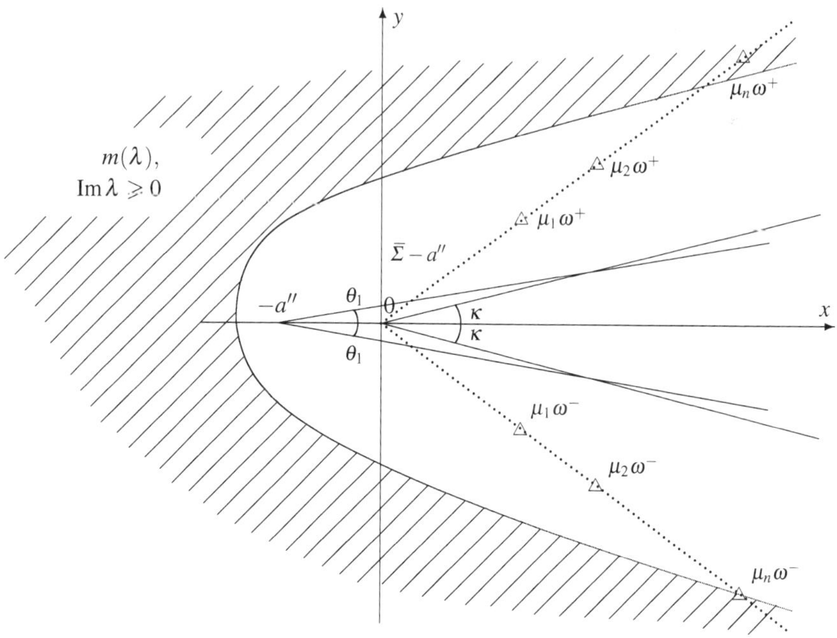

Figure 8

Recall that $\sigma(L)$ is inside some parabola $x = a'y^2 - b'$; $\lambda = x + \mathrm{i}y$, $a' > 0$. Thus, choosing a θ_1 such that $0 < \theta_1 < \min\left\{\arg\omega^+, \frac{\pi}{2}(2-\gamma)\right\}$, we have the estimate

$$\|(\lambda - L)^{-1}\| \leqslant \frac{\text{const}}{1+|\lambda|}, \quad \lambda \in \overline{\Sigma}_L - a'',$$

where $\overline{\Sigma}_L = \{\lambda \in \mathbb{C};\ \theta_1 \leqslant |\arg\lambda| \leqslant \pi\}$, and $a'' \in \mathbb{R}^1$. Let us introduce the map

$$m(\lambda) = (\lambda + \mathrm{i}R_0)^{\eta} e^{\mathrm{i}\kappa}, \quad \operatorname{Im}\lambda \geqslant 0,$$

where $R_0 > 0$. The constants η and κ are chosen so that

$$\max\left\{\gamma, 2 - \frac{2}{\pi}\arg\omega^+\right\} < \eta < 2 - \frac{2}{\pi}\theta_1, \quad \kappa = \frac{2-\eta}{2}\pi.$$

Thus we see that $\theta_1 < \kappa < \arg\omega^+$. By choosing an R_0 large enough in this map, the image of the upper half-plane by the map, that is the set $\{m(\lambda);\ \operatorname{Im}\lambda \geqslant 0\}$ is contained in the sector $\overline{\Sigma}_L - a''$. The relationship of $\{m(\lambda);\ \operatorname{Im}\lambda \geqslant 0\}$, $\sigma(L)$, and $\sigma(B)$ is illustrated in Figure 8. For simplicity of symbols, we write $f_k(m(\lambda);u)$ as $f(m(\lambda))$. Then the function $f(m(\lambda))$ is analytic in λ, $\operatorname{Im}\lambda \geqslant 0$, and

$$f(m(\sigma_j)) = 0, \quad \sigma_j = \mu_j^{1/\eta} e^{\mathrm{i}(\arg\omega^+ - \kappa)/\eta} - iR_0 \tag{3.20}$$

for $j \geqslant j_0$, where j_0 is the integer such that $\operatorname{Im}\sigma_{j_0} > 0$ or

$$\mu_{j_0}^{1/\eta} \sin\frac{\arg\omega^+ - \kappa}{\eta} - R_0 > 0.$$

We show that (3.20) implies that

$$f(m(\lambda)) \equiv 0, \quad \operatorname{Im}\lambda \geqslant 0. \tag{3.21}$$

Assuming the contrary, we derive a contradiction. We may assume that $f(m(0)) = f(-R_0^{\eta}) \neq 0$ by adjusting the number R_0 if necessary. Then Carleman's theorem (see Theorem 1.1) is applied to $\alpha f(m(\lambda))$, where $\alpha = 1/f(m(0))$: Let $R > 0$ be arbitrary but large enough. Suppose that $f(m(\lambda))$ has zeros $r_k e^{\mathrm{i}\theta_k}$, $1 \leqslant k \leqslant p$, inside the closed contour C_R consisting of the semicircle: $|\lambda| = R$, $0 < \arg\lambda < \pi$, and the segment: $|\lambda| \leqslant R$ on the real axis. Then, (1.5) reads

$$\begin{aligned}
&\sum_{k=1}^{p} m_k\left(\frac{1}{r_k} - \frac{r_k}{R^2}\right)\sin\theta_k \\
&= \frac{1}{\pi R}\int_0^{\pi} \log|\alpha f(m(Re^{\mathrm{i}\theta}))| \sin\theta\, d\theta \\
&\quad + \frac{1}{2\pi}\int_0^{R} \log|\alpha f(m(-x))\alpha f(m(x))| \left(\frac{1}{x^2} - \frac{1}{R^2}\right) dx \\
&\quad + \frac{1}{2}\operatorname{Im}\frac{d}{d\lambda}\alpha f(m(0)).
\end{aligned} \tag{3.22}$$

As for the first term of the right-hand side of (3.22), we note that

$$|f(m(Re^{i\theta}))| = \left|\left\langle (m(Re^{i\theta}) - L)^{-1}u, w\right\rangle_{\Gamma}\right| \leqslant \frac{\text{const}}{1 + |m(Re^{i\theta})|^{3/4-\varepsilon}} \leqslant \frac{\text{const}}{R^{(3/4-\varepsilon)\eta}}, \quad 0 < \varepsilon < \frac{1}{4}$$

in the case where $\alpha(\xi) \not\equiv 1$, and

$$|f(m(Re^{i\theta}))| = \left|\left\langle \frac{\partial}{\partial\nu}(m(Re^{i\theta}) - L)^{-1}u, w\right\rangle_{\Gamma}\right| \leqslant \frac{\text{const}}{1 + |m(Re^{i\theta})|^{1/4-\varepsilon}} \leqslant \frac{\text{const}}{R^{(1/4-\varepsilon)\eta}}, \quad 0 < \varepsilon < \frac{1}{4}$$

in the case where $\alpha(\xi) \equiv 1$. Thus the first term is bounded from above by

$$\frac{1}{\pi R}\int_0^{\pi} (\text{const} - (1/4 - \varepsilon)\eta \log R) \sin\theta \, d\theta \to 0 \quad \text{as } R \to \infty.$$

As for the second term, the estimate $\left|\log|\alpha f(m(-\lambda))\alpha f(m(\lambda))|\right| \leqslant \text{const}\, |\lambda|^2$ holds in a neighborhood of $\lambda = 0$. Thus decomposing the second term as $\int_0^{\delta} + \int_{\delta}^{R}$ for a small $\delta > 0$, we see that the second term is bounded from above by

$$\text{const} + \frac{1}{2\pi}\int_{\delta}^{R} (\text{const} - 2(1/4 - \varepsilon)\eta \log x)\left(\frac{1}{x^2} - \frac{1}{R^2}\right) dx \leqslant \text{const},$$

as $R \to \infty$. Thus the right-hand side of (3.22) remains bounded as $R \to \infty$.

Let us turn to the left-hand side. Let $N(x)$, $x \geqslant 0$, denote the number of $\mu_n^{1/\eta} < x$, that is, $N(x) = \#\{n \geqslant 0;\ \mu_n^{1/\eta} < x\}$. According to the growth rate assumption (2.5) on the sequence $\{\mu_i\}$, we easily find that

$$N(x) \geqslant \text{const}\, x^{\eta/\gamma} - 1.$$

We know from (3.20) that σ_j with $j \geqslant j_0$ are zeros of $f(m(\lambda))$. Thus

$$\sum_{\nu=1}^{p} m_\nu \left(\frac{1}{r_\nu} - \frac{r_\nu}{R^2}\right) \sin\theta_\nu \geqslant \sum_{\substack{j \geqslant j_0, \\ |\sigma_j| < R}} \left(\frac{1}{\mu_j^{1/\eta}} - \frac{\mu_j^{1/\eta}}{R^2}\right) \sin(\arg\sigma_j).$$

Since $\#\{j;\ |\sigma_j| < R\} \geqslant \#\{j;\ \mu_j^{1/\eta} < R\}$, the above right-hand side is obviously bounded from below by

$$\sum_{\substack{j \geqslant j_0, \\ \mu_j^{1/\eta} < R}} \left(\frac{1}{\mu_j^{1/\eta}} - \frac{\mu_j^{1/\eta}}{R^2}\right) \sin(\arg\sigma_{j_0}).$$

The last term $(\times 1/\sin(\arg\sigma_{j_0}))$ is calculated as follows:

$$
\begin{aligned}
&= \int_{\mu_{j_0}^{1/\eta}-\varepsilon}^{R} \left(\frac{1}{x}-\frac{x}{R^2}\right) dN(x) \\
&= \left(\frac{1}{x}-\frac{x}{R^2}\right) N(x)\Bigg|_{\mu_{j_0}^{1/\eta}-\varepsilon}^{R} + \int_{\mu_{j_0}^{1/\eta}-\varepsilon}^{R} \left(\frac{1}{x^2}+\frac{1}{R^2}\right) N(x)\,dx \\
&\geqslant \int_{\mu_{j_0}^{1/\eta}-\varepsilon}^{R} \left(\frac{1}{x^2}+\frac{1}{R^2}\right) \left(\mathrm{const}\, x^{\eta/\gamma}-1\right) dx - \mathrm{const} \\
&\geqslant \frac{\mathrm{const}}{(\eta/\gamma)^2-1} R^{\eta/\gamma-1} - \mathrm{const} \to \infty \quad \text{as } R \to \infty
\end{aligned}
$$

for a sufficiently small $\varepsilon > 0$, which is a contradiction. We have thus proven the relation (3.21). Going back to the original notations, we see that

$$
f_k(\lambda; u) = p_k\big((\lambda - L)^{-1} u\big) = 0, \quad 1 \leqslant k \leqslant N, \quad \lambda \in \rho(L). \tag{3.23}
$$

The idea leading to the observability matrices $\begin{pmatrix} \hat{W}_i & \hat{W}_i\Lambda_i & \dots & \hat{W}_i\Lambda_i^{m_i-1} \end{pmatrix}^{\mathrm{T}}$ in (3.5) is to note the simple algebraic relation:

$$
\begin{aligned}
(\lambda - L)^{-1} &= L_c(\lambda - L)^{-1} L_c^{-1} \\
&= -L_c^{-1} + (\lambda + c)(\lambda - L)^{-1} L_c^{-1},
\end{aligned} \tag{3.24}
$$

where $c > 0$ is large enough such that $-c \in \rho(L)$, and $L_c = L + c$. Following (3.24), let us introduce for each k, $1 \leqslant k \leqslant M$, a series of meromorphic functions $f_k^l(\lambda; u)$, $l = 0, 1, \dots$ as the recursion formula:

$$
f_k^0(\lambda; u) = f_k(\lambda; u), \qquad f_k^{l+1}(\lambda; u) = \frac{f_k^l(\lambda; u)}{\lambda + c}, \quad l = 0, 1, \dots. \tag{3.25}
$$

Then, we easily find that

$$
f_k^l(\lambda; u) = p_k\left((\lambda - L)^{-1} L_c^{-l} u\right) - \sum_{i=1}^{l} \frac{1}{(\lambda + c)^i}\, p_k\big(L_c^{-(l+1-i)} u\big), \tag{3.26}
$$

or more concretely

$$
f_k^l(\lambda; u) = \begin{cases}
\left\langle (\lambda - L)^{-1} L_c^{-l} u, w_k \right\rangle_\Gamma - \displaystyle\sum_{i=1}^{l} \frac{1}{(\lambda + c)^i} \left\langle L_c^{-(l+1-i)} u, w_k \right\rangle_\Gamma, \\
\quad \text{in the case where } \alpha(\xi) \not\equiv 1, \\
\left\langle \dfrac{\partial}{\partial \nu} (\lambda - L)^{-1} L_c^{-l} u, w_k \right\rangle_\Gamma - \displaystyle\sum_{i=1}^{l} \frac{1}{(\lambda + c)^i} \left\langle \dfrac{\partial}{\partial \nu} L_c^{-(l+1-i)} u, w_k \right\rangle_\Gamma, \\
\quad \text{in the case where } \alpha(\xi) \equiv 1.
\end{cases}
$$

By (3.23),

$$f_k^l(\lambda; u) = 0, \quad 1 \leqslant k \leqslant N, \quad l \geqslant 0, \quad \lambda \in \rho(L) \setminus \{-c\}. \tag{3.27}$$

Let us consider Laurent's expansion of $(\lambda - L)^{-1}$ in a neighborhood of the pole λ_i:

$$(\lambda - L)^{-1} = \sum_{j=1}^{l_i} \frac{A_{-j}}{(\lambda - \lambda_i)^j} + \sum_{j=0}^{\infty} (\lambda - \lambda_i)^j A_j, \quad \text{where}$$

$$l_i \leqslant m_i, \quad A_j = \frac{1}{2\pi \mathrm{i}} \int_{|\zeta - \lambda_i| = \delta} \frac{(\zeta - L)^{-1}}{(\zeta - \lambda_i)^{j+1}} d\zeta, \quad j = 0, \pm 1, \pm 2, \ldots.$$

Here, l_i denotes the *ascent* of $\lambda_i - L$. Note that $P_{\lambda_i} = A_{-1}$. It is clear that

$$LA_j u = \frac{1}{2\pi \mathrm{i}} \int_{|\zeta - \lambda_i| = \delta} \frac{-1 + \zeta(\zeta - L)^{-1}}{(\zeta - \lambda_i)^{j+1}} u \, d\zeta.$$

Thus the series $\sum_{j=0}^{\infty} (\lambda - \lambda_i)^j A_j u$ converges in the topology of $\mathscr{D}(L)$ and, thus, at least in the topology of $H^1(\Omega)$. In the specific case where $\alpha(\xi) \equiv 1$, note also that $\mathscr{D}(L) = H^2(\Omega) \cap H_0^1(\Omega) \hookrightarrow H^{3/2}(\Gamma)$. Thus, we see that

$$f_k(\lambda; u) = \sum_{j=1}^{l_i} \frac{1}{(\lambda - \lambda_i)^j} p_k(A_{-j} u) + \sum_{j=0}^{\infty} (\lambda - \lambda_i)^j p_k(A_j u) = 0$$

in a neighborhood of λ_i. Calculating the residue of $f_k(\lambda; u)$ at λ_i, we have

$$p_k(A_{-1} u) = p_k(P_{\lambda_i} u) = \sum_{j=1}^{m_i} p_k(\varphi_{ij}) u_{ij} = 0, \quad 1 \leqslant k \leqslant N, \quad i \geqslant 1. \tag{3.28}$$

Let us turn to $f_k^l(\lambda; u)$, $l \geqslant 1$. In view of (3.26) and (3.27), we similarly obtain the expansion in a neighborhood of each λ_i:

$$\begin{aligned} 0 &= f_k^l(\lambda; u) \\ &= \sum_{j=1}^{l_i} \frac{1}{(\lambda - \lambda_i)^j} p_k\left(A_{-j} L_c^{-l} u\right) + \sum_{j=0}^{\infty} (\lambda - \lambda_i)^j p_k\left(A_j L_c^{-l} u\right) \\ &\quad - \sum_{i=1}^{l} \frac{1}{(\lambda + c)^i} p_k\left(L_c^{-(l+1-i)} u\right). \end{aligned} \tag{3.29}$$

Here, note that $A_{-1} L_c^{-l} u = P_{\lambda_i} L_c^{-l} u = L_c^{-l} P_{\lambda_i} u$, and that the restriction of L_c^{-l} on $P_{\lambda_i} L^2(\Omega)$ is equivalent to the matrix $(\Lambda_i + c)^{-l}$:

$$A_{-1} L_c^{-l} u \quad \Leftrightarrow \quad (\Lambda_i + c)^{-l} \hat{u}_i, \quad \hat{u}_i = (u_{i1} \; u_{i2} \; \ldots \; u_{im_i})^{\mathrm{T}}.$$

Calculating the residue of $f_k^l(\lambda; u)$ at λ_i, we see that

$$\begin{aligned}0 = p_k(A_{-1}L_c^{-l}u) &= p_k\left(\sum_{j=1}^{m_i}\left((\Lambda_i+c)^{-l}\hat{u}_i\right)_j\varphi_{ij}\right)\\ &= \sum_{j=1}^{m_i}\left((\Lambda_i+c)^{-l}\hat{u}_i\right)_j p_k(\varphi_{ij})\\ &= \begin{pmatrix}p_k(\varphi_{i1}) & \dots & p_k(\varphi_{im_i})\end{pmatrix}(\Lambda_i+c)^{-l}\hat{u}_i\end{aligned}$$

for $i \geqslant 1$, $l \geqslant 1$, and $1 \leqslant k \leqslant N$. Thus, we obtain

$$\hat{W}_i(\Lambda_i+c)^{-l}\hat{u}_i = \mathbf{0}, \quad i \geqslant 1, \quad l \geqslant 1. \tag{3.30}$$

See (3.1) for the matrices $\hat{W}_i$. This, combined with (3.28), yields that

$$\begin{pmatrix}\hat{W}_i & \hat{W}_i(\Lambda_i+c)^{-1} & \dots & \hat{W}_i(\Lambda_i+c)^{-(m_i-1)}\end{pmatrix}^{\mathrm{T}}\hat{u}_i = \mathbf{0}, \quad i \geqslant 1. \tag{3.31}$$

It is clear that

$$\begin{aligned}&\operatorname{rank}\begin{pmatrix}\hat{W}_i & \hat{W}_i(\Lambda_i+c)^{-1} & \dots & \hat{W}_i(\Lambda_i+c)^{-(m_i-1)}\end{pmatrix}^{\mathrm{T}}\\ &= \operatorname{rank}\begin{pmatrix}\hat{W}_i & \hat{W}_i\Lambda_i & \dots & \hat{W}_i\Lambda_i^{m_i-1}\end{pmatrix}^{\mathrm{T}}, \quad i \geqslant 1.\end{aligned}$$

In view of the first rank conditions in (3.5), we find that $\hat{u}_i = \mathbf{0}$ for $1 \leqslant i \leqslant \nu$, that is, $P_\nu u = 0$. This is nothing but the relation (3.10), that is, what we hoped to prove.

Proof of Proposition 3.4. (i) Let T_λ, $\lambda \in \rho(L)$, be the operator defined by

$$\begin{aligned}z = T_\lambda u &= u - \sum_{k=1}^{M}\langle u, y_k\rangle N_\lambda g_k\\ &= u - \begin{pmatrix}N_\lambda g_1 & \dots & N_\lambda g_M\end{pmatrix}\langle u, \mathbf{y}\rangle, \quad u \in L^2(\Omega),\end{aligned} \tag{3.32}$$

where $\langle u, \mathbf{y}\rangle$ denotes the vector $(\langle u, y_1\rangle \ \dots \ \langle u, y_M\rangle)^{\mathrm{T}}$, and $N_\lambda g_k$, given in (2.6), are analytic in λ.

For the existence of the closure F, it is necessary and sufficient that

$$u_n \in \mathscr{D}(\hat{F}) \to 0 \quad \text{and } \hat{F}u_n \to y \quad \text{as } n \to \infty$$

implies that $y = 0$. For a sufficiently large $c > 0$, set $\lambda = -c$. If u is in $\mathscr{D}(\hat{F})$, then $z = T_{-c}u$ is in $\mathscr{D}(\hat{L})$ and

$$\hat{L}_c z = \mathscr{L}_c z = \mathscr{L}_c u = \hat{F}_c u.$$

Since $z_n = T_{-c}u_n \to 0$; $\hat{L}_c z_n = L_c z_n \to y$; and L is closed, we see that $y = 0$.

In order to consider the inverse of T_λ, let us introduce the matrix Φ_λ by

$$\Phi_\lambda = \left(\langle N_\lambda g_k, y_j\rangle; \begin{array}{ccc} j & \downarrow & 1, \dots, M \\ k & \rightarrow & 1, \dots, M\end{array}\right). \tag{3.33}$$

We show that Φ_λ goes to 0 as $\lambda \in \overline{\Sigma} - a \to \infty$. Abbreviating the subindices j and k in Φ_λ, suppose first that y is in $\mathscr{D}(\Omega) = C_0^\infty(\Omega)$. Choose a $c > 0$ large enough so that $-c$ is in $\rho(L)$. Then we see - via Green's formula - that

$$\begin{aligned}&\langle \mathscr{L} N_{-c} g, y\rangle - \langle N_{-c} g, \mathscr{L}^* y\rangle \\ &= - \langle (N_{-c} g)_\nu, y\rangle_\Gamma + \langle N_{-c} g, y_\nu\rangle_\Gamma + \langle (\boldsymbol{b}\cdot\boldsymbol{\nu}) N_{-c} g, y\rangle_\Gamma = 0.\end{aligned}$$

Thus,

$$\langle N_{-c} g, y\rangle = \frac{1}{-c} \langle N_{-c} g, \mathscr{L}^* y\rangle, \quad c > -\operatorname{Re}\lambda_1.$$

When $-c$ is replaced by $\lambda \in \rho(L)$ in the above relation, the both sides are analytic functions of λ. By analytic continuation, we obtain the relation

$$\langle N_\lambda g, y\rangle = \frac{1}{\lambda} \langle N_\lambda g, \mathscr{L}^* y\rangle, \quad \lambda \in \rho(L).$$

In view of the expression (2.6), the function $N_\lambda g$ is found bounded when $\lambda \to \infty$. Thus, $\langle N_\lambda g, y\rangle$ goes to 0 as $\lambda \to \infty$. For a general $y \in L^2(\Omega)$, we can also show the convergence of $\langle N_\lambda g, y\rangle$ to 0, approximating y arbitrarily by a sequence of elements of $\mathscr{D}(\Omega)$.

If a is chosen large enough, $(1 - \Phi_\lambda)^{-1}$ exists in $\overline{\Sigma} - a$, where $\overline{\Sigma} = \{\lambda \in \mathbb{C};\, \theta_0 \leqslant |\arg\lambda| \leqslant \pi\}$, $0 < \theta_0 < \pi/2$. Thus the bounded inverse T_λ^{-1} exists and it is expressed as

$$\begin{aligned}T_\lambda^{-1} z = z + (N_\lambda g_1 \ \ldots\ N_\lambda g_M)(1 - \Phi_\lambda)^{-1} \langle z, \boldsymbol{y}\rangle, \\ z \in L^2(\Omega), \quad \lambda \in \overline{\Sigma} - a.\end{aligned}$$

We have shown that both T_λ and T_λ^{-1} are analytic in $\overline{\Sigma} - a$.

For a given $c \geqslant a$ and $f \in L^2(\Omega)$, let us consider the boundary value problem:

$$(c + F)u = f.$$

By setting $z = L_c^{-1} f \in \mathscr{D}(L)$ and $u = T_{-c}^{-1} z = T_{-c}^{-1} L_c^{-1} f$, we find a sequence $\{z_n\} \subset \mathscr{D}(\hat{L})$ such that $z_n \to z$ and $\hat{L} z_n \to Lz$. Here we note that $\langle u_n, \boldsymbol{y}\rangle = (1 - \Phi_{-c})^{-1} \langle z_n, \boldsymbol{y}\rangle$, where $u_n = T_{-c}^{-1} z_n$. It is clear that $u_n \in \mathscr{D}(\hat{F})$. Thus,

$$u_n \to T_{-c}^{-1} z = u, \quad \text{and} \quad (c + \hat{F}) u_n = \mathscr{L}_c u_n = \hat{L}_c z_n \to L_c z.$$

We have shown that u is in $\mathscr{D}(F)$ and that $F_c u = L_c z = f$.

The uniqueness of solutions is shown as follows: Let $F_c u = 0$, and find a sequence $\{u_n\} \subset \mathscr{D}(\hat{F})$ such that $u_n \to u$ and $\hat{F} u_n \to Fu$. By setting $z_n = T_{-c} u_n \in \mathscr{D}(\hat{L})$, we see that

$$z_n \to T_{-c} u, \quad \text{and} \quad \hat{L}_c z_n = \mathscr{L}_c z_n = \hat{F}_c u_n \to F_c u.$$

Thus $T_{-c} u$ is in $\mathscr{D}(L)$ and $L_c T_{-c} u = F_c u = 0$. This means that $T_{-c} u = 0$ or $u = 0$.

We have shown that the bounded inverse: F_c^{-1} exists and is expressed as $T_{-c}^{-1}L_c^{-1}$, or

$$(-c-F)^{-1} = T_{-c}^{-1}(-c-L)^{-1}, \quad c \geqslant a.$$

The operator $T_\lambda^{-1}(\lambda - L)^{-1}$ is analytic in $\overline{\Sigma} - a$. Thus the resolvent $(-c-F)^{-1}$ on the real interval $(-\infty, -a]$ has an analytic extension as $T_\lambda^{-1}(\lambda - L)^{-1}$ in the sector $\overline{\Sigma} - a$. This extension is, however, nothing but the resolvent of F. We have finally found that $\overline{\Sigma} - a$ is contained in $\rho(F)$, and that

$$(\lambda - F)^{-1} = T_\lambda^{-1}(\lambda - L)^{-1} \text{ in } \overline{\Sigma} - a. \tag{3.34}$$

This also gives the decay estimate of the resolvent:

$$\left\|(\lambda - F)^{-1}\right\| \leqslant \frac{\text{const}}{1 + |\lambda|}, \quad \lambda \in \overline{\Sigma} - a. \tag{3.35}$$

Thus $-F$ generates an analytic semigroup e^{-tF}, $t > 0$.

Denseness of the domain $\mathscr{D}(F)$:

To see that $\mathscr{D}(F)$ is dense in $L^2(\Omega)$, we only have to show the implication:

$$\left\langle (\lambda - F)^{-1} f, \varphi \right\rangle = 0, \quad \forall f \in L^2(\Omega) \quad \Rightarrow \quad \varphi = 0.$$

By the relation (3.34),

$$\begin{aligned}
0 &= \langle T_\lambda^{-1}(\lambda - L)^{-1} f, \varphi \rangle \\
&= \left\langle (\lambda - L)^{-1} f, \varphi \right\rangle \\
&\quad + \left(\langle N_\lambda g_1, \varphi \rangle \ \dots \ \langle N_\lambda g_M, \varphi \rangle \right) (1 - \Phi_\lambda)^{-1} \left\langle (\lambda - L)^{-1} f, \mathbf{y} \right\rangle \\
&= \left\langle (\lambda - L)^{-1} f, \varphi \right\rangle + \sum_{k=1}^{M} a_k \left\langle (\lambda - L)^{-1} f, y_k \right\rangle \\
&= \left\langle f, (\overline{\lambda} - L^*)^{-1} \left(\varphi + \sum_{k=1}^{M} \overline{a_k} y_k \right) \right\rangle, \quad \forall f \in L^2(\Omega),
\end{aligned}$$

or $\varphi + \sum_{k=1}^{M} \overline{a_k} y_k = 0$. Here, we have set

$$\left(a_1 \ \dots \ a_M \right) = \left(\langle N_\lambda g_1, \varphi \rangle \ \dots \ \langle N_\lambda g_M, \varphi \rangle \right) (1 - \Phi_\lambda)^{-1}.$$

Thus we see that

$$0 = \left\langle N_\lambda g_j, \varphi + \sum_{k=1}^{M} \overline{a_k} y_k \right\rangle = \left\langle N_\lambda g_j, \varphi \right\rangle + \sum_{k=1}^{M} a_k \left\langle N_\lambda g_j, y_k \right\rangle, \quad 1 \leqslant j \leqslant M,$$

which readily implies that $a_k = 0$, $1 \leqslant k \leqslant M$, or $\varphi = 0$.

Let us consider the solution $u(t) = e^{-tF} u_0$ to the Cauchy problem:

$$\frac{du}{dt} + Fu = 0, \quad u(0) = u_0. \tag{3.36}$$

We have shown in (3.34) that $F_c^{-1} = T_{-c}^{-1}L_c^{-1}$ or $F_cT_{-c}^{-1} = L_c$ on $\mathscr{D}(L)$. By setting $z(t) = T_{-c}u(t)$, the function $z(t) \in \mathscr{D}(L)$ satisfies the equation

$$\frac{dz}{dt} + T_{-c}F_cT_{-c}^{-1}z = cz, \quad t > 0, \quad z(0) = T_{-c}u_0,$$

or

$$\begin{aligned} \frac{dz}{dt} + Lz &= \left(N_{-c}g_1 \ \dots \ N_{-c}g_M\right) \langle L_c z, \mathbf{y}\rangle \\ &= \left(N_{-c}g_1 \ \dots \ N_{-c}g_M\right) \left\langle L_c^{1-\beta}z, L_c^{*\beta}\mathbf{y}\right\rangle, \\ & \qquad t > 0, \quad z(0) = T_{-c}u_0. \end{aligned} \tag{3.37}$$

It is clear that eqn. (3.37) is well posed in $L^2(\Omega)$, and generates an analytic semigroup. Conversely, for the solution $z(t)$ to (3.37), $u(t) = T_{-c}^{-1}z(t)$ satisfies (3.36). Given an arbitrary $\varepsilon > 0$, let us consider the initial boundary value problem for $z^\varepsilon(t,x)$:

$$\begin{aligned} &\frac{\partial z^\varepsilon}{\partial t} + \mathscr{L}z^\varepsilon = \left(N_{-c}g_1 \ \dots \ N_{-c}g_M\right) \left\langle L_c^{1-\beta}z(t+\varepsilon), L_c^{*\beta}\mathbf{y}\right\rangle \quad \text{in } \mathbb{R}_+^1 \times \Omega, \\ &\tau z^\varepsilon = 0 \quad \text{on } \mathbb{R}_+^1 \times \Gamma, \\ &z^\varepsilon(0,x) = z(\varepsilon,x) \quad \text{in } \Omega. \end{aligned}$$

The function $L_c^{1-\beta}z(t+\varepsilon)$ on the right-hand side is analytic in $t \geqslant 0$, and $N_{-c}g_k$ are in $C^2(\Omega) \cap C^1(\overline{\Omega})$. In exactly the same way as in (2.10), the problem admits a unique genuine solution $z^\varepsilon(t,\cdot) \in \mathscr{D}(\hat{L})$ such that $\mathscr{L}z^\varepsilon(t,x)$ is bounded in $(t_1,t_2) \times \Omega$ for $0 < \forall t_1 < \forall t_2$; $\partial z^\varepsilon/\partial t = dz^\varepsilon/dt$; and consequently $z^\varepsilon(t,\cdot) = z(t+\varepsilon,\cdot)$, $t \geqslant 0$ (see [24]). Since $\varepsilon > 0$ is arbitrary, this means that the solution $z(t)$ to (3.37) is a genuine solution, and so is $u(t) = e^{-tF}u_0 = T_{-c}^{-1}z(t)$ with $u(t) \in \mathscr{D}(\hat{F})$, $t > 0$.

(ii) In order to achieve the stabilization, we consider (3.37) which is equivalent to (3.36). Assuming that y_k belong to $P_\nu^*L^2(\Omega) \subset \mathscr{D}(L^*)$, (3.37) is rewritten as

$$\frac{dz}{dt} + Lz - \sum_{k=1}^{M} \langle z, L_c^* y_k\rangle N_{-c}g_k = 0, \quad t > 0, \quad z(0) = Tu_0, \tag{3.37$'$}$$

so that

$$e^{-tF} = T_{-c}^{-1} \exp\left(-t\left(L - \textstyle\sum_{k=1}^{M} \langle \cdot, L_c^* y_k\rangle N_{-c}g_k\right)\right) T_{-c}, \quad t \geqslant 0.$$

The stabilization problem of (3.37′) is reduced to a simpler problem which is discussed in Corollary 2.2, Chapter 3, since y_k belong to $P_\nu^*L^2(\Omega)$: Let

$P_{\lambda_i} N_{-c} g_k = \sum_{1 \leqslant j \leqslant m_i} \zeta_{ij}^k \varphi_{ij}$, $1 \leqslant i \leqslant \nu$, and set (see (1.9) and (1.10), Chapter 3)

$$\tilde{G}_i = \left(\zeta_{ij}^k; \begin{array}{ccc} j & \downarrow & 1, \ldots, m_i \\ k & \rightarrow & 1, \ldots, M \end{array} \right) = \Pi_{\lambda_i}^{-1} \left(\langle N_{-c} g_k, \psi_{ij} \rangle; \begin{array}{ccc} j & \downarrow & 1, \ldots, m_i \\ k & \rightarrow & 1, \ldots, M \end{array} \right).$$

Owing to Corollary 2.2 of Chapter 3, if the controllability condition on the actuators $N_{-c} g_k$:

$$\operatorname{rank} \left(\tilde{G}_i \quad \Lambda_i \tilde{G}_i \quad \ldots \quad \Lambda_i^{m_i - 1} \tilde{G}_i \right) = m_i, \quad 1 \leqslant i \leqslant \nu \tag{3.38}$$

is satisfied, there exists suitable functions y_k or $L_c^* y_k \in P_\nu^* L^2(\Omega)$, $1 \leqslant k \leqslant M$, such that the decay estimate (3.15) is ensured. Here, y_k depends on $N_{-c} g_k$ and thus on c, too. In view of the definition of T_{-c} in (3.32), however, one might raise a question if the bounded inverse T_{-c}^{-1} would really exist. The inverse certainly exists, as long as $(1 - \Phi_{-c})^{-1}$ exists. Thus, if $\det(1 - \Phi_{-c}) = 0$, we only have to replace y_k by $(1 + \varepsilon)^{-1} y_k$ for a small $\varepsilon \neq 0$. By this change, the decay property (3.15) is *little* affected. The matrix Φ_{-c} is changed to $(1 + \varepsilon)^{-1} \Phi_{-c}$, and the polynomial, $\det(1 + \varepsilon - \Phi_{-c})$ of ε of order M is different from 0 in a neighborhood of $\varepsilon = 0$, so that the bounded inverse T_{-c}^{-1} exists for a very little modified y_k. Also note that r_1 cannot be generally replaced by $\operatorname{Re} \lambda_{\nu+1}$, due to possible algebraic multiplicities ($\geqslant 2$) of the eigenvalues on the vertical line: $\operatorname{Re} \lambda = \operatorname{Re} \lambda_{\nu+1}$. The proof of Proposition 3.4 is almost finished.

The condition (3.38) is stated in terms of $P_{\lambda_i} N_{-c} g_k$ in Ω. In the following we interpret (3.38) in terms of the actuators g_k on Γ. Green's formula implies that

$$\begin{aligned}
&\langle \mathscr{L}_c N_{-c} g_k, \psi_{ij} \rangle - \langle N_{-c} g_k, L_c^* \psi_{ij} \rangle \\
&= -\left\langle \frac{\partial N_{-c} g_k}{\partial \nu}, \psi_{ij} \right\rangle_\Gamma + \left\langle N_{-c} g_k, \frac{\partial \psi_{ij}}{\partial \nu} \right\rangle_\Gamma + \langle (\boldsymbol{b}(\xi) \cdot \boldsymbol{\nu}(\xi)) N_{-c} g_k, \psi_{ij} \rangle_\Gamma \\
&= -\left\langle g_k, (1 - \boldsymbol{b}(\xi) \cdot \boldsymbol{\nu}(\xi)) \psi_{ij} - \frac{\partial \psi_{ij}}{\partial \nu} \right\rangle_\Gamma \\
&= -\langle g_k, \sigma \psi_{ij} \rangle_\Gamma .
\end{aligned}$$

The restriction of L^* onto the invariant subspace $P_{\lambda_i}^* L^2(\Omega)$ has the matrix representation $\tilde{\Lambda}_i$ (see (1.8), Chapter 3). We calculate as

$$\begin{aligned}
&\left(\langle g_k, \sigma \psi_{i1} \rangle_\Gamma \quad \ldots \quad \langle g_k, \sigma \psi_{i m_i} \rangle_\Gamma \right)^{\mathrm{T}} \\
&= \left(\langle N_{-c} g_k, L_c^* \psi_{i1} \rangle \quad \ldots \quad \langle N_{-c} g_k, L_c^* \psi_{i m_i} \rangle \right)^{\mathrm{T}} \\
&= (\tilde{\Lambda}_i + c)^* \left(\langle N_{-c} g_k, \psi_{i1} \rangle \quad \ldots \quad \langle N_{-c} g_k, \psi_{i m_i} \rangle \right)^{\mathrm{T}}.
\end{aligned}$$

Thus,

$$\tilde{G}_i = \left((\tilde{\Lambda}_i^* + c)\Pi_{\lambda_i} \right)^{-1} \left(\langle g_k, \sigma\psi_{ij} \rangle_\Gamma ; \begin{array}{ccc} j & \downarrow & 1, \ldots, m_i \\ k & \to & 1, \ldots, M \end{array} \right)$$
$$= \Pi_{\lambda_i}^{-1} (\tilde{\Lambda}_i^* + c)^{-1} \hat{G}_i.$$

Recall the relation: $\Lambda_i = \Pi_{\lambda_i}^{-1} \tilde{\Lambda}_i^* \Pi_{\lambda_i}$ (see (1.11), Chapter 3). Then,

$$\begin{aligned}
&\left(\tilde{G}_i \quad (\Lambda_i + c)\tilde{G}_i \quad \ldots \quad (\Lambda_i + c)^{m_i - 1}\tilde{G}_i\right) \\
&= \left(\Pi_{\lambda_i}^{-1} (\tilde{\Lambda}_i^* + c)^{-1} \hat{G}_i \quad \Pi_{\lambda_i}^{-1} \hat{G}_i \quad \ldots \quad \Pi_{\lambda_i}^{-1} (\tilde{\Lambda}_i^* + c)^{m_i - 2} \hat{G}_i\right) \\
&= \Pi_{\lambda_i}^{-1} (\tilde{\Lambda}_i^* + c)^{-1} \left(\hat{G}_i \quad (\tilde{\Lambda}_i^* + c)\hat{G}_i \quad \ldots \quad (\tilde{\Lambda}_i^* + c)^{m_i - 1}\hat{G}_i\right).
\end{aligned}$$

Thus, the controllability condition (3.38) is rewritten as

$$\begin{aligned}
&\operatorname{rank} \left(\hat{G}_i \quad (\tilde{\Lambda}_i^* + c)\hat{G}_i \quad \ldots \quad (\tilde{\Lambda}_i^* + c)^{m_i - 1}\hat{G}_i\right) \\
&= \operatorname{rank} \left(\hat{G}_i \quad \tilde{\Lambda}_i^* \hat{G}_i \quad \ldots \quad (\tilde{\Lambda}_i^*)^{m_i - 1}\hat{G}_i\right) = m_i, \quad 1 \leqslant i \leqslant \nu.
\end{aligned} \tag{3.38$'$}$$

This is nothing but the second condition of (3.5).

Proof of Proposition 3.5 *and Lemma* 3.6.

Let us consider the perturbed operator F°. This operator is obtained as the closure of $\hat{F}^\circ$, which is defined by

$$\begin{aligned}
&\hat{F}^\circ u = \mathscr{L} u, \quad u \in \mathscr{D}(\hat{F}^\circ) \\
&\mathscr{D}(\hat{F}^\circ) = \left\{ u \in C^2(\Omega) \cap C^1(\overline{\Omega});\ \mathscr{L} u \in L^2(\Omega),\ \tilde{\tau}_f u = 0 \text{ on } \Gamma \right\}, \\
&\text{where} \quad \tilde{\tau}_f u = \tau u - \sum_{k=1}^{M} \langle u, \tilde{y}_k \rangle g_k, \quad \tilde{y}_k \in L^2(\Omega).
\end{aligned} \tag{3.14$'$}$$

By Proposition 3.4 we already know that, as long as $\sum_{k=1}^M \|\tilde{y}_k - y_k\|$ is close to 0, there is an $a \in \mathbb{R}^1$ such that the sector $\overline{\Sigma} - a$ is contained in $\rho(F^\circ)$ uniformly in $\tilde{y}_k$, $1 \leqslant k \leqslant M$. Choose any $c \geqslant a$ $(-c \in \overline{\Sigma} - a)$. In order to compare $(c + F^\circ)^{-1}$ with $(c + F)^{-1}$, let us first show Lemma 3.6, and obtain the expression of $(c + F^\circ)^{-1}$. On the analogy of $N_\lambda g$ (see (2.6)), we seek the solution to the boundary value problem (3.17):

$$(c + \mathscr{L})u = 0 \quad \text{in } \Omega, \quad \tau_f u = g \in C^{2+\omega}(\Gamma) \quad \text{on } \Gamma.$$

We may assume with no loss of generality that the set $\{y_1, \ldots, y_M\}$ is a linearly independent system. Choose ψ_k in $\mathscr{D}(\Omega)$ $(= C_0^\infty(\Omega))$ so that ψ_k are arbitrarily close to y_k in $L^2(\Omega)$. Then the matrix Ψ defined by

$$\Psi = \left(\langle \psi_j, y_k \rangle ; \begin{array}{ccc} j & \to & 1, \ldots, M \\ k & \downarrow & 1, \ldots, M \end{array} \right)$$

is non-singular. By setting $R_f g = Rg - (\psi_1 \ \ldots \ \psi_M)\Psi^{-1}\langle Rg, \mathbf{y}\rangle$, it is easily seen that $R_f g$ belongs to $C^{2+\omega}(\overline{\Omega})$ and satisfies the relation: $\tau_f R_f g = g$. Set

$$N_\lambda^f g = R_f g - (\lambda - F)^{-1}(\lambda - \mathscr{L})R_f g, \quad \lambda \in \rho(F). \tag{3.39}$$

Let c be greater than or equal to a. We note that $\mathscr{L}_c R_f g$ belongs to $C^\omega(\overline{\Omega})$. Then, $F_c^{-1}\mathscr{L}_c R_f g = T_{-c}^{-1} L_c^{-1}\mathscr{L}_c R_f g$ belongs to $\mathscr{D}(\hat{F})$. Thus $u = N_{-c}^f g$ solves (3.17) uniquely, and this proves Lemma 3.6 [2].

For a given $f \in L^2(\Omega)$, suppose for a moment that $u = (c + F^\circ)^{-1} f$ is in $\mathscr{D}(\hat{F}^\circ)$ and satisfies the equation:

$$\mathscr{L}_c u = f, \quad \tilde{\tau}_f u = 0, \quad \text{or} \quad \tau_f u = (g_1 \ \ldots \ g_M)\langle u, \tilde{\mathbf{y}} - \mathbf{y}\rangle,$$

where $\langle u, \tilde{\mathbf{y}} - \mathbf{y}\rangle = (\langle u, \tilde{y}_1 - y_1\rangle \ \ldots \ \langle u, \tilde{y}_M - y_M\rangle)^{\mathrm{T}}$. Then we have

$$\mathscr{L}_c\Big(u - (N_{-c}^f g_1 \ \ldots \ N_{-c}^f g_M)\langle u, \tilde{\mathbf{y}} - \mathbf{y}\rangle\Big) = f \quad \text{in } \Omega,$$
$$\tau_f\Big(u - (N_{-c}^f g_1 \ \ldots \ N_{-c}^f g_M)\langle u, \tilde{\mathbf{y}} - \mathbf{y}\rangle\Big) = 0 \quad \text{on } \Gamma.$$

This means that

$$u - (N_{-c}^f g_1 \ \ldots \ N_{-c}^f g_M)\langle u, \tilde{\mathbf{y}} - \mathbf{y}\rangle \quad \Big(\text{denoted as } \big(1 - N_{-c}^f \mathbf{g}\langle\cdot, \tilde{\mathbf{y}} - \mathbf{y}\rangle\big)u\Big)$$
$$= (c + F)^{-1} f.$$

We know from Proposition 3.4, (ii) that the set $(\overline{\Sigma} - a) \cup \{\lambda \in \mathbb{C};\ \mathrm{Re}\,\lambda \leqslant r_1\}$ is contained in $\rho(F)$. In view of (3.39), the functions $N_\lambda^f \mathbf{g} = \big(N_\lambda^f g_1 \ \ldots \ N_\lambda^f g_M\big)$ are analytic and bounded in $\lambda \in (\overline{\Sigma} - a) \cup \{\lambda \in \mathbb{C};\ \mathrm{Re}\,\lambda \leqslant r_1\}$. Thus the bounded inverse $\big(1 - N_\lambda^f \mathbf{g}\langle\cdot, \tilde{\mathbf{y}} - \mathbf{y}\rangle\big)^{-1}$ exists in $(\overline{\Sigma} - a) \cup \{\lambda \in \mathbb{C};\ \mathrm{Re}\,\lambda \leqslant r_1\}$ as long as $\|\tilde{\mathbf{y}} - \mathbf{y}\|$ is chosen small enough, and we have, as a necessary condition, the expression of the solution:

$$u = (c + F^\circ)^{-1} f = \big(1 - N_{-c}^f \mathbf{g}\langle\cdot, \tilde{\mathbf{y}} - \mathbf{y}\rangle\big)^{-1}(c + F)^{-1} f. \tag{3.40}$$

We show that, given any $f \in L^2(\Omega)$, the function u defined by (3.40) actually gives the solution $(c + F^\circ)^{-1} f$. Setting $v = F_c^{-1} f$, we find a sequence $\{v_n\} \subset \mathscr{D}(\hat{F})$ such that $v_n \to v$ and $f_n = \hat{F}_c v_n \to F_c v = f$. The functions $\varphi_n = \big(1 - N_{-c}^f \mathbf{g}\langle\cdot, \tilde{\mathbf{y}} - \mathbf{y}\rangle\big)^{-1} F_c^{-1} f_n$ satisfy the relations,

[2] It is easily seen that $T_{-c}^{-1} N_{-c} g = N_{-c} g + (N_{-c} g_1 \ldots N_{-c} g_M)(1 - \Phi_{-c})^{-1}\langle N_{-c} g, \mathbf{y}\rangle$ also gives the unique solution to (3.17). In other words, we have $N_{-c}^f g = T_{-c}^{-1} N_{-c} g$. However, the simpler expression $T_{-c}^{-1} N_{-c} g$ does not work in the following argument.

$\varphi_n - N_{-c}^f \boldsymbol{g} \langle \varphi_n, \tilde{\boldsymbol{y}} - \boldsymbol{y} \rangle = v_n$. Thus,

$$\mathscr{L}_c \varphi_n = \mathscr{L}_c N_{-c}^f \boldsymbol{g} \langle \varphi_n, \tilde{\boldsymbol{y}} - \boldsymbol{y} \rangle + \mathscr{L}_c v_n = f_n \quad \text{in } \Omega,$$
$$\tau_f \varphi_n = \tau_f N_{-c}^f \boldsymbol{g} \langle \varphi_n, \tilde{\boldsymbol{y}} - \boldsymbol{y} \rangle + \tau_f v_n = \left(g_1 \ \dots \ g_M \right) \langle \varphi_n, \tilde{\boldsymbol{y}} - \boldsymbol{y} \rangle \quad \text{on } \Gamma.$$

In other words, the functions φ_n belong to $\mathscr{D}(\hat{F}_c^\circ)$ and $\hat{F}_c^\circ \varphi_n = f_n$. Furthermore,

$$\varphi_n \to \left(1 - N_{-c}^f \boldsymbol{g} \langle \cdot, \tilde{\boldsymbol{y}} - \boldsymbol{y} \rangle \right)^{-1} F_c^{-1} f, \quad \text{and } \hat{F}_c^\circ \varphi_n \to f.$$

Thus, $\left(1 - N_{-c}^f \boldsymbol{g} \langle \cdot, \tilde{\boldsymbol{y}} - \boldsymbol{y} \rangle \right)^{-1} F_c^{-1} f$ belongs to $\mathscr{D}(F^\circ)$ and

$$F_c^\circ u = F_c^\circ \left(1 - N_{-c}^f \boldsymbol{g} \langle \cdot, \tilde{\boldsymbol{y}} - \boldsymbol{y} \rangle \right)^{-1} F_c^{-1} f = f.$$

Since $-c$ belongs to $\rho(F^\circ)$ (see the proof of Proposition 3.4), we have shown the correctness of the expression (3.40).

Recall that both the operators $\left(1 - N_\lambda^f \boldsymbol{g} \langle \cdot, \tilde{\boldsymbol{y}} - \boldsymbol{y} \rangle \right)^{-1}$ and $(\lambda - F)^{-1}$ are analytic in $\lambda \in (\overline{\Sigma} - a) \cup \{\lambda \in \mathbb{C};\ \mathrm{Re}\,\lambda \leqslant r\}$, $r < r_1$. We extend $(\lambda - F^\circ)^{-1}$ analytically via (3.40) to the set $(\overline{\Sigma} - a) \cup \{\lambda \in \mathbb{C};\ \mathrm{Re}\,\lambda \leqslant r\}$. The extension is nothing but the resolvent of F°. Thus we have shown the relation

$$(\lambda - F^\circ)^{-1} = \left(1 - N_\lambda^f \boldsymbol{g} \langle \cdot, \tilde{\boldsymbol{y}} - \boldsymbol{y} \rangle \right)^{-1} (\lambda - F)^{-1}, \quad \lambda \in (\overline{\Sigma} - a) \cup \{\lambda \in \mathbb{C};\ \mathrm{Re}\,\lambda \leqslant r\}, \tag{3.41}$$

from which we immediately obtain the estimate

$$\left\| e^{-tF^\circ} \right\| \leqslant \text{const}\, e^{-rt}, \quad t \geqslant 0,$$

or (3.16). The proof of Theorem 3.1 is thereby complete. □

We close Section 3 with the following remark: If an additional assumption is posed on the actuators g_k, a simpler approach is possible in the proof of Proposition 3.5. Let us see this briefly. Set $\Gamma_1 = \{\xi;\ \alpha(\xi) = 1\}$. Then,

Proposition 3.8. *Suppose that*

$$g_k(\xi) = 0 \quad \text{on } \Gamma_1, \quad \text{and} \quad \frac{g_k(\xi)}{1 - \alpha(\xi)} \in L^2(\Gamma \setminus \Gamma_1), \quad 1 \leqslant k \leqslant M. \tag{3.42}$$

Then the adjoint operator of F is expressed by [3]

$$
\begin{aligned}
F^*\varphi &= L^*\varphi - \sum_{k=1}^{M}\left\langle \varphi, \frac{g_k}{1-\alpha}\right\rangle_{\Gamma\setminus\Gamma_1} y_k \\
&= L^*\varphi - E_1\varphi, \qquad \varphi \in \mathscr{D}(F^*) = \mathscr{D}(L^*).
\end{aligned}
\tag{3.43}
$$

Proof. By Green's formula, we see that, for $u \in \mathscr{D}(\hat{F})$ and $\varphi \in \mathscr{D}(\hat{L}^*)$

$$
\begin{aligned}
\left\langle \hat{F}u, \varphi\right\rangle &= -\langle u_\nu, \varphi\rangle_\Gamma + \langle u, \varphi_\nu + (\boldsymbol{b}\cdot\boldsymbol{\nu})\varphi\rangle_\Gamma + \langle u, \mathscr{L}^*\varphi\rangle \\
&= -\sum_{k=1}^{M}\langle u, y_k\rangle \langle g_k, \sigma\phi\rangle_\Gamma + \langle u, \mathscr{L}^*\varphi\rangle \\
&= \left\langle u, \mathscr{L}^*\varphi - \sum_{k=1}^{M}\langle \sigma\varphi, g_k\rangle_\Gamma y_k\right\rangle = \left\langle u, \hat{F}^\dagger\varphi\right\rangle,
\end{aligned}
\tag{3.44}
$$

where $\sigma\varphi = (1-(\boldsymbol{b}\cdot\boldsymbol{\nu}))\varphi - \varphi_\nu$ (see (3.2)), and $\mathscr{D}(\hat{F}^\dagger) = \mathscr{D}(\hat{L}^*)$. By (3.42) we note that

$$
\langle \sigma\varphi, g_k\rangle_\Gamma = \langle \sigma\varphi, g_k\rangle_{\Gamma\setminus\Gamma_1} = \left\langle \frac{\varphi}{1-\alpha}, g_k\right\rangle_{\Gamma\setminus\Gamma_1} = \left\langle \varphi, \frac{g_k}{1-\alpha}\right\rangle_{\Gamma\setminus\Gamma_1}.
$$

Thus, $\hat{F}^\dagger$ is rewritten as

$$
\hat{F}^\dagger\varphi = \mathscr{L}^*\varphi - \sum_{k=1}^{M}\left\langle \varphi, \frac{g_k}{1-\alpha}\right\rangle_{\Gamma\setminus\Gamma_1} y_k, \quad \varphi \in \mathscr{D}(\hat{F}^\dagger).
$$

Set

$$
\begin{aligned}
F^\dagger\varphi &= L^*\varphi - \sum_{k=1}^{M}\left\langle \varphi, \frac{g_k}{1-\alpha}\right\rangle_{\Gamma\setminus\Gamma_1} y_k \\
&= L^*\varphi - E_1\varphi, \qquad \varphi \in \mathscr{D}(F^\dagger) = \mathscr{D}(L^*).
\end{aligned}
\tag{3.45}
$$

Then we see that $\hat{F}^\dagger \subset F^\dagger$. We recall the estimate: $\|L_c^* u\| \geqslant \text{const}\,\|u\|_{H^1(\Omega)}$ for $u \in \mathscr{D}(L^*)$. Then passage to the limit with respect to $u \in \mathscr{D}(\hat{F})$ and $\varphi \in \mathscr{D}(\hat{F}^\dagger)$ gives the refined version of (3.44):

$$
\langle Fu, \varphi\rangle = \left\langle u, F^\dagger\varphi\right\rangle, \quad u \in \mathscr{D}(F), \quad \varphi \in \mathscr{D}(F^\dagger). \tag{3.46}
$$

Thus we see that $F^\dagger \subset F^*$. We show that the bounded inverse $(\lambda - F^\dagger)^{-1}$ exists

[3]When $\alpha(\xi) \equiv 1$, we assume that g_k, $1 \leqslant k \leqslant M$, belong to $H^{3/2}(\Gamma)$. The adjoint operator F^* in this case is expressed by

$$
F^*\varphi = L^*\varphi - \sum_{k=1}^{M}\langle \varphi_\nu, g_k\rangle_\Gamma y_k, \quad \varphi \in \mathscr{D}(F^*),
$$

where $\mathscr{D}(F^*) = \mathscr{D}(L^*) = H^2(\Omega) \cap H_0^1(\Omega)$. The perturbing terms $\langle \varphi_\nu, g_k\rangle_\Gamma$ are subordinate to $L_c^{*\omega}$ with $\omega > 3/4$.

in the sector $\overline{\Sigma} - a$ if $a > 0$ is chosen large enough. Since L_c^* is m-accretive, we note a fairly rough relation: $\mathscr{D}\left(L_c^{*\omega}\right) \subset H^{\omega}(\Omega)$, $0 \leqslant \omega \leqslant 1$ (see, e.g., Subsection 3.3, Chapter 2). Then,

$$\begin{aligned}\left\|E_1(\lambda - L^*)^{-1}\varphi\right\| &\leqslant \text{const}\left\|(\lambda - L^*)^{-1}\varphi\right\|_{H^{\omega}(\Omega)}\\ &\leqslant \text{const}\left\|L_c^{*\omega}(\lambda - L^*)^{-1}\varphi\right\|\\ &\leqslant \frac{\text{const}}{1+|\lambda|^{1-\omega}}\|\varphi\|, \quad \frac{1}{2} < \omega < 1.\end{aligned}$$

Choosing an $a > 0$ large enough, we see that $\left\|E_1(\lambda - L^*)^{-1}\right\|$ is smaller than 1 in $\overline{\Sigma} - a$. Thus, the resolvent $(\lambda - F^{\dagger})^{-1}$ exists in $\overline{\Sigma} - a$, and it is expressed by $(\lambda - F^{\dagger})^{-1} = (\lambda - L^*)^{-1}\left(1 + E_1(\lambda - L^*)^{-1}\right)^{-1}$. Since both the resolvents $F_c^{\dagger\,-1}$ and $F_c^{*\,-1} = (F_c^{-1})^*$ exist for a sufficiently large $c > 0$, we find that $\mathscr{D}\left(F^{\dagger}\right)$ is equal to $\mathscr{D}(F^*)$, in other words, $F^{\dagger} = F^*$. □

Remark: An alternative and easier proof of Proposition 3.5 is possible. By Proposition 3.8 we see that

$$\begin{aligned}F^{\circ *}\varphi &= L^*\varphi - \sum_{k=1}^{M}\left\langle \varphi, \frac{g_k}{1-\alpha}\right\rangle_{\Gamma\setminus\Gamma_1}\tilde{y}_k\\ &= F^*\varphi + \sum_{k=1}^{M}\left\langle \varphi, \frac{g_k}{1-\alpha}\right\rangle_{\Gamma\setminus\Gamma_1}(y_k - \tilde{y}_k)\\ &= F^*\varphi + E_2\varphi, \quad \varphi \in \mathscr{D}(F^*).\end{aligned}$$

It is clear that

$$\|E_2\varphi\| \leqslant \text{const}\sum_{k=1}^{M}\|y_k - \tilde{y}_k\|\left\|L_c^*F^{*\,-1}\right\|\|F^*\varphi\|.$$

We already know that the set $(\overline{\Sigma} - a) \cup \{\lambda \in \mathbb{C};\ \text{Re}\,\lambda \leqslant r\}$ is contained in $\rho(F)$. Thus,

$$\|(\lambda - F^*)^{-1}\| \leqslant \frac{\text{const}}{1+|\lambda|}, \quad \lambda \in (\overline{\Sigma} - a) \cup \{\lambda \in \mathbb{C};\ \text{Re}\,\lambda \leqslant r\}.$$

Thus, as long as $\sum_{k=1}^{M}\|y_k - \tilde{y}_k\|$ is small enough, $\left\|E_2(\lambda - F^*)^{-1}\right\|$ is smaller than 1 for $\lambda \in (\overline{\Sigma} - a) \cup \{\lambda \in \mathbb{C};\ \text{Re}\,\lambda \leqslant r\}$, and thus

$$\begin{aligned}\left\|(\lambda - F^{\circ *})^{-1}\right\| &= \left\|(\lambda - F^*)^{-1}\left(1 - E_2(\lambda - F^*)^{-1}\right)^{-1}\right\|\\ &\leqslant \frac{\text{const}}{1+|\lambda|}, \quad \lambda \in (\overline{\Sigma} - a) \cup \{\lambda \in \mathbb{C};\ \text{Re}\,\lambda \leqslant r\}.\end{aligned}$$

This immediately implies that

$$\left\|e^{-tF^{\circ}}\right\| = \left\|\left(e^{-tF^{\circ}}\right)^*\right\| = \left\|e^{-tF^{\circ *}}\right\| \leqslant \text{const}\, e^{-rt}, \quad t \geqslant 0.$$

4.4 Another Construction of Stabilizing Compensators

In Section 3, the stabilizing compensator Σ_c in (2.7) is constructed, based on the controlled plant Σ_p for u with boundary inputs given in feedback form. We construct a compensator in this section, based on an equivalent controlled plant with only distributed inputs. It seems somewhat an easier approach to study the stabilization problem within this framework, since we are not bothered by nonhomogeneous boundary terms. The fundamental settings of the problem are the same as in the preceding sections, such as the operator B satisfying the separation condition (2.5) and the rank conditions (3.5) on the sensors and the actuators.

For u and $z_k \in L^2(\Omega)$, $1 \leqslant k \leqslant M$, let $\langle u, \boldsymbol{z}\rangle = \left(\langle u, z_1\rangle \ \dots \ \langle u, z_M\rangle\right)^{\mathrm{T}}$. For a $c > 0$ such that $-c \in \rho(L)$, let $\left(\underset{k\to}{\langle N_{-c}g_k, \boldsymbol{z}\rangle}\right)$ denote the $M \times M$ matrix defined by

$$\left(\underset{k\to}{\langle N_{-c}g_k, \boldsymbol{z}\rangle}\right) = \left(\langle N_{-c}g_1, \boldsymbol{z}\rangle \quad \dots \quad \langle N_{-c}g_M, \boldsymbol{z}\rangle\right)$$
$$= \left(\langle N_{-c}g_k, z_j\rangle\,;\ \begin{matrix} j & \downarrow & 1,\dots,M \\ k & \to & 1,\dots,M \end{matrix}\right).$$

Choose a set of $y_k \in P_\nu^* L^2(\Omega)$, say, y_{0k}, $1 \leqslant k \leqslant M$, stated in Proposition 3.4, (ii), and set $\boldsymbol{y}_0 = \left(y_{01} \ \dots \ y_{0M}\right)^{\mathrm{T}}$. By choosing $c > 0$ large enough, if necessary, we may assume with no loss of generality that (see (3.33))

$$\det\left(1 - \overline{\left(\underset{k\to}{\langle N_{-c}g_k, \boldsymbol{y}_0\rangle}\right)}\right) \neq 0.$$

The following Lemma is easily examined by direct computaitons, so that the proof is omitted:

Lemma 4.1. *The map $\boldsymbol{G}(\cdot)$ defined on $((L^2(\Omega))^M)$ as*

$$\boldsymbol{y} = \boldsymbol{G}(\boldsymbol{z}) = \left(1 + \overline{\left(\underset{k\to}{\langle N_{-c}g_k, \boldsymbol{z}\rangle}\right)}\right)^{-1} \boldsymbol{z}, \quad \boldsymbol{z} \in \left(L^2(\Omega)\right)^M \tag{4.1}$$

admits the inverse $\boldsymbol{G}^{-1}(\cdot)$ on $\left(L^2(\Omega)\right)^M$ as long as $\det\left(1 + \overline{\left(\underset{k\to}{\langle N_{-c}g_k, \boldsymbol{z}\rangle}\right)}\right) \neq 0.$

The inverse $\boldsymbol{G}^{-1}$ is given by

$$\begin{aligned} \boldsymbol{z} = \boldsymbol{G}^{-1}(\boldsymbol{y}) &= \left(1 - \overline{\left(\underset{k\to}{\langle N_{-c}g_k, \boldsymbol{y}\rangle}\right)}\right)^{-1} \boldsymbol{y}, \quad \text{and} \\ 1 - \overline{\left(\underset{k\to}{\langle N_{-c}g_k, \boldsymbol{y}\rangle}\right)} &= \left(1 + \overline{\left(\underset{k\to}{\langle N_{-c}g_k, \boldsymbol{z}\rangle}\right)}\right)^{-1}. \end{aligned} \tag{4.2}$$

By Lemma 4.1, let $\boldsymbol{z} = \left(z_1 \ \dots \ z_M\right)^{\mathrm{T}} \in \left(P_\nu^* L^2(\Omega)\right)^M$ be such that $\boldsymbol{z} = \boldsymbol{G}^{-1}(\boldsymbol{y}_0)$. By Proposition 3.3, we find suitable sequences of functions $X^*\rho_k$ which are arbitrarily close to z_k in the $L^2(\Omega)$-topology, $1 \leqslant k \leqslant M$. Thus, by setting $\boldsymbol{\rho} = \left(\rho_1 \ \dots \ \rho_M\right)^{\mathrm{T}} \in H^M$,

$$\|\boldsymbol{G}(X^*\boldsymbol{\rho}) - \boldsymbol{y}_0\| \to 0, \quad \text{and } \det\left(1 + \overline{\left(\underset{k\to}{\langle N_{-c}g_k, X^*\boldsymbol{\rho}\rangle}\right)}\right) \neq 0. \tag{4.3}$$

Choose the above $\boldsymbol{\rho}$ such that the perturbed operator (still denoted as F) with the parameters $\boldsymbol{y}_0$ replaced by $\boldsymbol{G}(X^*\boldsymbol{\rho}) \in \left(L^2(\Omega)\right)^M$ guarantees the decay (3.16). Noting that the set $\{\eta_{ij}^{\pm}\}$ forms an orthonormal basis for H, we can choose ρ_k, which are expressed as linear combinations of a finite number of $\eta_{ij}^{\pm}$, say, $1 \leqslant i \leqslant n$, $1 \leqslant j \leqslant n_i$. Based on the $\boldsymbol{\rho} \in \left(\mathscr{D}(B^*)\right)^M$, we define the matrices Θ, G_1, and G_2, respectively as:

$$\Theta = 1 + \overline{\left(\underset{k\to}{\langle N_{-c}g_k, X^*\boldsymbol{\rho}\rangle}\right)},$$

$$G_1 = \left(\underset{k\to}{\left\langle N_{-c}g_k, \Theta^{-1}X^*\boldsymbol{\rho}\right\rangle}\right) = \overline{\Theta}^{-1}\left(\underset{k\to}{\langle N_{-c}g_k, X^*\boldsymbol{\rho}\rangle}\right), \quad \text{and} \tag{4.4}$$

$$G_2 = \left(\underset{k\to}{\langle \xi_k, \boldsymbol{\rho}\rangle_H}\right) = \left(\langle \xi_k, \rho_j\rangle_H\,;\ \begin{matrix} j & \downarrow & 1,\dots,M \\ k & \to & 1,\dots,N \end{matrix}\right).$$

Derivation of the differential equation with distributed feedback.

Let us derive a system of differential equations with state $(q(t), v(t))$, which is fundamental in our stabilization scheme as well as well-posedness. Let $f_k(t)$, $1 \leqslant k \leqslant M$, be input functions, not specified at this point, and consider the differential equation,

$$\begin{cases} \dfrac{du}{dt} + \mathscr{L}u = 0 \quad \text{in } \mathbb{R}_+^1 \times \Omega, \\ \tau u = \displaystyle\sum_{k=1}^{M} f_k(t) g_k \quad \text{on } \mathbb{R}_+^1 \times \Gamma, \\ u(0,\cdot) = u_0 \quad \text{in } \Omega. \end{cases} \tag{4.5}$$

Assuming for a moment that f_k are of class C^1 and setting

$$q(t) = u(t) - \sum_{k=1}^{M} f_k(t) N_{-c} g_k,$$

we obtain the equation for $q(t)$:

$$\begin{aligned} \frac{dq}{dt} + Lq &= \sum_{k=1}^{M} \left(cf_k(t) - f_k'(t) \right) N_{-c} g_k, \\ q(0) = q_0 &= u_0 - \sum_{k=1}^{M} f_k(0) N_{-c} g_k. \end{aligned} \tag{4.6}$$

A new feature is that we construct a dynamic compensator, based not on (4.5) but on (4.6) with state $q(t) \in \mathscr{D}(L)$. The system of differential equations in $L^2(\Omega) \times H$ is then described as

$$\begin{cases} \dfrac{dq}{dt} + Lq = \displaystyle\sum_{k=1}^{M} \left(cf_k(t) - f_k'(t) \right) N_{-c} g_k, \\ \dfrac{dv}{dt} + Bv = -Cq + \displaystyle\sum_{k=1}^{M} \left(cf_k(t) - f_k'(t) \right) X N_{-c} g_k, \\ q(0) = q_0 \in L^2(\Omega), \quad v(0) = v_0 \in H, \end{cases} \tag{4.7}$$

where the operators B and C are given by (2.2) and (3.8), respectively. Whatever the inputs $f_k(t)$ may be, Proposition 3.2 implies that

$$\frac{d}{dt}(Xq - v) + B(Xq - v) = 0, \quad t > 0,$$

or $Xq(t) - v(t) = e^{-tB}(Xq_0 - v_0)$, $t \geqslant 0$: Ensuring this relation is the role of the compensator. By the decay property (2.3) of e^{-tB},

$$\left\| Xq(t) - v(t) \right\|_H \leqslant e^{-a\mu_1 t} \left\| Xq_0 - v_0 \right\|_H, \quad t \geqslant 0. \tag{4.8}$$

Let $\boldsymbol{g}(q, v)$ be the vector-valued function defined by

$$\boldsymbol{g}(q, v) = \overline{\Theta}^{-1} \left(-G_2 \boldsymbol{p}(q) + \langle v, B_c^* \boldsymbol{\rho} \rangle_H \right) = \begin{pmatrix} g_1(q, v) \\ \vdots \\ g_M(q, v) \end{pmatrix}, \tag{4.9}$$

where $\boldsymbol{p}(q) = \begin{pmatrix} p_1(q) & p_2(q) & \dots & p_N(q) \end{pmatrix}^{\mathrm{T}}$, and the matrices Θ and G_2 are defined in (4.4). Replacing $cf_k(t) - f_k'(t)$ by $g_k(q, v)$ in (4.7), we obtain the system of differential equations with state (q, v):

$$\begin{cases} \dfrac{dq}{dt} + Lq = \left(N_{-c} g_1 \; \dots \; N_{-c} g_M \right) \boldsymbol{g}(q, v), \\ \dfrac{dv}{dt} + Bv = -Cq + \left(X N_{-c} g_1 \; \dots \; X N_{-c} g_M \right) \boldsymbol{g}(q, v), \\ q(0) = q_0 \in L^2(\Omega), \quad v(0) = v_0 \in H. \end{cases} \tag{4.10}$$

Equation (4.10), our basic equation, is clearly well posed in $L^2(\Omega)\times H$, and the decay estimate (4.8) holds. As in Theorem 2.1, $q(t)$ belongs to $\mathscr{D}(\hat{L})$ for each $t>0$.

In (4.10), set

$$u(t) = q(t) + \sum_{k=1}^{M} f_k(t) N_{-c} g_k = q(t) + \big(N_{-c}g_1 \ \dots \ N_{-c}g_M\big)\boldsymbol{f}(t), \qquad \text{where} \quad f_k(t) = \langle v(t), \rho_k\rangle_H, \quad 1 \leqslant k \leqslant M. \tag{4.11}$$

Then $u(t)$ belongs to $C^2(\Omega)\cap C^1(\overline{\Omega})$ for each $t>0$. In view of (4.8), we calculate as

$$\begin{aligned}
|\boldsymbol{f}(t) - \langle q(t), X^*\boldsymbol{\rho}\rangle| &= \left|\boldsymbol{f}(t) - \langle u(t), X^*\boldsymbol{\rho}\rangle + \sum_{k=1}^{M} f_k(t)\langle N_{-c}g_k, X^*\boldsymbol{\rho}\rangle\right| \\
&= \left|\boldsymbol{f}(t) - \langle u(t), \Theta\boldsymbol{y}\rangle + \overline{\Theta}G_1\boldsymbol{f}(t)\right| \\
&= \left|(1+\overline{\Theta}G_1)\boldsymbol{f}(t) - \overline{\Theta}\langle u(t), \boldsymbol{y}\rangle\right| \\
&\leqslant \text{const}\, e^{-a\mu_1 t}\left(\|q_0\| + \|v_0\|_H\right), \quad t \geqslant 0,
\end{aligned}$$

where $\boldsymbol{y} = \boldsymbol{G}(X^*\boldsymbol{\rho}) = \Theta^{-1}X^*\boldsymbol{\rho}$. Since $\overline{\Theta} = 1 + \overline{\Theta}G_1$, we find that

$$|\boldsymbol{f}(t) - \langle u(t), \boldsymbol{y}\rangle| \leqslant \text{const}\, e^{-a\mu_1 t}\left(\|q_0\| + \|v_0\|_H\right), \quad t \geqslant 0. \tag{4.12}$$

Recall that $\boldsymbol{\rho}$ belongs to $\left(\mathscr{D}(B^*)\right)^M$. Then we similarly obtain the estimate

$$|\boldsymbol{f}'(t) - \langle u_t(t), \boldsymbol{y}\rangle| \leqslant \text{const}\, e^{-a\mu_1 t}\left(\|q_0\| + \|v_0\|_H\right), \quad t > 0. \tag{4.13}$$

In view of the equation for v in (4.10), we calculate as

$$\begin{aligned}
\langle v_t, \boldsymbol{\rho}\rangle_H + \langle Bv, \boldsymbol{\rho}\rangle_H &= \sum_{k=1}^{N} p_k(q)\langle \xi_k, \boldsymbol{\rho}\rangle_H \\
&\quad + \left(\langle XN_{-c}g_1, \boldsymbol{\rho}\rangle_H \ \dots \ \langle XN_{-c}g_M, \boldsymbol{\rho}\rangle_H\right)\boldsymbol{g}(q,v),
\end{aligned}$$

or

$$\boldsymbol{f}'(t) + \langle v, B^*\boldsymbol{\rho}\rangle_H = G_2\,\boldsymbol{p}(q) + \left(\underset{k\to}{\langle N_{-c}g_k, X^*\boldsymbol{\rho}\rangle}\right)\boldsymbol{g}(q,v),$$

from which we find that

$$c\boldsymbol{f}(t) - \boldsymbol{f}'(t) = \boldsymbol{g}(q,v), \quad t > 0. \tag{4.14}$$

Thus (4.10) is rewritten as

$$\begin{cases}
\dfrac{dq}{dt} + Lq = (N_{-c}g_1 \ \dots \ N_{-c}g_M)\,(c\boldsymbol{f}(t) - \boldsymbol{f}'(t)), \\
\dfrac{dv}{dt} + Bv = -Cq + (XN_{-c}g_1 \ \dots \ XN_{-c}g_M)\,(c\boldsymbol{f}(t) - \boldsymbol{f}'(t)), \\
q(0) = q_0 \in L^2(\Omega), \quad v(0) = v_0 \in H.
\end{cases} \tag{4.10$'$}$$

Then $u(t)$ defined by (4.11) satisfies the differential equation:

$$\begin{cases} \dfrac{du}{dt} + \mathscr{L}u = 0 \quad \text{in } \mathbb{R}^1_+ \times \Omega, \\ \tau u = \big(g_1 \ \dots \ g_M\big)\boldsymbol{f}(t) \quad \text{on } \mathbb{R}^1_+ \times \Gamma, \\ u(0,\cdot) = u_0 = q_0 + \big(N_{-c}g_1 \ \dots \ N_{-c}g_M\big)\langle v_0, \boldsymbol{\rho}\rangle_H \quad \text{in } \Omega. \end{cases} \tag{4.5'}$$

The behavior of $u(t)$ on Γ is written as $\tau_f u = \big(g_1 \ \dots \ g_M\big)(\boldsymbol{f}(t) - \langle u(t), \boldsymbol{y}\rangle)$. Set

$$p(t) = u(t) - \left(N^f_{-c}g_1 \ \dots \ N^f_{-c}g_M\right)\boldsymbol{\varepsilon}(t), \quad \text{with} \quad \boldsymbol{\varepsilon}(t) = \boldsymbol{f}(t) - \langle u(t), \boldsymbol{y}\rangle,$$

where $N^f_{-c}g_i$ are introduced in Lemma 3.6. The function $p(t)$, $t > 0$, belongs to $\mathscr{D}(\hat{F})$ and satisfies the equation

$$\begin{aligned} &\frac{dp}{dt} + Fp = \left(N^f_{-c}g_1 \ \dots \ N^f_{-c}g_M\right)(c\boldsymbol{\varepsilon}(t) - \boldsymbol{\varepsilon}'(t)), \\ &p(0) = u_0 - \left(N^f_{-c}g_1 \ \dots \ N^f_{-c}g_M\right)\left(\langle v_0, \boldsymbol{\rho}\rangle_H - \langle u_0, \boldsymbol{y}\rangle\right). \end{aligned}$$

By (4.12) and (4.13), we already know that

$$|\boldsymbol{\varepsilon}(t)|, \ |\boldsymbol{\varepsilon}'(t)| \leqslant \text{const}\, e^{-a\mu_1 t}\left(\|u_0\| + \|v_0\|_H\right), \quad t > 0.$$

In view of the decay (3.16), we obtain the decay estimate

$$\|p(t)\|, \ \|u(t)\|, \ \text{and} \quad |\boldsymbol{f}(t)| \leqslant \text{const}\, e^{-rt}\left(\|u_0\| + \|v_0\|_H\right), \quad t \geqslant 0.$$

This immediately gives the decay estimate for every solution $(q(t), v(t))$ to (4.10):

$$\|q(t)\| + \|v(t)\|_H \leqslant \text{const}\, e^{-rt}\left(\|q_0\| + \|v_0\|_H\right), \quad t \geqslant 0. \tag{4.15}$$

Reduction to a finite-dimensional compensator.

We go back to eqn. (4.10) satisfying the decay (4.15). Recall that the vector $\boldsymbol{\rho}$ is chosen in the subspace $\left(P^H_n H\right)^M$. In (4.10), set $v_1(t) = P^H_n v(t)$. Note that $\boldsymbol{g}(q,v) = \boldsymbol{g}(q,v_1)$ (see (4.9)). Applying P^H_n to the both sides of the equation for v, we obtain the system of differential equations

$$\begin{cases} \dfrac{dq}{dt} + Lq = (N_{-c}g_1 \ \dots \ N_{-c}g_M)\,\boldsymbol{g}(q,v_1), \\ \dfrac{dv_1}{dt} + B_1 v_1 = -P^H_n Cq + (P^H_n XN_{-c}g_1 \ \dots \ P^H_n XN_{-c}g_M)\,\boldsymbol{g}(q,v_1), \\ q(0) = q_0 \in L^2(\Omega), \quad v_1(0) = P^H_n v_0 \in P^H_n H. \end{cases} \tag{4.16}$$

In (4.16), B_1 denotes the restriction of B onto the invariant subspace $P_n^H H$, i.e., $B_1 = B|_{P_n^H H}$. Just as in (4.10), eqn. (4.16) with state (q, v_1) is well posed in $L^2(\Omega) \times P_n^H H$. The semigroup generated by (4.16) is analytic in $t > 0$. Solution $q(t,\cdot) \in \mathscr{D}(L)$ actually belongs to $\mathscr{D}(\hat{L})$ for each $t > 0$. Every solution $(q(t), v_1(t))$ to (4.16) with initial value $(q_0, v_0) \in L^2(\Omega) \times P_n^H H$ is derived from the solution $(\tilde{q}(t), \tilde{v}(t))$ to (4.10) with the same initial value, and is expressed by $(q(t), v_1(t)) = \left(\tilde{q}(t), P_n^H \tilde{v}(t)\right)$. Thus every solution $(q(t), v_1(t))$ to (4.16) satisfies the decay estimate

$$\|q(t)\| + \|v_1(t)\|_H \leqslant \text{const}\, e^{-rt}\left(\|q_0\| + \|v_0\|_H\right), \quad t \geqslant 0. \tag{4.15$'$}$$

The equation for v_1 in (4.16) means a finite-dimensioanl compensator in the subspace $P_n^H H$. In (4.11) note that $\boldsymbol{f}(t) = \langle v(t), \boldsymbol{\rho}\rangle_H = \langle v_1(t), \boldsymbol{\rho}\rangle_H$ satisfies the relation

$$c\boldsymbol{f}(t) - \boldsymbol{f}'(t) = \boldsymbol{g}(q, v_1), \quad t > 0.$$

Thus (4.16) is rewritten as

$$\begin{cases} \dfrac{dq}{dt} + Lq = (N_{-c}g_1 \; \dots \; N_{-c}g_M)\,(c\boldsymbol{f}(t) - \boldsymbol{f}'(t)), \\ \dfrac{dv_1}{dt} + B_1 v_1 = -P_n^H Cq + \left(P_n^H XN_{-c}g_1 \; \dots \; P_n^H XN_{-c}g_M\right)(c\boldsymbol{f}(t) - \boldsymbol{f}'(t)), \\ q(0) = q_0 \in L^2(\Omega), \quad v_1(0) \in P_n^H H. \end{cases} \tag{4.16$'$}$$

We rewrite (4.16) in terms of $(u(t), v_1(t))$, where $u(t)$ is defined by (4.11) with v replaced by v_1. In view of (4.16′), we easily obtain

$$\begin{cases} \dfrac{du}{dt} + \mathscr{L}u = 0 \quad \text{in } \mathbb{R}^1_+ \times \Omega, \\ \quad \tau u = \left(g_1 \; \dots \; g_M\right)\langle v_1, \boldsymbol{\rho}\rangle_H \quad \text{on } \mathbb{R}^1_+ \times \Gamma, \\ \dfrac{dv_1}{dt} + B_1 v_1 = -P_n^H Cu + \left(P_n^H CN_{-c}g_1 \; \dots \; P_n^H CN_{-c}g_M\right)\langle v_1, \boldsymbol{\rho}\rangle_H \\ \qquad\qquad + \left(P_n^H XN_{-c}g_1 \; \dots \; P_n^H XN_{-c}g_M\right)\tilde{\boldsymbol{g}}(u, v_1) \quad \text{in } \mathbb{R}^1_+, \\ u(0,\cdot) = u_0 \in L^2(\Omega), \quad v_1(0) = v_{10} \in P_n^H H. \end{cases} \tag{4.17}$$

where $u_0 = q_0 + (N_{-c}g_1 \; \dots \; N_{-c}g_M)\langle v_{10}, \boldsymbol{\rho}\rangle_H$, and

$$\begin{aligned} \tilde{\boldsymbol{g}}(u, v_1) &= \boldsymbol{g}(q, v_1) \\ &= \overline{\Theta}^{-1}\left(-G_2\,\boldsymbol{p}(u) + G_2 G_3 \langle v_1, \boldsymbol{\rho}\rangle_H + \langle v_1, B_c^* \boldsymbol{\rho}\rangle_H\right), \\ G_3 &= \left(\boldsymbol{p}\underset{k\to}{(N_{-c}g_k)}\right) = \left(p_j(N_{-c}g_k); \; \begin{matrix} j & \downarrow & 1, \dots, N \\ k & \to & 1, \dots, M \end{matrix}\right). \end{aligned} \tag{4.18}$$

In the equation for v_1, we get together the operator B_1 and the terms which include the inner products $\langle v_1, \boldsymbol{\rho}\rangle_H$ and $\langle v_1, B^*\boldsymbol{\rho}\rangle_H$ in a lump. The resultant operator is denoted by the same symbol B_1 with no confusion. We can then finally obtain the desired control system (1.3). □

4.5 Alternative Framework of Stabilization

The concepts of compensator design in Sections 3 and 4 are the same, while the designs are based on different (but equivalent) descriptions of the controlled plant Σ_p (see (2.7) and (4.10)). In this section, we introduce an alternative framework of feedback. This influences the determination of the dimension of a stabilizing compensator Σ_c.

Let us recall that the controlled plant Σ_p is characterized by the pair of differential operators $(\mathscr{L}, \tau)$ in (1.1), and the setting of the operator B in (2.2) and the separation condition (2.5) are unchanged. With the preparation in the preceding sections, let us describe the control system which finally leads to eqn. (1.3). Our control system is formulated as the differential equation in the product space $L^2(\Omega) \times H$:

$$\begin{cases} u_t + \mathscr{L}u = 0, \qquad \tau u = -\sum_{k=1}^{M} \langle v, \rho_k\rangle_H \, g_k(\xi), \quad u(0,\cdot) = u_0, \\ v_t + Bv = \sum_{k=1}^{N} p_k(u - Xv)\xi_k, \quad v(0) = v_0. \end{cases} \tag{5.1}$$

Here, we recall that given the observation weights $w_k \in L^2(\Gamma)$, $1 \leqslant k \leqslant N$, $p_k(\cdot)$ are defined by (1.4), that is,

$$p_k(u) = \begin{cases} \langle u, w_k\rangle_\Gamma, & \text{in the case where } \alpha(\xi) \not\equiv 1, \\ \left\langle \dfrac{\partial u}{\partial \nu}, w_k \right\rangle_\Gamma, & \text{in the case where } \alpha(\xi) \equiv 1. \end{cases} \tag{5.2}$$

As for the actuators of Σ_p, we assume that g_k belongs to $C^{2+\omega}(\Gamma)$, $1 \leqslant k \leqslant M$, and that they are given in the form: $g_k(\xi) = (1-\alpha(\xi))\hat{g}_k(\xi)$ with $\hat{g}_k \in C^{2+\omega}(\Gamma)$ in the case where $\alpha(\xi) \not\equiv 1$. Thus no control is assumed in this case, on the Dirichlet part of the boundary, $\Gamma_1 = \{\xi;\ \alpha(\xi) = 1\}$. In (5.1), X will denote a unique solution in $\mathscr{L}(H; L^2(\Omega))$ to the operator equation (5.5) below, and ξ_k the vectors such that $X^*\xi_k$ approximate specific functions belonging to a finite-dimensional generalized eigenspace associated with L.

In engineering implementation of the controller design, it is one of the most important issues to seek the best possible parameters, such as the number of sensors and actuators and the dimension of controllers. The dimension of the compensator Σ_c in the preceding sections is determined only by the actuators g_k

(see Section 1). The alternative stabilization framework of this section enables us to construct a compensator Σ_c with *lower* dimension: In fact, we show that the dimension of Σ_c here is determined only by the sensors w_k. Thus, we can choose the minimum of these dimensions for the stabilization. The alternative control scheme ensures the exponential decay of $\|u - Xv\|$, instead of the decay of $\|Xu - v\|_H$ in our previous results: Assuming the existing decay of $\|Xu - v\|_H$, for example, one might make possible use of the inverse X^{-1} to derive the decay of $\|u - X^{-1}v\|$. This attempt, however, faces a serious difficulty: For the existence of X^{-1}, one has to assume the *complete observability* condition on the sensors. In addition, the inverse X^{-1}, even if it exists, must be necessarily *unbounded*, due to compactness property of X. We remark that the situation is fairly different from the one in the case where the so called *identity* compensators are employed. This type of compensators can be applied *only* when the pair $(\mathscr{L}, \tau)$ admits the associated Riesz basis. The dimension of the compensator in this scheme must be determined both by the sensors and the actuators, where we make use of the fact that both the sensors and the actuators can be arbitrarily approximated by vectors belonging to a finite dimensional subspace -via the Riesz basis.

In studying the boundary observation/boundary control scheme, we may assume with no loss of generality that the point 0 belongs to $\rho(L)$. Let us first consider the case where $\alpha(\xi) \not\equiv 1$. Given a function $\hat{g} \in C^{2+\omega}(\Gamma)$, consider the boundary value problem:

$$\mathscr{L}q = 0, \quad \text{in } \Omega, \qquad \tau q = (1 - \alpha(\xi))\hat{g}(\xi) = g(\xi) \quad \text{on } \Gamma. \tag{5.3$_1$}$$

The problem admits a unique solution $q \in C^2(\Omega) \cap C^1(\overline{\Omega})$. The solution q has a non-unique expression as $q = Rg - L^{-1}\mathscr{L}Rg = R(1-\alpha)\hat{g} - L^{-1}\mathscr{L}R(1-\alpha)\hat{g}$, where R denotes the operator of prolongation appearing in (2.6) such that

$$Rg \in C^{2+\omega}(\overline{\Omega}), \quad Rg\big|_\Gamma = \frac{\partial}{\partial \nu}Rg\bigg|_\Gamma = g.$$

In the case where $\alpha(\xi) \equiv 1$, the boundary value problem:

$$\mathscr{L}q = 0, \quad \text{in } \Omega, \qquad \tau q = q\big|_\Gamma = g(\xi) \quad \text{on } \Gamma \tag{5.3$_2$}$$

similarly admits a unique solution $q \in C^2(\Omega) \cap C^1(\overline{\Omega})$. Our first result is the following:

Lemma 5.1. *For any $\varphi \in \mathscr{D}(L^*)$, we have the relation*

$$\begin{cases} \langle L^*\varphi, q\rangle = \langle \varphi, \hat{g}\rangle_\Gamma, & \text{in the case where } \alpha(\xi) \not\equiv 1, \\ \langle L^*\varphi, q\rangle = -\left\langle \dfrac{\partial \varphi}{\partial \nu}, g\right\rangle_\Gamma, & \text{in the case where } \alpha(\xi) \equiv 1. \end{cases} \tag{5.4}$$

Proof. Consider first the case where $\alpha(\xi) \not\equiv 1$. For any $\varphi \in \mathscr{D}(\hat{L}^*) \subset C^2(\Omega) \cap C^1(\overline{\Omega})$, we calculate through Green's formula as

$$\begin{aligned}\langle L^* \varphi, q \rangle &= -\left\langle \frac{\partial \varphi}{\partial \nu}, q \right\rangle_\Gamma + \left\langle \varphi, \frac{\partial q}{\partial \nu} \right\rangle_\Gamma - \langle (\boldsymbol{b} \cdot \boldsymbol{\nu}) \varphi, q \rangle_\Gamma + \langle \varphi, \mathscr{L} q \rangle \\ &= -\left\langle \frac{\partial \varphi}{\partial \nu}, q \right\rangle_\Gamma + \left\langle \varphi, \frac{\partial q}{\partial \nu} \right\rangle_\Gamma - \langle (\boldsymbol{b} \cdot \boldsymbol{\nu}) \varphi, q \rangle_\Gamma,\end{aligned}$$

and

$$\tau^* \varphi = \alpha \varphi + (1 - \alpha) \left(\frac{\partial \varphi}{\partial \nu} + (\boldsymbol{b} \cdot \boldsymbol{\nu}) \varphi \right) = 0.$$

Since $\tau q = (1 - \alpha) \hat{g}$, we see that

$$-\frac{\partial \varphi}{\partial \nu} q + \varphi \frac{\partial q}{\partial \nu} - (\boldsymbol{b} \cdot \boldsymbol{\nu}) \varphi q = \varphi \hat{g} \quad \text{on } \Gamma.$$

Thus,

$$\langle L^* \varphi, q \rangle = \langle \varphi, \hat{g} \rangle_\Gamma, \quad \forall \varphi \in \mathscr{D}(\hat{L}^*).$$

Recall that $\|L_c^* \varphi\| \geqslant \text{const} \|\varphi\|_{H^1(\Omega)}$, and thus $\mathscr{D}(L^*)$ is contained in $H^1(\Omega)$ (see, e.g., Subsection 3.3, Chapter 2). For any $\varphi \in \mathscr{D}(L^*)$, there is a sequence $\{\varphi_n\} \subset \mathscr{D}(\hat{L}^*)$, such that $\varphi_n \to \varphi$ and $\hat{L}^* \varphi_n \to L^* \varphi$ in the $L^2(\Omega)$-topology. Thus, passage to the limit regarding φ shows the first relation of (5.4). The other case (the Dirichlet boundary) is similarly handled by noting that $\mathscr{D}(L^*) = H^2(\Omega) \cap H_0^1(\Omega)$. □

Let us go back to (5.1). The operator $X \in \mathscr{L}(H; L^2(\Omega))$ is a possible solution to the operator equation:

$$L \left(X v + \sum_{k=1}^{M} \langle v, \rho_k \rangle_H q_k \right) = X B v, \quad v \in \mathscr{D}(B), \tag{5.5}$$

where q_k denotes the unique solution to the boundary value problem (5.3_1) or (5.3_2). The operator equation (5.5) is, in a sense, a dual of Sylvester's equation (3.8). Sylvester's equation has a long history, and the readers can refer, e.g., to [6] and a number of references therein. As far as the author knows, however, the equation in the form of (5.5) has never appeared, since it reflects the non-homogeneous boundary input. The existence and uniqueness of the solution X to (5.5) will be discussed later in the proof of Theorem 5.2. Let us recall the matrices $\hat{W}_i$ and $\hat{G}_i$ in (3.1):

$$\begin{aligned}\hat{W}_i &= \left(\rho_k(\varphi_{ij}); \begin{array}{ccc} j & \to & 1, \dots, m_i \\ k & \downarrow & 1, \dots, N \end{array} \right), \quad \text{and} \\ \hat{G}_i &= \left(\langle g_k, \sigma \psi_{ij} \rangle_\Gamma; \begin{array}{ccc} j & \downarrow & 1, \dots, m_i \\ k & \to & 1, \dots, M \end{array} \right),\end{aligned} \tag{5.6}$$

respectively, where the boundary operator σ is defined in (3.2):

$$\sigma\psi_{ij} = (1 - \boldsymbol{b}\cdot\boldsymbol{\nu})\psi_{ij} - \frac{\partial\psi_{ij}}{\partial\nu}.$$

Note that, due to the setting: $g_k(\xi) = (1 - \alpha(\xi))\hat{g}_k(\xi)$, the inner products $\langle g_k, \sigma\psi_{ij}\rangle_\Gamma$ in $\hat{G}_i$ contain only integrations on $\Gamma \setminus \Gamma_1$, $\Gamma_1 = \{\xi \in \Gamma;\, \alpha(\xi) = 1\}$ in the case where $\alpha(\xi) \not\equiv 1$. According to the orthonormal basis $\{\eta_{ij}^{\pm}\}$ for H, let us express the sensors ρ_k in (5.1) as

$$\rho_k = \sum_{i,j} \left(\rho_{ij}^k \eta_{ij}^+ + \overline{\rho_{ij}^k}\eta_{ij}^-\right),$$

and define $n_i \times M$ matrices R_i as

$$R_i = \left(\rho_{ij}^k;\ \begin{matrix} j & \downarrow & 1,\ldots,n_i \\ k & \rightarrow & 1,\ldots,M \end{matrix}\right), \quad i \geqslant 1. \tag{5.7}$$

Compare it with Ξ_i in (3.4). With the alternative setting (5.1) of the feedback scheme then, our stabilization result is stated as follows:

Theorem 5.2. *Assume the basic condition* (2.1) *on L and the separation condition* (2.5) *on B. Then,*

(i) *there exists a unique solution* $X \in \mathscr{L}(H; L^2(\Omega))$ *to the operator equation* (5.5). *Eqn.* (3.1) *is thus well posed in* $L^2(\Omega) \times H$.

Let any r, $0 < r < \mathrm{Re}\,\lambda_{\nu+1}$, *be given. Suppose that* w_k, $g_k = (1-\alpha)\hat{g}_k$, *and* ρ_k *satisfy the rank conditions*

$$\begin{aligned} &\mathrm{rank}\left(\hat{W}_i \quad \hat{W}_i\Lambda_i \quad \ldots \quad \hat{W}_i\Lambda_i^{m_i-1}\right)^{\mathrm{T}} = m_i, \quad 1 \leqslant i \leqslant \nu, \\ &\mathrm{rank}\left(\hat{G}_i \quad \tilde{\Lambda}_i^*\hat{G}_i \quad \ldots \quad (\tilde{\Lambda}_i^*)^{m_i-1}\hat{G}_i\right) = m_i, \quad 1 \leqslant i \leqslant \nu, \quad \textit{and} \\ &\mathrm{rank}\, R_i = M, \quad i \geqslant 1, \end{aligned} \tag{5.8}$$

respectively. Then we find a suitable integer n and $\xi_k \in P_n^H H$, $1 \leqslant k \leqslant N$, *such that every solution* $(u(t,\cdot), v(t))$ *to* (5.1) *satisfies the decay estimate*

$$\|u(t,\cdot)\| + \|v(t)\|_H \leqslant \mathrm{const}\, e^{-rt}\left(\|u_0\| + \|v_0\|_H\right), \quad t \geqslant 0. \tag{5.9}$$

(ii) *As long as the initial data* v_0 *stays in* $P_n^H H$, (5.1) *means a differential equation in* $L^2(\Omega) \times P_n^H H$, *which is finally described as eqn.* (1.3).

Remark: In the condition on $\hat{G}_i$, no information on g_k on Γ_1 is required in the case of $\alpha(\xi) \not\equiv 1$. On the other hand, the terms $\int_{\Gamma_1} g_k(\xi)\,\frac{\partial\psi_{ij}}{\partial\nu}\,d\xi$ enter the $\hat{G}_i$ in Theorem 3.1.

Proof of Theorem 5.2:

Step I (*Operator equation and well-posedness of eqn.* (5.1)):

(i) In view of the separation condition (2.5) on $\sigma(L)$ and $\sigma(B)$, we can define the operator X in H as

$$\begin{aligned} Xv = &\sum_{i,j}\sum_{k=1}^{M} v_{ij}^{+}\overline{\rho_{ij}^{k}} L(\mu_i\omega^{+} - L)^{-1} q_k \\ &+ \sum_{i,j}\sum_{k=1}^{M} v_{ij}^{-}\rho_{ij}^{k} L(\mu_i\omega^{-} - L)^{-1} q_k \end{aligned} \tag{5.10}$$

for $v = \sum_{i,j}\left(v_{ij}^{+}\eta_{ij}^{+} + v_{ij}^{-}\eta_{ij}^{-}\right) \in H$. Then the operator X clearly belongs to $\mathscr{L}(H; L^2(\Omega))$. For any $v \in \mathscr{D}(B)$, we calculate as

$$\begin{aligned} Xv = &\sum_{i,j}\sum_{k=1}^{M} v_{ij}^{+}\overline{\rho_{ij}^{k}}(-q_k + \mu_i\omega^{+}(\mu_i\omega^{+} - L)^{-1} q_k) \\ &+ \sum_{i,j}\sum_{k=1}^{M} v_{ij}^{-}\rho_{ij}^{k}(-q_k + \mu_i\omega^{-}(\mu_i\omega^{-} - L)^{-1} q_k) \\ = &-\sum_{k=1}^{M}\sum_{i,j}\left(v_{ij}^{+}\overline{\rho_{ij}^{k}} + v_{ij}^{-}\rho_{ij}^{k}\right) q_k + \sum_{i,j}\sum_{k=1}^{M} v_{ij}^{+}\overline{\rho_{ij}^{k}}\mu_i\omega^{+}(\mu_i\omega^{+} - L)^{-1} q_k \\ &+ \sum_{i,j}\sum_{k=1}^{M} v_{ij}^{-}\rho_{ij}^{k}\mu_i\omega^{-}(\mu_i\omega^{-} - L)^{-1} q_k \\ = &-\sum_{k=1}^{M}\langle v, \rho_k\rangle_H q_k + \sum_{i,j}\sum_{k=1}^{M} v_{ij}^{+}\overline{\rho_{ij}^{k}}\mu_i\omega^{+}(\mu_i\omega^{+} - L)^{-1} q_k \\ &+ \sum_{i,j}\sum_{k=1}^{M} v_{ij}^{-}\rho_{ij}^{k}\mu_i\omega^{-}(\mu_i\omega^{-} - L)^{-1} q_k. \end{aligned}$$

Note that the above second and the third terms belong to $\mathscr{D}(L)$. In fact, it is correct for any finite summation of the second term, and then,

$$\begin{aligned} &L\left(\sum_{i,j\,(i\leqslant n)}\sum_{k=1}^{M} v_{ij}^{+}\overline{\rho_{ij}^{k}}\mu_i\omega^{+}(\mu_i\omega^{+} - L)^{-1} q_k\right) \\ &= \sum_{i,j\,(i\leqslant n)}\sum_{k=1}^{M} v_{ij}^{+}\overline{\rho_{ij}^{k}}\mu_i\omega^{+} L(\mu_i\omega^{+} - L)^{-1} q_k. \end{aligned}$$

Since $\left\|L(\mu_i\omega^{+} - L)^{-1} q_k\right\|$ are uniformly bounded, and $\sum_{i,j}\left|v_{ij}^{+}\mu_i \cdot \overline{\rho_{ij}^{k}}\right| < \infty$, the right-hand side converges as $n \to \infty$. Thus we see that the second term, and similarly the third term, too, belong to $\mathscr{D}(L)$. It means that

$$Xv + \sum_{k=1}^{M}\langle v, \rho_k\rangle_H q_k$$

belongs to $\mathscr{D}(L)$, and that

$$
\begin{aligned}
L\left(Xv+\sum_{k=1}^{M}\langle v,\rho_k\rangle_H q_k\right) &= \sum_{i,j}\sum_{k=1}^{M} v_{ij}^{+}\overline{\rho_{ij}^{k}}\mu_i\omega^{+}L(\mu_i\omega^{+}-L)^{-1}q_k \\
&\quad+\sum_{i,j}\sum_{k=1}^{M} v_{ij}^{-}\rho_{ij}^{k}\mu_i\omega^{-}L(\mu_i\omega^{-}-L)^{-1}q_k.
\end{aligned}
\tag{5.11}
$$

In view of the definition (5.10) of X and the relation: $(Bv)_{ij}^{\pm}=\mu_i\omega^{\pm}v_{ij}^{\pm}$, the right-hand side of (5.11) is nothing but XBv.

Supposing two operator solutions to (5.5), say X_1 and X_2, and taking the difference, we have the relation: $L(X_1-X_2)v=(X_1-X_2)Bv$ for any $v\in\mathscr{D}(B)$. Setting $v=\eta_{ij}^{\pm}$ then, we see that

$$
(\mu_i\omega^{\pm}-L)(X_1-X_2)\eta_{ij}^{\pm}=0 \quad\Rightarrow\quad (X_1-X_2)\eta_{ij}^{\pm}=0 \quad \text{for } \forall\eta_{ij}^{\pm}.
$$

Thus we conclude that $X_1=X_2$.

The proof of well-posedness of eqn. (5.1) in $L^2(\Omega)\times H$ is carried out by introducing a new state $z=u+\sum_{k=1}^{M}\langle v,\rho_k\rangle_H q_k$, and then changing the problem into the equivalent one for the equation with state (z,v): The equation for (z,v) has the operator $\begin{pmatrix} L & 0\\ 0 & B\end{pmatrix}$ as the principal operator. The proof is similar to the proof of Theorem 2.1, and thus omitted.

Step II (*Stabilization*):

Let us turn to the stabilization problem of (5.1). Note that the solution $u(t,\cdot)$ is in $C^2(\Omega)\cap C^1(\overline{\Omega})$ for $t>0$, the proof of which is almost the same as the proof of Theorem 2.1. Since $\tau\left(u+\sum_{k=1}^{M}\langle v,\rho_k\rangle_H q_k\right)=0$ on Γ, in addition, we see that $u+\sum_{k=1}^{M}\langle v,\rho_k\rangle_H q_k$ belongs to $\mathscr{D}(L)$. By applying X to the equation for v, and taking the relation (5.5) into account, eqn. (5.1) is transformed into

$$
\begin{cases}
u_t+L\left(u+\sum_{k=1}^{M}\langle v,\rho_k\rangle_H q_k\right)=0,\\
Xv_t+L\left(Xv+\sum_{k=1}^{M}\langle v,\rho_k\rangle_H q_k\right)=\sum_{k=1}^{N}p_k(u-Xv)X\xi_k,
\end{cases}
\tag{5.12}
$$

both equations of which are described in $L^2(\Omega)$. Taking the difference of these equations, we have the equation for $u-Xv$:

$$
\begin{aligned}
&(u-Xv)_t+L(u-Xv)+\sum_{k=1}^{N}p_k(u-Xv)X\xi_k=0,\\
&\text{or } u(t)-Xv(t)=\exp\left(-t\left(L+\textstyle\sum_{k=1}^{N}p_k(\cdot)X\xi_k\right)\right)(u_0-Xv_0).
\end{aligned}
\tag{5.13}
$$

Since the observability condition on w_k (the first condition of (5.8)) is satisfied, there is a set of $f_k \in P_\nu L^2(\Omega)$ such that

$$\left\| \exp\left(-t\left(L + \sum_{k=1}^{N} p_k(\cdot) f_k\right)\right) \right\| \leqslant \text{const}\, e^{-r_1 t}, \quad t \geqslant 0, \tag{5.14}$$

where $r < r_1 < \operatorname{Re} \lambda_{\nu+1}$: The situation is the same as in Theorem 2.1, Chapter 3. The only technical difference is that the outputs on Γ in this case are unbounded operators. The problem is thus how we could approximate these $f_k \in P_\nu L^2(\Omega)$ by functions of the form $X\xi_k$. The following result is a counterpart of Proposition 3.3, and forms the key result in Theorem 5.1. The proof will be given later in *Step* III of this section.

Proposition 5.3. *Under the second and the third conditions of* (5.9) *on* g_k *and* ρ_k, *respectively, and the separation condition* (2.5) *on L and B, we have the inclution relation:*

$$P_\nu L^2(\Omega) \subset \overline{XH}. \tag{5.15}$$

By Proposition 5.3, we find vectors ξ_k, $1 \leqslant k \leqslant N$, such that $X\xi_k$ approximate $f_k \in P_\nu L^2(\Omega)$ in (5.14) arbitrarily in the $L^2(\Omega)$-topology. Thus we obtain the decay estimate:

$$\left\| \exp\left(-t\left(L + \sum_{k=1}^{N} p_k(\cdot) X\xi_k\right)\right) \right\| \leqslant \text{const}\, e^{-rt}, \quad t \geqslant 0,$$

which ensures -via (5.13)- the decay estimate:

$$\|u(t) - Xv(t)\| \leqslant \text{const}\, e^{-rt} \|u(0) - Xv(0)\|, \quad t \geqslant 0. \tag{5.16}$$

In view of the equation for $v(t)$ and the decay (2.3) for e^{-tB}, we have

$$v(t) = e^{-tB} v(0) + \int_0^t e^{-(t-s)B} \sum_{k=1}^{N} p_k(u(s) - Xv(s))\xi_k \, ds,$$

and thus,

$$\|v(t)\|_H \leqslant \text{const}\, e^{-rt} \left(\|v(0)\|_H + \|u(0)\| \right), \quad t \geqslant 0.$$

By (5.16), we have the decay estimate for $\|u(t,\cdot)\|$ similar to the above, which establishes the stabilization of eqn. (5.1).

Proof of (ii). Since the set $\{\eta_{ij}^{\pm}\}$ forms a basis for H, we can assume that the vectors q_k stay in the finite dimensional subspace $P_n^H H$ for a suitable n. Suppose that $v(0)$ is in $P_n^H H$, and set $v_1(t) = P_n^H v(t)$ and $v_2(t) = \left(1 - P_n^H\right) v(t)$. It is clear that $v_2(t)$ satisfies the equation:

$$\frac{dv_2}{dt} + Bv_2 = 0, \quad v_2(0) = 0, \quad t \geqslant 0.$$

This means that $v_2(t) = 0$ for $t \geqslant 0$, and thus $v(t) = v_1(t) \in P_n^H H$ for $t \geqslant 0$. Note that the terms

$$p_k(u - Xv) = \begin{cases} \langle u - Xv,\, w_k\rangle_\Gamma, & \text{in the case where } \alpha(\xi) \not\equiv 1, \\ \langle (u - Xv)_\nu,\, w_k\rangle_\Gamma, & \text{in the case where } \alpha(\xi) \equiv 1 \end{cases}$$

appear in (5.1). In view of the expression (5.10) of X, the terms $p_k(Xv) = p_k(Xv_1)$ are bounded functionals.

Remark: In the approach through the framework of (2.7), we have to introduce the projection $(u(t), P_n^H v(t))$ of $(u(t), v(t))$ to obtain a finite-dimensional compensator. In other words, (2.7) is not the final form of the feedback scheme. On the other hand, the approach in this section, (5.1) contains a finite-dimensional compensator as an internal structure.

Step III (*Proof of Proposition* 5.3):

In order to prove the relation (5.15), we only have to show its geometrical dual:

$$\ker X^* \subset \left\{u \in L^2(\Omega);\ P_\nu^* u = 0\right\}. \tag{5.17}$$

The proof is similar to the proof of Proposition 3.3. Consider first the case where $\alpha(\xi) \not\equiv 1$. In view of the definition (5.10), the operator $X^* \in \mathscr{L}(L^2(\Omega); H)$ is calculated as

$$\begin{aligned} X^* u &= \sum_{i,j}\sum_{k=1}^{M} \rho_{ij}^k \left\langle L^*(\mu_i\omega^- - L^*)^{-1}u,\, q_k\right\rangle \eta_{ij}^+ \\ &\quad + \sum_{i,j}\sum_{k=1}^{M} \overline{\rho_{ij}^k} \left\langle L^*(\mu_i\omega^+ - L^*)^{-1}u,\, q_k\right\rangle \eta_{ij}^- \\ &= \sum_{i,j}\sum_{k=1}^{M} \rho_{ij}^k \left\langle (\mu_i\omega^- - L^*)^{-1}u,\, \hat{g}_k\right\rangle_\Gamma \eta_{ij}^+ \\ &\quad + \sum_{i,j}\sum_{k=1}^{M} \overline{\rho_{ij}^k} \left\langle (\mu_i\omega^+ - L^*)^{-1}u,\, \hat{g}_k\right\rangle_\Gamma \eta_{ij}^-, \qquad u \in L^2(\Omega). \end{aligned} \tag{5.18$_1$}$$

Here we have used the relation in Lemma 5.1. Second, in the case where $\alpha(\xi) \equiv 1$, we similarly obtain -via Lemma 5.1

$$\begin{aligned} X^* u = &-\sum_{i,j}\sum_{k=1}^{M} \rho_{ij}^k \left\langle \frac{\partial}{\partial\nu}(\mu_i\omega^- - L^*)^{-1}u,\, g_k\right\rangle_\Gamma \eta_{ij}^+ \\ &-\sum_{i,j}\sum_{k=1}^{M} \overline{\rho_{ij}^k} \left\langle \frac{\partial}{\partial\nu}(\mu_i\omega^+ - L^*)^{-1}u,\, g_k\right\rangle_\Gamma \eta_{ij}^-, \qquad u \in L^2(\Omega). \end{aligned} \tag{5.18$_2$}$$

Let u be in $\ker X^*$. Since the set $\{\eta_{ij}^{\pm}\}$ forms an orthonormal basis for H, we see that

$$\begin{cases} \sum_{k=1}^{M} \rho_{ij}^k \langle (\mu_i\omega^- - L^*)^{-1}u, \hat{g}_k \rangle_\Gamma = \sum_{k=1}^{M} \overline{\rho_{ij}^k} \langle (\mu_i\omega^+ - L^*)^{-1}u, \hat{g}_k \rangle_\Gamma = 0 \\ \qquad \text{for } i \geqslant 1, \ 1 \leqslant j \leqslant n_i, \quad \text{in the case where } \alpha(\xi) \not\equiv 1, \\ \sum_{k=1}^{M} \rho_{ij}^k \left\langle \frac{\partial}{\partial \nu}(\mu_i\omega^- - L^*)^{-1}u, g_k \right\rangle_\Gamma = \sum_{k=1}^{M} \overline{\rho_{ij}^k} \left\langle \frac{\partial}{\partial \nu}(\mu_i\omega^+ - L^*)^{-1}u, g_k \right\rangle_\Gamma \\ \qquad = 0 \qquad \text{for } i \geqslant 1, \ 1 \leqslant j \leqslant n_i, \quad \text{in the case where } \alpha(\xi) \equiv 1. \end{cases}$$

In other words, we obtain the relations (see (5.7) for the matrices R_i):

$$\begin{cases} R_i \langle (\mu_i\omega^- - L^*)^{-1}u, \hat{\boldsymbol{g}} \rangle_\Gamma = \overline{R_i} \langle (\mu_i\omega^+ - L^*)^{-1}u, \hat{\boldsymbol{g}} \rangle_\Gamma = \mathbf{0} \quad \text{for } i \geqslant 1, \\ \qquad \text{in the case where } \alpha(\xi) \not\equiv 1, \\ R_i \left\langle \frac{\partial}{\partial \nu}(\mu_i\omega^- - L^*)^{-1}u, \boldsymbol{g} \right\rangle_\Gamma = \overline{R_i} \left\langle \frac{\partial}{\partial \nu}(\mu_i\omega^+ - L^*)^{-1}u, \boldsymbol{g} \right\rangle_\Gamma = \mathbf{0} \\ \qquad \text{for } i \geqslant 1, \quad \text{in the case where } \alpha(\xi) \equiv 1, \end{cases}$$

where

$$\langle (\mu_i\omega^- - L^*)^{-1}u, \hat{\boldsymbol{g}} \rangle_\Gamma = \Big(\langle (\mu_i\omega^- - L^*)^{-1}u, \hat{g}_k \rangle_\Gamma ; \, k \downarrow 1, \dots, M \Big),$$

$$\left\langle \frac{\partial}{\partial \nu}(\mu_i\omega^- - L^*)^{-1}u, \boldsymbol{g} \right\rangle_\Gamma = \left(\left\langle \frac{\partial}{\partial \nu}(\mu_i\omega^- - L^*)^{-1}u, g_k \right\rangle_\Gamma ; \, k \downarrow 1, \dots, M \right).$$

Since rank $R_i = M$, $i \geqslant 1$, by (5.8), we see that

$$\begin{cases} \langle (\mu_i\omega^\pm - L^*)^{-1}u, \hat{\boldsymbol{g}} \rangle_\Gamma = \mathbf{0} \quad \text{for } i \geqslant 1, \text{ in the case where } \alpha(\xi) \not\equiv 1, \\ \left\langle \frac{\partial}{\partial \nu}(\mu_i\omega^\pm - L^*)^{-1}u, \boldsymbol{g} \right\rangle_\Gamma = \mathbf{0} \quad \text{for } i \geqslant 1, \text{ in the case where } \alpha(\xi) \equiv 1. \end{cases} \tag{5.19}$$

Setting, for each k, $1 \leqslant k \leqslant M$,

$$f_k(\lambda; u) = \begin{cases} \langle (\lambda - L^*)^{-1}u, \hat{g}_k \rangle_\Gamma, & \text{in the case where } \alpha(\xi) \not\equiv 1, \\ \left\langle \frac{\partial}{\partial \nu}(\lambda - L^*)^{-1}u, g_k \right\rangle_\Gamma, & \text{in the case where } \alpha(\xi) \equiv 1, \end{cases} \tag{5.20}$$

the relation (5.19) is rewritten as

$$f_k(\mu_i\omega^\pm; u) = 0, \quad i \geqslant 1, \ 1 \leqslant k \leqslant M. \tag{5.21}$$

As in (3.24), we note the simple algebraic relation:

$$\begin{aligned} (\lambda - L^*)^{-1} &= L_c^*(\lambda - L^*)^{-1} L_c^{*-1} \\ &= -L_c^{*-1} + (\lambda + c)(\lambda - L^*)^{-1} L_c^{*-1}, \end{aligned} \tag{5.22}$$

where $c > 0$ is large enough such that $-c \in \rho(L^*)$, and $L_c^* = L^* + c$. Following (5.22), let us introduce for each k, $1 \leqslant k \leqslant M$, a series of meromorphic functions $f_k^l(\lambda; u)$, $l = 0, 1, \ldots$ by the recursion formula:

$$f_k^0(\lambda; u) = f_k(\lambda; u), \qquad f_k^{l+1}(\lambda; u) = \frac{f_k^l(\lambda; u)}{\lambda + c}, \quad l = 0, 1, \ldots. \tag{5.23}$$

Then, just as in (3.26), we find that

$$f_k^l(\lambda; u) = \begin{cases} \left\langle (\lambda - L^*)^{-1} L_c^{*-l} u, \hat{g}_k \right\rangle_\Gamma - \sum_{i=1}^{l} \frac{1}{(\lambda + c)^i} \left\langle L_c^{*-(l+1-i)} u, \hat{g}_k \right\rangle_\Gamma, \\ \quad \text{in the case where } \alpha(\xi) \not\equiv 1, \\ \left\langle \frac{\partial}{\partial \nu} (\lambda - L^*)^{-1} L_c^{*-l} u, g_k \right\rangle_\Gamma - \sum_{i=1}^{l} \frac{1}{(\lambda + c)^i} \left\langle \frac{\partial}{\partial \nu} L_c^{*-(l+1-i)} u, g_k \right\rangle_\Gamma, \\ \quad \text{in the case where } \alpha(\xi) \equiv 1, \end{cases} \tag{5.24}$$

and, by (5.21),

$$f_k^l(\mu_i \omega^\pm; u) = 0, \quad i \geqslant 1, \quad 1 \leqslant k \leqslant M, \quad l \geqslant 0. \tag{5.25}$$

Recall the growth rate condition of the sequence $\{\mu_n\}$ in (2.5). Just as in the proof of Proposition 3.3, (5.25) implies — via Theorem 1.1 (Carleman's theorem) — that

$$f_k^l(\lambda; u) = 0, \quad 1 \leqslant k \leqslant M, \quad l \geqslant 0. \quad \lambda \in \rho(L) \setminus \{-c\}. \tag{5.26}$$

Thus we see that, when $l \geqslant 1$,

$$\begin{cases} \left\langle (\lambda - L^*)^{-1} L_c^{*-l} u, \hat{g}_k \right\rangle_\Gamma - \sum_{i=1}^{l} \frac{1}{(\lambda + c)^i} \left\langle L_c^{*-(l+1-i)} u, \hat{g}_k \right\rangle_\Gamma \equiv 0, \\ \quad \text{in the case where } \alpha(\xi) \not\equiv 1, \\ \left\langle \frac{\partial}{\partial \nu} (\lambda - L^*)^{-1} L_c^{*-l} u, g_k \right\rangle_\Gamma - \sum_{i=1}^{l} \frac{1}{(\lambda + c)^i} \left\langle \frac{\partial}{\partial \nu} L_c^{*-(l+1-i)} u, g_k \right\rangle_\Gamma \equiv 0, \\ \quad \text{in the case where } \alpha(\xi) \equiv 1. \end{cases} \tag{5.27}$$

Like Laurent's expansion of $(\lambda - L)^{-1}$, Laurent's expansion of the resolvent $(\lambda - L^*)^{-1}$ in a neighborhood of the pole $\overline{\lambda_i} \in \sigma(L^*)$ is :

$$(\lambda - L^*)^{-1} = \sum_{j=1}^{l_i} \frac{A_{-j}^*}{(\lambda - \overline{\lambda_i})^j} + \sum_{j=0}^{\infty} (\lambda - \overline{\lambda_i})^j A_j^*, \quad \text{where}$$

$$A_j^* = \frac{1}{2\pi \mathrm{i}} \int_{|\zeta - \overline{\lambda_i}| = \delta} \frac{(\zeta - L^*)^{-1}}{(\zeta - \overline{\lambda_i})^{j+1}} d\zeta, \quad j = 0, \pm 1, \pm 2, \ldots.$$

Here, l_i denotes the *ascent* of $\lambda_i - L$, and thus of $\overline{\lambda_i} - L^*$. Recall that the adjoint $P_{\lambda_i}^*$ of P_{λ_i} is the projector corresponding to the eigenvalue $\overline{\lambda_i}$ of L^* (see (1.10) and (1.11), Chapter 3). Set $P_{\lambda_i}^* u = \sum_{j=1}^{m_i} u_{ij}\psi_{ij} = A_{-1}^* u$. Note that $\sum_{j=0}^{\infty}(\lambda - \overline{\lambda_i})^j A_j^* u$ generally converges in the topology of $\mathscr{D}(L)$ and, thus, at least in the topology of $H^1(\Omega)$. In the specific case where $\alpha(\xi) \equiv 1$, note also that $\mathscr{D}(L) = H^2(\Omega) \cap H_0^1(\Omega) \hookrightarrow H^{3/2}(\Gamma)$. Thus, we see that

$$
\begin{aligned}
0 &= f_k(\lambda; u) \\
&= \begin{cases}
\displaystyle\sum_{j=1}^{l_i} \frac{\left\langle A_{-j}^* u, \hat{g}_k \right\rangle_\Gamma}{(\lambda - \overline{\lambda_i})^j} + \sum_{j=0}^{\infty} (\lambda - \overline{\lambda_i})^j \left\langle A_j^* u, \hat{g}_k \right\rangle_\Gamma, \\
\qquad \text{in the case where } \alpha(\xi) \not\equiv 1, \text{ and} \\
\displaystyle\sum_{j=1}^{l_i} \frac{1}{(\lambda - \overline{\lambda_i})^j} \left\langle \frac{\partial}{\partial \nu} A_{-j}^* u, g_k \right\rangle_\Gamma + \sum_{j=0}^{\infty} (\lambda - \overline{\lambda_i})^j \left\langle \frac{\partial}{\partial \nu} A_j^* u, g_k \right\rangle_\Gamma, \\
\qquad \text{in the case where } \alpha(\xi) \equiv 1.
\end{cases}
\end{aligned}
\tag{5.28}
$$

Calculating the residue of $f_k(\lambda; u)$ at $\overline{\lambda_i}$, we see that

$$
\begin{cases}
\left\langle A_{-1}^* u, \hat{g}_k \right\rangle_\Gamma = \left\langle P_{\lambda_i}^* u, \hat{g}_k \right\rangle_\Gamma \\
\qquad = \displaystyle\sum_{j=1}^{m_i} \left\langle \psi_{ij}, \hat{g}_k \right\rangle_\Gamma u_{ij} = 0, \quad \text{in the case where } \alpha(\xi) \not\equiv 1, \text{ and} \\
\left\langle \dfrac{\partial}{\partial \nu} A_{-1}^* u, g_k \right\rangle_\Gamma = \left\langle \dfrac{\partial}{\partial \nu} P_{\lambda_i}^* u, g_k \right\rangle_\Gamma \\
\qquad = \displaystyle\sum_{j=1}^{m_i} \left\langle \frac{\partial \psi_{ij}}{\partial \nu}, g_k \right\rangle_\Gamma u_{ij}, \quad \text{in the case where } \alpha(\xi) \equiv 1
\end{cases}
$$

for $i \geqslant 1$, and $1 \leqslant k \leqslant M$. In the former case, however, recalling that $\psi_{ij}(\xi) = 0$ and $g_k(\xi) = (1 - \alpha(\xi))\hat{g}_k(\xi) = 0$ on $\Gamma_1 = \{\xi \in \Gamma;\ \alpha(\xi) = 1\}$, we calculate as

$$
\begin{aligned}
\left\langle \psi_{ij}, \hat{g}_k \right\rangle_\Gamma &= \left\langle \psi_{ij}, \frac{g_k}{1 - \alpha} \right\rangle_{\Gamma \setminus \Gamma_1} = \left\langle \frac{\psi_{ij}}{1 - \alpha}, g_k \right\rangle_{\Gamma \setminus \Gamma_1} \\
&= \left\langle (1 - \boldsymbol{b} \cdot \boldsymbol{\nu})\psi_{ij} - \frac{\partial \psi_{ij}}{\partial \nu}, g_k \right\rangle_{\Gamma \setminus \Gamma_1} = \left\langle \sigma \psi_{ij}, g_k \right\rangle_\Gamma .
\end{aligned}
$$

Thus, the above relations are expressed in a unified manner as

$$
\begin{aligned}
&\sum_{j=1}^{m_i} \left\langle \sigma \psi_{ij}, g_k \right\rangle_\Gamma u_{ij} = \left(\langle \sigma \psi_{i1}, g_k \rangle_\Gamma \quad \ldots \quad \langle \sigma \psi_{im_i}, g_k \rangle_\Gamma \right) \boldsymbol{u}_i = 0, \\
&\qquad 1 \leqslant k \leqslant M, \quad i \geqslant 1, \quad \text{where } \boldsymbol{u}_i = \left(u_{i1}\ u_{i2} \ldots u_{im_i} \right)^{\mathrm{T}}.
\end{aligned}
\tag{5.29}
$$

Let us turn to $f_k^l(\lambda; u)$, $l \geqslant 1$. In view of Laurent's expansion of $(\lambda - L^*)^{-1}$, we similarly obtain the expansion in a neighborhood of each $\overline{\lambda_i}$:

$$0 = f_k^l(\lambda; u)$$
$$= \begin{cases} \sum_{j=1}^{l_i} \frac{\left\langle A_{-j}^* L_c^{*-l} u, \hat{g}_k \right\rangle_\Gamma}{(\lambda - \overline{\lambda_i})^j} + \sum_{j=0}^{\infty} (\lambda - \overline{\lambda_i})^j \left\langle A_j^* L_c^{*-l} u, \hat{g}_k \right\rangle_\Gamma \\ \quad - \sum_{i=1}^{l} \frac{1}{(\lambda + c)^i} \left\langle L_c^{*-(l+1-i)} u, \hat{g}_k \right\rangle_\Gamma, \quad \text{in the case where } \alpha(\xi) \not\equiv 1, \text{ and} \\ \sum_{j=1}^{l_i} \frac{1}{(\lambda - \overline{\lambda_i})^j} \left\langle \frac{\partial}{\partial \nu} A_{-j}^* L_c^{*-l} u, g_k \right\rangle_\Gamma + \sum_{j=0}^{\infty} (\lambda - \overline{\lambda_i})^j \left\langle \frac{\partial}{\partial \nu} A_j^* L_c^{*-l} u, g_k \right\rangle_\Gamma \\ \quad - \sum_{i=1}^{l} \frac{1}{(\lambda + c)^i} \left\langle \frac{\partial}{\partial \nu} L_c^{*-(l+1-i)} u, g_k \right\rangle_\Gamma, \quad \text{in the case where } \alpha(\xi) \equiv 1, \end{cases} \tag{5.30}$$

Note that $A_{-1}^* L_c^{*-l} u = P_{\overline{\lambda_i}}^* L_c^{*-l} u = L_c^{*-l} P_{\overline{\lambda_i}}^* u$, and that the restriction of L_c^{*-l} on $P_{\overline{\lambda_i}}^* L^2(\Omega)$ is equivalent to the matrix $(\tilde{\Lambda}_i + c)^{-l}$: $A_{-1}^* L_c^{*-l} u \Leftrightarrow (\tilde{\Lambda}_i + c)^{-l} \boldsymbol{u}_i$. Calculating the residue of $f_k^l(\lambda; u)$ at $\overline{\lambda_i}$, we see that

$$\begin{cases} \left\langle A_{-1}^* L_c^{*-l} u, \hat{g}_k \right\rangle_\Gamma = 0, \quad \text{in the case where } \alpha(\xi) \not\equiv 1, \text{ and} \\ \left\langle \frac{\partial}{\partial \nu} A_{-1}^* L_c^{*-l} u, g_k \right\rangle_\Gamma = 0, \quad \text{in the case where } \alpha(\xi) \equiv 1 \end{cases}$$

for $i \geqslant 1$, $l \geqslant 1$, and $1 \leqslant k \leqslant M$. In the former case, we further calculate as

$$\begin{aligned} 0 &= \left\langle A_{-1}^* L_c^{*-l} u, \hat{g}_k \right\rangle_\Gamma = \left\langle \sum_{j=1}^{m_i} \left((\tilde{\Lambda}_i + c)^{-l} \boldsymbol{u}_i \right)_j \psi_{ij}, \hat{g}_k \right\rangle_\Gamma \\ &= \sum_{j=1}^{m_i} \left((\tilde{\Lambda}_i + c)^{-l} \boldsymbol{u}_i \right)_j \langle \psi_{ij}, \hat{g}_k \rangle_\Gamma \\ &= \left(\langle \psi_{i1}, \hat{g}_k \rangle_\Gamma \ \ \dots \ \ \langle \psi_{im_i}, \hat{g}_k \rangle_\Gamma \right) (\tilde{\Lambda}_i + c)^{-l} \boldsymbol{u}_i, \\ &= \left(\langle \sigma \psi_{i1}, g_k \rangle_\Gamma \ \ \dots \ \ \langle \sigma \psi_{im_i}, g_k \rangle_\Gamma \right) (\tilde{\Lambda}_i + c)^{-l} \boldsymbol{u}_i, \quad 1 \leqslant k \leqslant M. \end{aligned}$$

In the latter case, we similarly obtain

$$\begin{aligned} 0 &= \left\langle \frac{\partial}{\partial \nu} A_{-1}^* L_c^{*-l} u, g_k \right\rangle_\Gamma \\ &= \left(\left\langle \frac{\partial \psi_{i1}}{\partial \nu}, g_k \right\rangle_\Gamma \ \ \dots \ \ \left\langle \frac{\partial \psi_{im_i}}{\partial \nu}, g_k \right\rangle_\Gamma \right) (\tilde{\Lambda}_i + c)^{-l} \boldsymbol{u}_i \\ &= - \left(\langle \sigma \psi_{i1}, g_k \rangle_\Gamma \ \ \dots \ \ \langle \sigma \psi_{im_i}, g_k \rangle_\Gamma \right) (\tilde{\Lambda}_i + c)^{-l} \boldsymbol{u}_i, \quad 1 \leqslant k \leqslant M. \end{aligned}$$

Thus, in each case we obtain

$$\begin{aligned}
&\left(\langle \sigma\psi_{ij}, g_k\rangle_\Gamma; \begin{array}{ccc} j & \to & 1,\dots,m_i \\ k & \downarrow & 1,\dots,M \end{array}\right)(\tilde{\Lambda}_i + c)^{-l}\boldsymbol{u}_i \\
&= \hat{G}_i^*(\tilde{\Lambda}_i + c)^{-l}\boldsymbol{u}_i = \boldsymbol{0}, \quad i \geqslant 1, \quad l \geqslant 1.
\end{aligned} \tag{5.31}$$

This, combined with (5.29), yields that

$$\begin{pmatrix} \hat{G}_i^* & \hat{G}_i^*(\tilde{\Lambda}_i + c)^{-1} & \dots & \hat{G}_i^*(\tilde{\Lambda}_i + c)^{-(m_i-1)} \end{pmatrix}^{\mathrm{T}} \boldsymbol{u}_i = \boldsymbol{0}, \quad i \geqslant 1. \tag{5.32}$$

As we have seen in Section 3, it is clear that the rank of the above coefficient matrix is equal to

$$\begin{aligned}
&\operatorname{rank}\begin{pmatrix} \hat{G}_i^* & \hat{G}_i^*\tilde{\Lambda}_i & \dots & \hat{G}_i^*\tilde{\Lambda}_i^{m_i-1} \end{pmatrix}^{\mathrm{T}} \\
&= \operatorname{rank}\begin{pmatrix} \hat{G}_i & \tilde{\Lambda}_i^*\hat{G}_i & \dots & (\tilde{\Lambda}_i^*)^{m_i-1}\hat{G}_i \end{pmatrix}, \quad i \geqslant 1.
\end{aligned}$$

In view of the second rank conditions in (5.8), we find that $\boldsymbol{u}_i = \boldsymbol{0}$ for $1 \leqslant i \leqslant \nu$, that is, $P_\nu^* u = 0$. This is nothing but the relation (5.17), that is, what we hoped to prove. The proof of Proposition 5.3, and thus the proof of Theorem 5.2 is thereby complete. □

4.6 The Robin Boundary and Fractional Powers

We have so far studied the feedback control systems (2.7), (4.10) and (5.1) within the algebraic framework. We show in this section that fractional powers L_c^ω of the operator $L_c = L + c$ are a useful tool, if the boundary operator τ in (1.1) is of the Robin type, that is, the case where $0 \leqslant \alpha(\xi) < 1$. This method is also applied to other control systems, as long as the boundary conditions are essentially derived from the Robin boundary. An advantage is that the domain $\mathscr{D}(L_c^\omega)$, $0 \leqslant \omega \leqslant 1$, is characterized in terms of fractional Sobolev spaces. By a transformation of the state $u(t,\cdot)$ via the fractional power, the original system with boundary inputs is transformed into an equivalent system with distributed inputs only, so that the standard semigroup theory is effectively applied. This transformation of u works just like an integral transformation cancelling nonhomogeneous boundary terms.

It is generally a difficult problem to characterize the domain $\mathscr{D}(L_c^\omega)$ in terms of Sobolev spaces, if the boundary operator is described by τ in (1.1) with $0 \leqslant \alpha(\xi) \leqslant 1$, or $\Gamma_1 = \{\xi \in \Gamma;\ \alpha(\xi) = 1\} \neq \varnothing$. A serious reason for this is that the Dirichlet boundary is locally *continuously* connected with the Neumann boundary. In addition, this method has the following disadvantage: in the case of the Dirichlet boundary, i.e., the case where $\alpha(\xi) \equiv 1$, the characterization of $\mathscr{D}(L_c^\omega)$ is well known (see Subsection 3.4, Chapter 2). Nevertheless, this method *cannot* be applied to linear systems with the Dirichlet boundary, and, in fact, faces a fatal difficulty of well-posedness of the system (see the remark at the end of this section). We thus limit ourselves to linear systems with the Robin boundary.

Let us consider again eqn. (2.7) with the Robin boundary, so that $\mathscr{D}(L) = \{u \in H^2(\Omega);\ \tau u = 0\}$ simply. We observe Theorem 3.1 by an approach via L_c^ω. Given the weights $w_k \in L^2(\Gamma)$, the outputs of Σ_p are given by $p_k(u) = \langle u, w_k\rangle_\Gamma$, $1 \leqslant k \leqslant N$. By assuming that the actuators g_k, $1 \leqslant k \leqslant M$, belong to $H^{1/2}(\Gamma)$, the boundary control system is then described again as

$$\begin{cases} \dfrac{\partial u}{\partial t} + \mathscr{L}u = 0 \quad \text{in } \mathbb{R}^1_+ \times \Omega, \\ \tau u = \displaystyle\sum_{k=1}^{M} \langle v, \rho_k\rangle_H\, g_k \quad \text{on } \mathbb{R}^1_+ \times \Gamma, \\ \dfrac{dv}{dt} + Bv = \displaystyle\sum_{k=1}^{N} \langle u, w_k\rangle_\Gamma\, \xi_k + \sum_{k=1}^{M} \langle v, \rho_k\rangle_H\, \zeta_k \quad \text{in } \mathbb{R}^1_+ \times H, \\ u(0,\cdot) = u_0(\cdot) \in L^2(\Omega), \qquad v(0) = v_0 \in H. \end{cases} \tag{6.1}$$

Let $c > 0$ be chosen large enough, so that $\sigma(L_c)$ is contained in the right-half plane $\mathbb{C}_+$. Thus, the fractional powers L_c^ω and $L_c^{*\omega}$, $\omega \in \mathbb{R}^1$, are well defined. It is well known that (see Subsection 3.4, Chapter 2)

$$\mathscr{D}(L_c^\omega) = \mathscr{D}\left(L_c^{*\omega}\right) = H^{2\omega}(\Omega), \quad 0 \leqslant \omega < \frac{3}{4}. \tag{6.2}$$

In the case where $\frac{3}{4} \leqslant \omega \leqslant 1$, the homogeneous boundary condition enters the right-hand side, $H^{2\omega}(\Omega)$. For each k, let $u = N_{-c}g_k \in H^2(\Omega)$ be a unique solution to the boundary value problem

$$(c + \mathscr{L})u = 0 \quad \text{in } \Omega, \qquad \tau u = g_k \in H^{1/2}(\Gamma) \quad \text{on } \Gamma.$$

Note that $N_{-c}g_k \in \mathscr{D}(L_c^\omega)$, $\omega < \frac{3}{4}$. Given an ε, $0 < \varepsilon < \frac{1}{2}$, set in (6.1)

$$x(t) = L_c^{-\alpha}u(t,\cdot), \quad t > 0, \quad \alpha = \frac{1}{4} + \varepsilon. \tag{6.3}$$

The above transform of the state works just like an *integral transform* which makes the state u smoother in space variables. Since $u(t,\cdot)$ belongs to $H^2(\Omega) \subset H^{3/2-2\varepsilon}(\Omega) = \mathscr{D}\left(L_c^{3/4-\varepsilon}\right)$, we note that $x(t)$, $t > 0$, belongs to $\mathscr{D}(L)$. The equation for $u(t,\cdot)$ is rewriten as

$$\frac{du}{dt} + L_c\left(u - \sum_{k=1}^{M} \langle v, \rho_k\rangle_H N_{-c}g_k\right) = cu.$$

Applying $L_c^{-\alpha}$ to the both sides, we obtain

$$\frac{dx}{dt} + L_c^{3/4-\varepsilon}\left(u - \sum_{k=1}^{M} \langle v, \rho_k \rangle_H N_{-c} g_k\right) = cx, \quad \text{or}$$

$$\frac{dx}{dt} + Lx - \sum_{k=1}^{M} \langle v, \rho_k \rangle_H L_c^{3/4-\varepsilon} N_{-c} g_k = 0, \quad t > 0, \quad x(0) = x_0 = L_c^{-\alpha} u_0.$$

Thus, the control system is transformed into the system with state (x, v) in $L^2(\Omega) \times H$:

$$\begin{cases} \dfrac{dx}{dt} + Lx = \displaystyle\sum_{k=1}^{M} \langle v, \rho_k \rangle_H L_c^{3/4-\varepsilon} N_{-c} g_k, & \text{in } \mathbb{R}_+^1 \times L^2(\Omega), \\ \dfrac{dv}{dt} + Bv = \displaystyle\sum_{k=1}^{N} \langle L_c^{\alpha} x, w_k \rangle_\Gamma \xi_k + \sum_{k=1}^{M} \langle v, \rho_k \rangle_H \zeta_k & \text{in } \mathbb{R}_+^1 \times H, \\ x(0) = x_0 \in L^2(\Omega), \qquad v(0) = v_0 \in H. \end{cases} \tag{6.4}$$

By the trace theorem (see (2.9_1), Chapter 2) and (6.2), we note that, for an arbitrarily small $\delta > 0$,

$$\begin{aligned} \left|\langle L_c^{\alpha} x, w_k \rangle_\Gamma\right| &\leqslant \text{const} \left\|L_c^{\alpha} x\right\|_{H^{1/2+2\delta}(\Omega)} \left\|w_k\right\|_{L^2(\Gamma)} \\ &\leqslant \text{const} \left\|L_c^{1/2+\varepsilon+\delta} x\right\|. \end{aligned}$$

Thus, the terms $\langle L_c^{\alpha} x, w_k \rangle_\Gamma$ in (6.4) are subordinate to L. Consequently, eqn. (6.4) is well posed in $L^2(\Omega) \times H$, and the coefficient operator is the infinitesimal generator of an analytic semigroup.

The operator $X \in \mathscr{L}(L^2(\Omega); H)$ is the unique solution to Sylvester's equation (3.8). Setting $\zeta_k = L_c^{3/4-\varepsilon} N_{-c} g_k$, $1 \leqslant k \leqslant M$, in (6.4), we see that

$$\begin{aligned} &\frac{d}{dt}(Xx - v) + B(Xx - v) = 0, \quad \text{or} \\ &Xx(t) - v(t) = e^{-tB}(Xx_0 - v_0), \quad t \geqslant 0, \end{aligned}$$

and thus (compare it with (3.12))

$$\|Xx(t) - v(t)\| \leqslant \text{const}\, e^{-a\mu_1 t} \|Xx_0 - v_0\|, \quad t \geqslant 0. \tag{6.5}$$

The equation for x is rewritten as

$$\begin{aligned} &\frac{dx}{dt} + \left(L - \sum_{k=1}^{M} \langle \cdot, X^* \rho_k \rangle L_c^{3/4-\varepsilon} N_{-c} g_k\right) x \\ &= \sum_{k=1}^{M} \left\langle e^{-tB}(v_0 - Xx_0), \rho_k \right\rangle_H L_c^{3/4-\varepsilon} N_{-c} g_k. \end{aligned} \tag{6.6}$$

What is the controllability condition on the actuators $L_c^{3/4-\varepsilon}N_{-c}g_k$, $1 \leqslant k \leqslant M$? The situation is similar to the problem in (3.37′) of Section 4.3, Chapter 4, where the actuators are instead $N_{-c}g_k$. Following (2.8), Chapter 3, let $P_{\lambda_i}L_c^{3/4-\varepsilon}N_{-c}g_k = \sum_{1\leqslant j\leqslant m_i} g_{ij}^k \varphi_{ij}$. By (1.10), Chapter 3,

$$\begin{aligned}
\begin{pmatrix} g_{i1}^k & \cdots & g_{im_i}^k \end{pmatrix}^{\mathrm{T}} &= \Pi_{\lambda_i}^{-1}\left(\left\langle L_c^{3/4-\varepsilon}N_{-c}g_k, \psi_{i1}\right\rangle \quad \cdots \quad \left\langle L_c^{3/4-\varepsilon}N_{-c}g_k, \psi_{im_i}\right\rangle\right)^{\mathrm{T}} \\
&= \Pi_{\lambda_i}^{-1}\left(\left\langle N_{-c}g_k, L_c^{*3/4-\varepsilon}\psi_{i1}\right\rangle \quad \cdots \quad \left\langle N_{-c}g_k, L_c^{*3/4-\varepsilon}\psi_{im_i}\right\rangle\right)^{\mathrm{T}} \\
&= \Pi_{\lambda_i}^{-1}(\tilde{\Lambda}_i + c)^{3/4-\varepsilon^*}\left(\langle N_{-c}g_k, \psi_{i1}\rangle \quad \cdots \quad \langle N_{-c}g_k, \psi_{im_i}\rangle\right)^{\mathrm{T}}.
\end{aligned}$$

As we have seen in Section 3, we know via Green's formula that

$$\begin{aligned}
&\left(\langle N_{-c}g_k, \psi_{i1}\rangle \quad \cdots \quad \langle N_{-c}g_k, \psi_{im_i}\rangle\right)^{\mathrm{T}} \\
&= (\tilde{\Lambda}_i + c)^{-1^*}\left(\langle g_k, \sigma\psi_{i1}\rangle_\Gamma \quad \cdots \quad \langle g_k, \sigma\psi_{im_i}\rangle_\Gamma\right)^{\mathrm{T}}.
\end{aligned}$$

Thus,

$$\begin{pmatrix} g_{i1}^k & \cdots & g_{im_i}^k \end{pmatrix}^{\mathrm{T}} = \Pi_{\lambda_i}^{-1}(\tilde{\Lambda}_i + c)^{*^{-\alpha}}\left(\langle g_k, \sigma\psi_{i1}\rangle_\Gamma \quad \cdots \quad \langle g_k, \sigma\psi_{im_i}\rangle_\Gamma\right)^{\mathrm{T}},$$

where $\alpha = \frac{1}{4} + \varepsilon$. In other words,

$$\begin{aligned}
\tilde{G}_i &= \begin{pmatrix} g_{ij}^k; & \begin{matrix} j & \downarrow & 1,\ldots,m_i \\ k & \rightarrow & 1,\ldots,M \end{matrix} \end{pmatrix} \\
&= \Pi_{\lambda_i}^{-1}(\tilde{\Lambda}_i + c)^{*^{-\alpha}}\begin{pmatrix} \langle g_k, \sigma\psi_{ij}\rangle_\Gamma; & \begin{matrix} j & \downarrow & 1,\ldots,m_i \\ k & \rightarrow & 1,\ldots,M \end{matrix} \end{pmatrix} \\
&= \Pi_{\lambda_i}^{-1}\left(\tilde{\Lambda}_i^* + c\right)^{-\alpha}\hat{G}_i.
\end{aligned}$$

At this stage, we assume the rank condition (3.5). Recalling that $\Lambda_i = \Pi_{\lambda_i}^{-1}\tilde{\Lambda}_i^*\Pi_{\lambda_i}$, we calculate as

$$\begin{aligned}
&\left(\tilde{G}_i \quad (\Lambda_i + c)\tilde{G}_i \quad \cdots \quad (\Lambda_i + c)^{m_i-1}\tilde{G}_i\right) \\
&= \left(\Pi_{\lambda_i}^{-1}\left(\tilde{\Lambda}_i^* + c\right)^{-\alpha}\hat{G}_i \quad \Pi_{\lambda_i}^{-1}\left(\tilde{\Lambda}_i^* + c\right)^{1-\alpha}\hat{G}_i \quad \cdots \quad \Pi_{\lambda_i}^{-1}\left(\tilde{\Lambda}_i^* + c\right)^{m_i-1-\alpha}\hat{G}_i\right) \\
&= \Pi_{\lambda_i}^{-1}\left(\tilde{\Lambda}_i^* + c\right)^{-\alpha}\left(\hat{G}_i \quad (\tilde{\Lambda}_i^* + c)\hat{G}_i \quad \cdots \quad (\tilde{\Lambda}_i^* + c)^{m_i-1}\hat{G}_i\right).
\end{aligned}$$

Thus, the controllability condition on $L_c^{3/4-\varepsilon}N_{-c}g_k$ becomes

$$\begin{aligned}
&\operatorname{rank}\left(\tilde{G}_i \quad \Lambda_i\tilde{G}_i \quad \cdots \quad \Lambda_i^{m_i-1}\tilde{G}_i\right) \\
&= \operatorname{rank}\left(\hat{G}_i \quad (\tilde{\Lambda}_i^* + c)\hat{G}_i \quad \cdots \quad \left(\tilde{\Lambda}_i^* + c\right)^{m_i-1}\hat{G}_i\right) \\
&= \operatorname{rank}\left(\hat{G}_i \quad \tilde{\Lambda}_i^*\hat{G}_i \quad \cdots \quad (\tilde{\Lambda}_i^*)^{m_i-1}\hat{G}_i\right) = m_i, \quad 1 \leqslant i \leqslant \nu.
\end{aligned}$$

This is nothing but the second condition of (3.5). Thus, there exist functions $y_k \in P_\nu^* L^2(\Omega)$, $1 \leqslant k \leqslant M$, such that

$$\left\| \exp\left(-t\left(L - \sum_{k=1}^{M} \langle \cdot, y_k \rangle L_c^{3/4-\varepsilon} N_{-c} g_k\right)\right)\right\| \leqslant \mathrm{const}\, e^{-r_1 t}, \quad t \geqslant 0,$$

where $r < r_1 < \mathrm{Re}\,\lambda_{\nu+1}$. Owing to the first and the third conditions of (3.5), Proposition 3.3 ensures the inclusion relation (3.10). Thus, there exist suitable vectors $\rho_k \in H$, $1 \leqslant k \leqslant M$, such that

$$\left\| \exp\left(-t\left(L - \sum_{k=1}^{M} \langle \cdot, X^* \rho_k \rangle L_c^{3/4-\varepsilon} N_{-c} g_k\right)\right)\right\| \leqslant \mathrm{const}\, e^{-rt}, \quad t \geqslant 0.$$

In view of (6.6), we immediately obtain the decay estimate

$$\|x(t)\| + \|v(t)\|_H \leqslant \mathrm{const}\, e^{-rt}\left(\|x_0\| + \|v_0\|_H\right), \quad t \geqslant 0, \tag{6.7}$$

which establishes the stabilization of (6.4). As before, note that ρ_k can be constructed in a finite-dimensional subspace $P_n^H H$ for a suitable n. The reduction procedure of (6.4) into the system with a finite-dimensional compensator is the same as *Step* IV of Theorem 3.1: Setting $v_1(t) = P_n^H v(t)$, we obtain the equation for (x, v_1):

$$\begin{cases} \dfrac{dx}{dt} + Lx = \displaystyle\sum_{k=1}^{M} \langle v_1, \rho_k \rangle_H L_c^{3/4-\varepsilon} N_{-c} g_k, \quad \text{in } \mathbb{R}_+^1 \times L^2(\Omega), \\ \dfrac{dv_1}{dt} + B_1 v = \displaystyle\sum_{k=1}^{N} \langle L_c^\alpha x, w_k \rangle_\Gamma P_n^H \xi_k + \sum_{k=1}^{M} \langle v_1, \rho_k \rangle_H P_n^H \zeta_k \quad \text{in } \mathbb{R}_+^1 \times H, \\ x(0) = x_0 \in L^2(\Omega), \qquad v_1(0) = P_n^H v_0 = v_{10} \in P_n^H H, \end{cases} \tag{6.4'}$$

which is well posed in $L^2(\Omega) \times P_n^H H$, and the decay estimate (6.7) with $v(t)$ and $v_0 \in H$ replaced by $v_1(t)$ and $v_{10} \in P_n^H H$, respectively.

Let us go back to the original state $u(t,\cdot) = L_c^\alpha x(t)$. By (6.4′),

$$\begin{aligned} u(t,\cdot) &= L_c^\alpha x(t) \\ &= e^{-tL} u_0 + \int_0^t L_c^\alpha e^{-(t-s)L} \sum_{k=1}^{M} \langle v_1(s), \rho_k \rangle_H L_c^{3/4-\varepsilon} N_{-c} g_k \, ds. \end{aligned}$$

Thus, $\|u(t,\cdot)\|$ has no singularity at $t = 0$, and in fact satisfies an estimate:

$$\|u(t,\cdot)\| \leqslant \mathrm{const}\left(\|u_0\| + \|v_{10}\|_H\right)$$

in a neighborhood of $t = 0$. As for the decay of $\|u(t,\cdot)\|$ for $t > 0$, let $e^{-t\mathscr{N}}$ be the analytic semigroup generated by (6.4′), where $\mathscr{D}(\mathscr{N}) = \mathscr{D}(L) \times P_n^H H$ and $0 \in \rho(\mathscr{N})$. Then, the decay estimate for, e.g., $t \geqslant 1$ is obtained by the expression,

$$\begin{pmatrix} u(t,\cdot) \\ 0 \end{pmatrix} = \begin{pmatrix} L_c^\alpha & 0 \\ 0 & 0 \end{pmatrix} \mathscr{N}^{-1} \mathscr{N} e^{-t\mathscr{N}} \begin{pmatrix} L_c^{-\alpha} u_0 \\ v_{10} \end{pmatrix}, \quad t > 0. \tag{6.8}$$

It is easy to obtain the equation for $(u(t,\cdot), v_1(t))$ in the form of (1.3).

We close this section with the following remarks:

Remark 1: In the case of the Dirichlet boundary, the output of the controlled plant Σ_p is given by the second of (1.4), namely,

$$\left\langle \frac{\partial u}{\partial \nu}, w_k \right\rangle_\Gamma, \quad 1 \leqslant k \leqslant N. \tag{6.9}$$

The boundary condition in (6.1) is: $\tau u = u|_\Gamma = \sum_{1\leqslant k\leqslant M} \langle v, \rho_k \rangle_H g_k$. The fractional structure of L_c with this boundary operator is fairly different from the former case: The relation corresponding to (6.2) is that (see Subsection 3.4, Chapter 2)

$$\mathscr{D}(L_c^\omega) = \mathscr{D}(L_c^{*\omega}) = H^{2\omega}(\Omega), \quad 0 \leqslant \omega < \frac{1}{4}.$$

In the case where $\frac{1}{4} \leqslant \omega \leqslant 1$, the homogeneous boundary condition enters the right-hand side, $H^{2\omega}(\Omega)$. On the analogy of (6.3), we might set

$$x(t) = L_c^{-3/4-\varepsilon} u(t,\cdot), \quad t > 0.$$

Then, we similarly obtain the equation for $x(t)$:

$$\frac{dx}{dt} + Lx = \sum_{k=1}^{M} \langle v, \rho_k \rangle_H L_c^{3/4-\varepsilon} D_{-c} g_k,$$

where $D_{-c} g_k \in H^2(\Omega)$, $1 \leqslant k \leqslant M$, denote unique solutions to the boundary value problems: $(c + \mathscr{L}) D_{-c} g_k = 0$ in Ω, $\tau D_{-c} g_k = D_{-c} g_k|_\Gamma = g_k$ on Γ. Then the output (6.9) of Σ_p is rewritten as

$$\left\langle \frac{\partial u}{\partial \nu}, w_k \right\rangle_\Gamma = \left\langle \frac{\partial}{\partial \nu} L_c^{3/4+\varepsilon} x, w_k \right\rangle_\Gamma. \tag{6.9'}$$

A difficulty arises at this stage: Due to the strong unboundedness of the right-hand side, the above functionals on x is no more subordinate to the operator L.

Remark 2: We remark another approach to transform boundary inputs into distributed inputs. It is based on the formulation of the equation for u in weak form. According to this formulation, the equation for u is regarded as the one in a space of linear forms, and L is interpreted as the extended and generalized operator. In [60], this formulation is extensively studied in studying optimal control problems, etc. Let us review briefly this approach in two ways. When u belongs to $\mathscr{D}(L)$ and ψ to $\mathscr{D}(L^*)$, Green's formula implies the well known relation: $\langle Lu, \psi \rangle = \langle u, L^* \psi \rangle$, the right-hand side of which is an anti-linear form on $\mathscr{D}(L^*)$ in which the graph norm is equipped. Thus we see that there is a unique map $A_1 : L^2(\Omega) \to \mathscr{D}(L^*)'$ such that $\langle u, L^* \psi \rangle = \langle A_1 u, \psi \rangle$, where the bracket $\langle \cdot, \cdot \rangle$ is understood as the one between the pair of spaces: $\mathscr{D}(L^*)'$ and

$\mathscr{D}(L^*)$. This allows us to extend L defined on $\mathscr{D}(L)$ to the operator A_1 on $L^2(\Omega)$ by the above formula. Identifying $L^2(\Omega)$ as its dual, we obtain: $\mathscr{D}(L^*) \subset L^2(\Omega) \subset \mathscr{D}(L^*)'$ with continuous, dense injections. For each $\psi \in \mathscr{D}(L^*)$ and $\varphi_k = N_{-c}h_k$ in (1.4), Green's formula implies that

$$0 = \langle \mathscr{L}_c N_{-c} g_k, \psi \rangle = -\langle g_k, \sigma\psi \rangle_\Gamma + \langle N_{-c} g_k, L_c^* \psi \rangle .$$

Thus $\langle g_k, \sigma\psi \rangle_\Gamma$ defines an anti-linear form on $\mathscr{D}(L^*)$. According to the extended L, i.e., A_1, this anti-linear form is rewritten as

$$\langle g_k, \sigma\psi \rangle_\Gamma = \langle N_{-c} g_k, L_c^* \psi \rangle = \langle L_c N_{-c} g_k, \psi \rangle, \quad L_c N_{-c} g_k \in \mathscr{D}(L^*).$$

We formulate the equation for u in weak form as follows: In (1.3), when the solution $u(t,\cdot)$, $t > 0$, belongs to $H^2(\Omega)$ and ψ to $\mathscr{D}(L^*)$, we calculate the term $\langle u, \psi \rangle$ - via Green's formula as

$$\begin{aligned} c\,\langle u, \psi \rangle &= \frac{d}{dt} \langle u, \psi \rangle + \langle \mathscr{L}_c u, \psi \rangle \\ &= \frac{d}{dt} \langle u, \psi \rangle - \langle \tau u, \sigma\psi \rangle_\Gamma + \langle u, L_c^* \psi \rangle \\ &= \frac{d}{dt} \langle u, \psi \rangle - \sum_{k=1}^{M} \langle v, \rho_k \rangle_{\mathbb{R}^\ell} \langle g_k, \sigma\psi \rangle_\Gamma + \langle u, L_c^* \psi \rangle \\ &= \frac{d}{dt} \langle u, \psi \rangle - \sum_{k=1}^{M} \langle v, \rho_k \rangle_{\mathbb{R}^\ell} \langle L_c N_{-c} g_k, \psi \rangle + \langle L_c u, \psi \rangle . \end{aligned}$$

Thus u satisfies the equation in $\mathscr{D}(L^*)'$:

$$\frac{du}{dt} + Lu = \sum_{k=1}^{M} \langle v, \rho_k \rangle_{\mathbb{R}^\ell} L_c N_{-c} g_k, \tag{6.10}$$

which turns out a counterpart of the equation for x in (6.4). We stress in (6.10) that L is regarded as the extended operator A_1. Thus, $\mathscr{D}(L)$ is equal to $L^2(\Omega)$ in (6.10). An advantage of the form of (6.10) is that it allows the boundary operator τ in our problem. On the other hand, the regularity problem remains: examining if the solution u would be actually an $H^2(\Omega)$-function satisfying the original boundary condition. In addition, a serious difficulty arises: The output $\langle u, w_k \rangle_\Gamma$ of the controlled plant Σ_p is no more subordinate to the extended L.

A formulation somewhat stronger than (6.10) is possible in the dual space of the Hilbert space $H^1_\alpha(\Omega)$. Here,

$$\begin{aligned} &H^1_\alpha(\Omega) \\ &= \left\{ u \in H^1(\Omega);\ u = 0 \text{ on } \Gamma_1, \quad \left(\frac{\alpha(\xi)}{1-\alpha(\xi)} \right)^{1/2} u \in L^2(\Gamma \setminus \Gamma_1) \right\}, \end{aligned} \tag{6.11}$$

and $\Gamma_1 = \{\xi \in \Gamma;\ \alpha(\xi) = 1\} \neq \varnothing$. The space appeared in (3.27), Chapter 2. The sesqui-linear form associated with the pair $(\mathscr{L}, \tau)$ is defined by

$$B(u, \psi) = \left\langle \frac{\alpha(\xi)}{1-\alpha(\xi)} u,\ \psi \right\rangle_{\Gamma \setminus \Gamma_1} + \sum_{i,j=1}^{m} \left\langle a_{ij}(x) \frac{\partial u}{\partial x_j},\ \frac{\partial \psi}{\partial x_i} \right\rangle + \sum_{i=1}^{m} \left\langle b_i(x) \frac{\partial u}{\partial x_i},\ \psi \right\rangle + \langle c(x) u,\ \psi \rangle .$$

When u and ψ belong to $\mathscr{D}(L)$ and $H^1_\alpha(\Omega)$, respectively, we see that $\langle Lu,\ \psi \rangle = B(u, \psi)$. Since $B(u, \psi)$ is an anti-linear form in $\psi \in H^1_\alpha(\Omega)$, there is a unique map $A_2 :\ H^1_\alpha(\Omega) \to H^1_\alpha(\Omega)'$ such that $B(u, \psi) = \langle A_2 u,\ \psi \rangle$, where the bracket is understood as the one between the pair of spaces: $H^1_\alpha(\Omega)'$ and $H^1_\alpha(\Omega)$. Thus L defined on $\mathscr{D}(L)$ is extended to A_2 on $H^1_\alpha(\Omega)$ by the above formula. By assuming an additional condition: $g_k|_{\Gamma_1} = 0,\ 1 \leqslant k \leqslant M$, solutions $u(t,\cdot)$ belongs to $H^1_\alpha(\Omega)$. Then u satisfies the equation in $H^1_\alpha(\Omega)'$ with unbounded controls:

$$\frac{du}{dt} + Lu = \sum_{k=1}^{M} \langle v,\ \rho_k \rangle_{\mathbb{R}^\ell} J g_k, \quad J g_k \in \mathscr{D}(L^*)', \tag{6.12}$$

where $L = A_2$ and $\mathscr{D}(L) = H^1_\alpha(\Omega)$. An advantage of the form of (6.12) is that, since solutions are sought in $H^1_\alpha(\Omega)$, the outputs $\langle u,\ w_k \rangle_\Gamma$ are subordinate to L. However, we have to require the superfluous assumptions: $g_k|_{\Gamma_1} = 0,\ 1 \leqslant k \leqslant M$.

4.7 Some Related Topics

4.7.1 On the growth rate of $\sigma(B)$

In constructing the spectrum $\sigma(B) = \{\mu_i \omega^{\pm};\ i \geqslant 1\}$ of the compensator Σ_c, it is assumed in (2.5) that

$$0 < \exists \gamma < 2; \quad \mu_i \leqslant \text{const}\, i^\gamma, \quad i \geqslant 1.$$

It is natural to expect that the critical point of the growth rate γ might be 2. Hereafter we show that the choice $\gamma = 2$ is actually possible in some situation. In (6.1), let L be a self-adjoint operator equipped with the homogeneous Robin boundary condition. We need to show an inclusion relation similar to (3.10) with $\gamma = 2$. By introducing a new state $x(t) = L_c^{-\alpha} u(t,\cdot)$, $\alpha = 1/4 + \varepsilon$, $0 < \varepsilon < 1/4$ (see (6.3)), the control system (6.1) is transformed into (6.4) with state $(x,\ v)$. By further assuming that the observation weight w_k belongs to $H^{1/2}(\Gamma)$ for each k, we find a unique solution $u \in H^2(\Omega)$ to the boundary value problem:

$$(c + \mathscr{L}) u = 0 \quad \text{in } \Omega, \qquad \tau u = w_k \quad \text{on } \Gamma.$$

The solution is denoted as $u = h_k = N_{-c}w_k$. Then, by Green's formula,

$$\langle u, w_k\rangle_\Gamma = \langle L_c^\alpha x, w_k\rangle_\Gamma = \left\langle L_c^{2\alpha}x, L_c^{3/4-\varepsilon}h_k\right\rangle.$$

The unique solution $X \in \mathscr{L}(L^2(\Omega); H)$ to Sylvester's equation (3.8) is expressed by (3.9), where $f_k(\lambda; u) = \left\langle L_c^{2\alpha}(\lambda - L)^{-1}u, L_c^{3/4-\varepsilon}h_k\right\rangle$. Operator X with restricted domain $\mathscr{D}(L_c^{2\alpha})$ is denoted as $X_\alpha \in \mathscr{L}(\mathscr{D}(L_c^{2\alpha}); H)$. Then, we have a version of Proposition 3.3:

$$P_\nu L^2(\Omega) \subset \overline{X_\alpha^* H}, \tag{7.1}$$

where the overline means the closure in $\mathscr{D}(L_c^{2\alpha})$.

Proof. We only have to show that

$$\ker X_\alpha \subset \left\{u \in \mathscr{D}(L_c^{2\alpha});\ \langle u, \varphi_{ij}\rangle = 0,\ 1 \leqslant i \leqslant \nu,\ 1 \leqslant j \leqslant m_i\right\}.$$

Let $u \in \ker X_\alpha\ (\subset \mathscr{D}(L_c^{2\alpha}))$. By the last assumption of (3.5), or $\operatorname{rank}\Xi_i = N$, $i \geqslant 1$, we have

$$f_k(\mu_i\omega^\pm; u) = \left\langle(\mu_i\omega^\pm - L)^{-1}L_c^{2\alpha}u, L_c^{3/4-\varepsilon}h_k\right\rangle = 0, \qquad i \geqslant 1, \quad 1 \leqslant k \leqslant N. \tag{7.2}$$

Here we note the identity,

$$2\lambda\left(\lambda^2 - L_c\right)^{-1} = \left(\lambda + L_c^{1/2}\right)^{-1} + \left(\lambda - L_c^{1/2}\right)^{-1}$$

for $\lambda \in \rho\left(\pm L_c^{1/2}\right)$, and set for each k

$$\begin{aligned}\varphi(\lambda) &= \left\langle(\lambda - \mathrm{i}L_c^{1/2})^{-1}L_c^{2\alpha}u, L_c^{3/4-\varepsilon}h_k\right\rangle + \left\langle(\lambda + \mathrm{i}L_c^{1/2})^{-1}L_c^{2\alpha}u, L_c^{3/4-\varepsilon}h_k\right\rangle \\ &= 2\lambda\left\langle(\lambda^2 + L_c)^{-1}L_c^{2\alpha}u, L_c^{3/4-\varepsilon}h_k\right\rangle.\end{aligned} \tag{7.3}$$

Then, by (7.2),

$$\varphi\left(\pm\mathrm{i}\left(\mu_i\omega^\pm + c\right)^{1/2}\right) = 0, \quad i \geqslant 1. \tag{7.4}$$

Since $L_c^{1/2}$ is positive-definite, the two Cauchy problems in $L^2(\Omega)$

$$\frac{dx}{dt} = \mathrm{i}L_c^{1/2}x, \quad \text{and} \quad \frac{dx}{dt} = -\mathrm{i}L_c^{1/2}x$$

are well posed on $\mathscr{D}(L_c^{1/2})$, and generate strongly continuous semigroups, i.e., C_0-semigroups, $\exp(\mathrm{i}tL_c^{1/2})$ and $\exp(-\mathrm{i}tL_c^{1/2})$, respectively. Note that

$$\left\|e^{\pm\mathrm{i}tL_c^{1/2}}\right\|_{\mathscr{L}(L^2(\Omega))} = 1, \quad t \in \mathbb{R}^1.$$

It is clear that

$$\int_0^\infty e^{-\lambda t} e^{\pm \mathrm{i} t L_c^{1/2}} u\,dt = \left(\lambda \mp \mathrm{i} L_c^{1/2}\right)^{-1} u, \quad u \in L^2(\Omega), \quad \mathrm{Re}\,\lambda > 0.$$

Thus,

$$\varphi(\lambda) = \int_0^\infty e^{-\lambda t} \left\langle \left(e^{\mathrm{i} t L_c^{1/2}} + e^{-\mathrm{i} t L_c^{1/2}} \right) L_c^{2\alpha} u, L_c^{3/4-\varepsilon} h_k \right\rangle dt, \quad \mathrm{Re}\,\lambda > 0.$$

In (7.4), $\arg \mathrm{i}(\mu_i \omega^- + c)^{1/2}$ is monotone decreasing in i and approaches $\frac{\pi}{2} - \frac{1}{2}\theta_0$, $\theta_0 = \mathrm{Tan}^{-1}\left(\sqrt{1-a^2}/a\right)$. Choose a $\delta > 0$ small enough. Then we see that

$$\begin{aligned} &\int_0^\infty e^{-(\mathrm{i}(\mu_i\omega^- + c)^{1/2} - \delta)t} e^{-\delta t} \left\langle \left(e^{\mathrm{i} t L_c^{1/2}} + e^{-\mathrm{i} t L_c^{1/2}} \right) L_c^{2\alpha} u, L_c^{3/4-\varepsilon} h_k \right\rangle dt \\ &= 0, \quad i \geqslant 1. \end{aligned} \tag{7.5}$$

Here recall the celebrated Szász's theorem [64] (see also [31, 54, 72]). It is stated as follows:

Let $\mathrm{Re}\,\lambda_i > -\frac{1}{2}$, $i \geqslant 1$. *The set of functions* $\left\{\exp\left(-\left(\lambda_i + \frac{1}{2}\right)t\right); i \geqslant 1\right\}$ *is closed in* $L^2(0,\infty)$ *if and only if*

$$\sum_{i=1}^\infty \frac{1 + 2\,\mathrm{Re}\,\lambda_i}{1 + |\lambda_i|^2} = \infty. \tag{7.6}$$

We examine the condition (7.6) in the case where $\lambda_i + \frac{1}{2} = \mathrm{i}(\mu_i\omega^- + c)^{1/2} - \delta$. Let $\mu_i\omega^- + c = r_i \exp(-\mathrm{i}\theta_i)$ with $\theta_i \nearrow \theta_0 = \mathrm{Tan}^{-1}\left(\sqrt{1-a^2}/a\right)$. Then, by choosing a small constant, we have

$$\frac{1 + 2\,\mathrm{Re}\,\lambda_i}{1 + |\lambda_i|^2} > \frac{\mathrm{const}}{r_i^{1/2}}$$

for i large enough. In addition, $r_i - \mu_i \to c\cos\theta_0$ as $i \to \infty$. Thus,

$$\sum_{i \geqslant i_0} \frac{1 + 2\,\mathrm{Re}\,\lambda_i}{1 + |\lambda_i|^2} \geqslant \mathrm{const} \sum_{i \geqslant i_0} \frac{1}{i} = \infty$$

for some integer i_0. and (7.6) is satisfied. Thus, we conclude from (7.5) that

$$e^{-\delta t} \left\langle \left(e^{\mathrm{i} t L_c^{1/2}} + e^{-\mathrm{i} t L_c^{1/2}} \right) L_c^{2\alpha} u, L_c^{3/4-\varepsilon} h_k \right\rangle = 0, \quad t \geqslant 0,$$

and that $\varphi(\lambda) = 0$ for $\mathrm{Re}\,\lambda > 0$. In view of (7.3), the meromorphic function $f_k(\lambda; u) = \left\langle (\lambda - L)^{-1} L_c^{2\alpha} u, L_c^{3/4-\varepsilon} h_k \right\rangle$ is identically equal to 0 except at possible poles λ_i, $i \geqslant 1$. Calculating the residue at each λ_i, we have

$$\sum_{j=1}^{m_i} \langle u, \varphi_{ij} \rangle \langle \varphi_{ij}, h_k \rangle = (\lambda_i + c) \sum_{j=1}^{m_i} \langle u, \varphi_{ij} \rangle \langle \varphi_{ij}, w_k \rangle_\Gamma = 0, \quad i \geqslant 1.$$

The first rank conditions on w_k in (3.5) turn out to be rank $\hat{W}_i = m_i$, $1 \leqslant i \leqslant \nu$, since there arises no generalized eigenfunctions of L (see the arguments in Chapter 5). Thus, we have (7.1).

The rest of the stabilization procedure is almost the same as in Theorem 3.1. The difference is just of technical nature. For example, eqn. (6.6) for x is rewritten as

$$\frac{dx}{dt} + \left(L - \sum_{k=1}^{M} \langle \cdot, X_\alpha^* \rho_k \rangle_{\mathscr{D}(L_c^{2\alpha})} L_c^{3/4-\varepsilon} N_{-c} g_k \right) x = \sum_{k=1}^{M} \left\langle e^{-tB}(v_0 - X x_0), \rho_k \right\rangle_H L_c^{3/4-\varepsilon} N_{-c} g_k.$$

The above coefficient operator is then rewritten as

$$A = L - \sum_{k=1}^{M} \left\langle L_c^{2\alpha} \cdot, L_c^{2\alpha} X_\alpha^* \rho_k \right\rangle L_c^{3/4-\varepsilon} N_{-c} g_k.$$

The second rank conditions on g_k in (3.5) are in this case equivalent to the conditions, rank $\hat{G}_i = m_i$, $1 \leqslant i \leqslant \nu$. There exist functions $y_k \in P_\nu L^2(\Omega)$, $1 \leqslant k \leqslant M$, such that

$$\left\| e^{-t\hat{A}} \right\| \leqslant \text{const}\, e^{-r_1 t}, \quad t \geqslant 0,$$
$$\text{where } \hat{A} = L - \sum_{k=1}^{M} \left\langle L_c^{2\alpha} \cdot, L_c^{2\alpha} y_k \right\rangle L_c^{3/4-\varepsilon} N_{-c} g_k. \tag{7.7}$$

We remark that, by choosing an $\ell > 0$ large enough, $\hat{A}_\ell = \hat{A} + \ell$ is m-accretive. Thus, we see that

$$\mathscr{D}(\hat{A}_\ell^\beta) = \mathscr{D}(L_c^\beta), \quad 0 \leqslant \beta \leqslant 1.$$

Set $\delta_k = L_c^{2\alpha}(y_k - X_\alpha^* \rho_k)$. Since each y_k is approximated by vectors of the form $X_\alpha^* \rho_k$ arbitrarily in the topology of $\mathscr{D}(L_c^{2\alpha})$ (see (7.1)), we see that $\delta_k \to 0$ in $L^2(\Omega)$. Since

$$A = \hat{A} + \sum_{k=1}^{M} \left\langle L_c^{2\alpha} \cdot, L_c^{2\alpha} \delta_k \right\rangle L_c^{3/4-\varepsilon} N_{-c} g_k = \tilde{A} + D_\delta;$$

the resolvent $(\lambda - \hat{A})^{-1}$ exists in the set $(\overline{\Sigma} - a) \cup \{\lambda;\ \mathrm{Re}\,\lambda \leqslant r_1\}$; and $D_\delta \hat{A}_\ell^{-2\alpha} \to 0$, we see that

$$(\lambda - A)^{-1} = (\lambda - \hat{A})^{-1} \left(1 - D_\delta \hat{A}_\ell^{-2\alpha} \hat{A}_\ell^{2\alpha} (\lambda - \hat{A})^{-1}\right)^{-1}$$

also exists in the same set as long as $\|\delta_k\|$ are small enough. We have thus obtained the desired decay estimate

$$\left\| e^{-tA} \right\| \leqslant \text{const}\, e^{-r_1 t}, \quad t \geqslant 0,$$

which immediately leads to the decay of $x(t)$ and $v(t)$.

4.7.2 On fractional powers of elliptic operators characterized by feedback boundary conditions

In Proposition 3.4, we have defined the closed operator F through $\hat{F}$ in (3.14). Specifically the boundary operator τ_f in F is characterized in feedback form. We seek the domain of the fractional powers of F or its right-shift $F_c = F + c$, $c > 0$. The result will be applied in Section 2, Chapter 5. In doing so, the boundary operator τ is limited to the two cases: One is the Dirichlet boundary (case I) or $\alpha(\xi) \equiv 1$, and the other the Robin boundary (case II) or $0 \leqslant \alpha(\xi) < 1$. To clarify these two cases, let us set

$$\begin{aligned} &L_i u = \mathscr{L} u, \quad u \in \mathscr{D}(L_i) = \{ u \in H^2(\Omega);\ \tau_i u = 0 \}, \quad i = 1, 2, \\ &\text{where } \tau_1 u = u|_\Gamma, \quad \text{and } \tau_2 u = \frac{\partial u}{\partial \nu} + \sigma(\xi) u. \quad \left(\sigma(\xi) = \frac{\alpha(\xi)}{1 - \alpha(\xi)} \right) \end{aligned} \tag{7.8}$$

Then, we define

$$\begin{aligned} &F_i u = \mathscr{L} u, \quad u \in \mathscr{D}(F_i) = \left\{ u \in H^2(\Omega);\ \tau_i^f u = 0 \right\}, \\ &\text{where } \tau_i^f u = \tau_i u - \sum_{k=1}^{M} \langle u, w_k \rangle g_k, \quad i = 1, 2. \end{aligned} \tag{7.9}$$

We have shown in (3.35) that both F_1 and F_2 are sectorial operators with dense domains (see Proposition 3.4). Thus fractional powers of $F_{ic} = F_i + c$, $i = 1, 2$ are well defined, where $c > 0$ is chosen large enough. A particular difference between F_1 and F_2 lies in *accretiveness*. In fact, it is easily shown that F_2 or its right shift $F_{2c} = F_2 + c$, if necessary, is m-accretive, while F_1 is not. Thus, different approaches are necessary for F_1 and F_2. The boundary conditions $\tau_i^f u = 0$ enter the domains, if $\theta \geqslant 1/4$ in the case I, and $\theta \geqslant 3/4$ in the case II.

Let $\zeta(x)$, $x \in \mathbb{R}^m$, denote the distance from x to the boundary Γ. It is assumed that $\sigma(\xi)$ in τ_2 has a suitable smooth extension to $\overline{\Omega}$. In view of (2.10), Chapter 2, let R_1 and R_2 denote non-unique prolongation operators from functions on Γ to functions on $\overline{\Omega}$ such that

$$R_1 \in \mathscr{L}(H^{3/2}(\Gamma);\ H^2(\Omega)); \quad R_1 g|_\Gamma = g, \quad \left. \frac{\partial}{\partial \nu} R_1 g \right|_\Gamma = 0, \quad \forall g \in H^{3/2}(\Gamma),$$

and

$$R_2 \in \mathscr{L}(H^{1/2}(\Gamma);\ H^2(\Omega)); \quad R_2 g|_\Gamma = 0, \quad \left. \frac{\partial}{\partial \nu} R_2 g \right|_\Gamma = g, \quad \forall g \in H^{1/2}(\Gamma),$$

respectively. The following results show a complete characterization of the domains of fractional powers F_{ic}^θ for $0 \leqslant \theta \leqslant 1$, where $c > 0$ is chosen large enough, so that $\sigma(F_{ic}) \subset \mathbb{C}_+$. What is necessary in Chapter 5 is, however, a domain domain where no such boundary condition enters. Thus, we limit ourselves to the proof of $\mathscr{D}(F_{1c}^{\theta})$ and $\mathscr{D}(F_{2c}^{\theta})$ for $0 \leqslant \theta < 1/4$ and $0 \leqslant \theta < 3/4$, respectively. The readers refer to [41] for a complete proof.

Theorem 7.1. (*Case* I. *The Dirichlet boundary condition*). *Suppose that* $w_k \in H^{2\varepsilon}(\Omega)$, $\varepsilon > 0$, *and* $g_k \in H^{3/2}(\Gamma)$, $1 \leqslant k \leqslant M$. *Then,*

(i) $\mathscr{D}(F_{1c}^{\theta}) = H^{2\theta}(\Omega)$, $0 \leqslant \theta < \frac{1}{4}$;

(ii) $\mathscr{D}(F_{1c}^{1/4}) = \left\{ u \in H^{1/2}(\Omega);\ \int_{\Omega} \frac{1}{\zeta(x)} \left| u - \sum_{k=1}^{M} \langle u, w_k \rangle R_1 g_k \right|^2 dx < \infty \right\}$;

and

(iii) $\mathscr{D}(F_{1c}^{\theta}) = \left\{ u \in H^{2\theta}(\Omega);\ \tau_1^f u = 0 \text{ on } \Gamma \right\}$, $\frac{1}{4} < \theta \leqslant 1$.

Moreover, we have the interpolation relation

$$\mathscr{D}(F_{1c}^{\theta}) = \left[\mathscr{D}(F_1), L^2(\Omega) \right]_{1-\theta}, \quad 0 \leqslant \theta \leqslant 1. \tag{7.10}$$

(*Case* II. *The Robin boundary condition*). *Suppose that* $w_k \in L^2(\Omega)$, *and* $g_k \in H^{1/2}(\Gamma)$, $1 \leqslant k \leqslant M$. *Then,*

(i) $\mathscr{D}(F_{2c}^{\theta}) = H^{2\theta}(\Omega)$, $0 \leqslant \theta < \frac{3}{4}$;

(ii) $\mathscr{D}(F_{2c}^{3/4})$
$= \left\{ u \in H^{3/2}(\Omega);\ \int_{\Omega} \frac{1}{\zeta(x)} \left| \tau_{\Omega} u - \sum_{k=1}^{M} \langle u, w_k \rangle \tau_{\Omega} R_2 g_k \right|^2 dx < \infty \right\}$;

and

(iii) $\mathscr{D}(F_{2c}^{\theta}) = \left\{ u \in H^{2\theta}(\Omega);\ \tau_2^f u = 0 \text{ on } \Gamma \right\}$, $\frac{3}{4} < \theta \leqslant 1$.

where $\tau_{\Omega} u = \dfrac{\partial u}{\partial \zeta} + \sigma(x) u$.

Remark: In the case II, the above results (i) – (iii) also hold, when the boundary operator τ_2^f in (7.9) is replaced by

$$\tau_2^f u = \tau_2 u - \sum_{k=1}^{M} \langle u, w_k \rangle_{\Gamma}\, g_k,$$

where $w_k \in L^2(\Gamma)$, $1 \leqslant k \leqslant M$ (see [41] for the proof).

Proof. (Case I) The proof is divided into several steps.

First step (*Operator* T_1). A difficulty is that F_1 is *no more* an accretive operator. So, our strategy is to introduce, instead, another operator K defined below in the second step. Let T_1 be the operator defined by

$$v = T_1 u = u - \sum_{k=1}^{M} \langle u, w_k \rangle R_1 g_k. \tag{7.11}$$

Operator T_1 clearly belongs to $\mathscr{L}(L^2(\Omega); L^2(\Omega)) \cap \mathscr{L}(\mathscr{D}(F_1); \mathscr{D}(L_1))$, where both $\mathscr{D}(F_1)$ and $\mathscr{D}(L_1)$ are equipped with the graph topology. Let us examine its inverse T_1^{-1}. Set $T_1 u = 0$. Then $\langle u, w_j \rangle = \sum_{k=1}^{M} \langle u, w_k \rangle \langle R_1 g_k, w_j \rangle$, $1 \leqslant j \leqslant M$, or $\langle u, \mathbf{w} \rangle = \Phi \langle u, \mathbf{w} \rangle$, where Φ is an $M \times M$ matrix defined by

$$\Phi = \left(\langle R_1 g_k, w_j \rangle ; \begin{array}{ccc} j & \downarrow & 1, \dots, M \\ k & \rightarrow & 1, \dots, M \end{array} \right).$$

Since R_1 admits a great deal of freedom of choice, we may assume with no loss of generality that $\det(1 - \Phi) \neq 0$. Thus, we see that $\langle u, \mathbf{w} \rangle = \mathbf{0}$, or $u = 0$. Thus T_1 is injective, and the inverse T_1^{-1} is given by

$$u = T_1^{-1} v = v + \sum_{k=1}^{M} \left((1 - \Phi)^{-1} \langle v, \mathbf{w} \rangle \right)_k R_1 g_k. \tag{7.12}$$

Note that $\langle u, \mathbf{w} \rangle = (1 - \Phi)^{-1} \langle v, \mathbf{w} \rangle$. Thus T_1^{-1} maps $\mathscr{D}(L_1)$ onto $\mathscr{D}(F_1)$ and belongs to $\mathscr{L}(\mathscr{D}(L_1); \mathscr{D}(F_1))$. The well known interpolation theory [32] implies that

$$\begin{aligned} &T_1 \in \mathscr{L}\left([\mathscr{D}(F_1), L^2(\Omega)]_{1-\theta}; \mathscr{D}(L_{1c}^{\theta}) \right), \quad \text{and} \\ &T_1^{-1} \in \mathscr{L}\left(\mathscr{D}(L_{1c}^{\theta}); [\mathscr{D}(F_1), L^2(\Omega)]_{1-\theta} \right), \quad 0 \leqslant \theta \leqslant 1. \end{aligned} \tag{7.13}$$

Here we have used the fact that $[\mathscr{D}(L_1), L^2(\Omega)]_{1-\theta}$ is equal to $\mathscr{D}(L_{1c}^{\theta})$ due to the m-accretiveness of L_{1c}.

Second step (*Operator K*). Let us introduce the operator K as

$$K = T_1 F_1 T_1^{-1}, \quad \mathscr{D}(K) = \mathscr{D}(L_1) = H^2(\Omega) \cap H_0^1(\Omega). \tag{7.14}$$

The operator K plays a role of connecting F_1 with L_1 (see the diagram at the end of the Second step). If λ is in $\rho(F_1)$, then $\lambda - K$ has a bounded inverse, and

$$(\lambda - K)^{-1} = T_1 (\lambda - F_1)^{-1} T_1^{-1} \in \mathscr{L}(L^2(\Omega)).$$

In view of the decay estimate (3.35), the sector $\overline{\Sigma} - a$ is contained in $\rho(K)$ and

$$\left\| (\lambda - K)^{-1} \right\| \leqslant \frac{\text{const}}{1 + |\lambda|}, \quad \lambda \in \overline{\Sigma} - a.$$

Thus, if c is large enough, fractional powers of $K_c = K + c$ are well defined. The operator $K_c^{-\theta}$ is by definition calculated as follows:

$$\begin{aligned} K_c^{-\theta} &= \frac{-1}{2\pi \mathrm{i}} \int_C \lambda^{-\theta} (\lambda - K_c)^{-1} \, d\lambda = \frac{-1}{2\pi \mathrm{i}} \int_C \lambda^{-\theta} T_1 (\lambda - F_{1c})^{-1} T_1^{-1} \, d\lambda \\ &= T_1 F_{1c}^{-\theta} T_1^{-1}, \quad \theta \geqslant 0, \end{aligned} \tag{7.15}$$

where C denotes the boundary of a suitable right-shift of the sector $\overline{\Sigma}$, oriented according to increasing $\operatorname{Im} \lambda$. The operator K enjoys nice properties. For example, relation (7.15) immediately implies that

$$\begin{aligned} T_1 &\in \mathscr{L}\left(\mathscr{D}(F_{1c}^{\theta}); \mathscr{D}(K_c^{\theta})\right) \quad \text{and} \\ T_1^{-1} &\in \mathscr{L}\left(\mathscr{D}(K_c^{\theta}); \mathscr{D}(F_{1c}^{\theta})\right), \quad 0 \leqslant \theta \leqslant 1. \end{aligned} \tag{7.16}$$

The following proposition forms a key result of the theorem.

Proposition 7.2. (i) *If c is large enough, the equivalence relation*

$$\mathscr{D}\left(K_c^{\theta}\right) = \mathscr{D}\left(L_{1c}^{\theta}\right), \quad 0 \leqslant \theta \leqslant 1 \tag{7.17}$$

holds algebraically and topologically.

Proof. Let us find a concrete form of the operator $K = T_1 F_1 T_1^{-1}$. By the assumption, w_k belong to $H^{2\varepsilon}(\Omega) = \mathscr{D}\left(L_{1c}^{*\,\varepsilon}\right)$. Then K is written as

$$\begin{aligned} Ku &= L_1 u - \sum_{k=1}^{M} \langle L_1 u, w_k \rangle \, R_1 g_k + \sum_{k=1}^{M} \left((1-\Phi)^{-1} \langle u, \mathbf{w} \rangle\right)_k T_1 \mathscr{L} R_1 g_k \\ &= L_1 u - \sum_{k=1}^{M} \left\langle L_{1c}^{1-\varepsilon} u, L_{1c}^{*\,1-\varepsilon} w_k \right\rangle R_1 g_k \\ &\quad + \sum_{k=1}^{M} \left((1-\Phi)^{-1} \langle u, \mathbf{w} \rangle\right)_k T_1 \mathscr{L} R_1 g_k + c \sum_{k=1}^{M} \langle u, w_k \rangle \, R_1 g_k \\ &= L_1 u + Du, \quad u \in \mathscr{D}(K). \end{aligned}$$

The operator D is subordinate to $L_{1c}^{1-\varepsilon}$, that is, $\|Du\| \leqslant \text{const} \left\|L_{1c}^{1-\varepsilon} u\right\|$ for $u \in \mathscr{D}(L_{1c}^{1-\varepsilon})$. Since $\mathscr{D}(K_c)$ is equal to $\mathscr{D}(L_{1c})$ anyway, we see that the relations

$$\mathscr{D}(K_c^{\beta}) \subset \mathscr{D}(L_{1c}^{\alpha}), \quad \text{and} \quad \mathscr{D}(L_{1c}^{\beta}) \subset \mathscr{D}(K_c^{\alpha}), \quad 0 \leqslant \alpha < \beta \tag{7.18}$$

hold in general algebraically and topologically [26]. Note that

$$\begin{aligned} K_c^{-\omega} - L_{1c}^{-\omega} &= \frac{-1}{2\pi \mathrm{i}} \int_C \lambda^{-\omega} (\lambda - K_c)^{-1} D (\lambda - L_{1c})^{-1} \, d\lambda \\ &= \frac{-1}{2\pi \mathrm{i}} \int_C \lambda^{-\omega} (\lambda - L_{1c})^{-1} D (\lambda - K_c)^{-1} \, d\lambda, \quad 0 \leqslant \omega \leqslant 1, \end{aligned}$$

where C denotes a contour of a suitable right-shift of $\partial\overline{\Sigma}$ oriented according to increasing Im λ. For any given $u \in \mathscr{D}(K_c^\theta)$, $0 \leqslant \theta \leqslant 1$, set $\varphi = K_c^\theta u \in L^2(\Omega)$. Thus,

$$K_c^{-\theta}\varphi = L_{1c}^{-\theta}\varphi - \frac{1}{\pi \mathrm{i}}\int_C \lambda^{-\theta}(\lambda - L_{1c})^{-1}D(\lambda - K_c)^{-1}\varphi\, d\lambda.$$

According to (7.18) and the moment inequality for K_c, the integrand is estimated as follows:

$$\begin{aligned}\left\|D(\lambda - K_c)^{-1}\varphi\right\| &\leqslant \text{const}\left\|L_{1c}^{1-\varepsilon}(\lambda - K_c)^{-1}\varphi\right\| \\ &\leqslant \text{const}\left\|L_{1c}^{1-\varepsilon}K_c^{-\eta}\right\|\left\|K_c^{\eta}(\lambda - K_c)^{-1}\varphi\right\| \\ &\leqslant \frac{\text{const}}{(1+|\lambda|)^{1-\eta}}\left\|\varphi\right\|,\end{aligned}$$

where $1-\varepsilon < \eta < 1$. Thus we see that

$$\left\|\lambda^{-\theta}L_{1c}^{\theta}\lambda^{-\theta}(\lambda - L_{1c})^{-1}D(\lambda - K_c)^{-1}\varphi\right\| \leqslant \frac{\text{const}}{(1+|\lambda|)^{2-\eta}}\left\|\varphi\right\|,$$

the last term of which is integrable on C. This means that $\mathscr{D}(K_c^\theta)$ is contained in $\mathscr{D}(L_{1c}^\theta)$, and that

$$\begin{aligned}L_{1c}^{\theta}u &= \varphi - \frac{1}{2\pi \mathrm{i}}\int_C \lambda^{-\theta}L_{1c}^{\theta}(\lambda - L_{1c})^{-1}D(\lambda - K_c)^{-1}\varphi\, d\lambda, \\ \text{and} \quad &\left\|L_{1c}^{\theta}u\right\| \leqslant \text{const}\left\|\varphi\right\| = \text{const}\left\|K_c^{\theta}u\right\|.\end{aligned}$$

The converse relation,

$$\mathscr{D}\left(L_{1c}^{\theta}\right) \subset \mathscr{D}\left(K_c^{\theta}\right) \quad \text{and} \quad \left\|K_c^{\theta}u\right\| \leqslant \text{const}\left\|L_{1c}^{\theta}u\right\|$$

is similarly proven. □

By (7.16) and (7.17), we obtain the assertion:

Proposition 7.3. *The operator T_1 is a continuous bijection from $\mathscr{D}\left(F_{1c}^{\theta}\right)$ onto $\mathscr{D}\left(L_{1c}^{\theta}\right)$ for each $0 \leqslant \theta \leqslant 1$, and thus,*

$$\begin{aligned}T_1 &\in \mathscr{L}\left(\mathscr{D}(F_{1c}^{\theta});\, \mathscr{D}(L_{1c}^{\theta})\right) \quad \text{and} \\ T_1^{-1} &\in \mathscr{L}\left(\mathscr{D}(L_{1c}^{\theta});\, \mathscr{D}(F_{1c}^{\theta})\right), \quad 0 \leqslant \theta \leqslant 1.\end{aligned} \tag{7.19}$$

Although the m-accretiveness of M_{1c} is *never* expected and thus a generalization of Heinz's inequality [25] cannot be applied, relation (7.13) combined with Proposition 7.3 yields the last assertion (7.10) of the theorem:

$\mathscr{D}(F_{1c}^{\theta}) = [\mathscr{D}(F_1), L^2(\Omega)]_{1-\theta}$, $0 \leqslant \theta \leqslant 1$. The above relations are are summarized as the following diagram:

$$[\mathscr{D}(F_1); L^2(\Omega)]_{1-\theta} \underset{T_1^{-1}}{\overset{T_1}{\rightleftarrows}} \mathscr{D}(L_{1c}^{\theta}) = \mathscr{D}(K_c^{\theta}) \underset{T_1}{\overset{T_1^{-1}}{\rightleftarrows}} \mathscr{D}(F_{1c}^{\theta})$$

Third step (*Proof of* (i)). Both operators T_1 and T_1^{-1} belong to $\mathscr{L}(H^{2\theta}(\Omega))$, $0 \leqslant \theta \leqslant 1$. Since T_1 also belongs to $\mathscr{L}(\mathscr{D}(F_{1c}^{\theta});\ H^{2\theta}(\Omega))$ and T_1^{-1} to $\mathscr{L}(H^{2\theta}(\Omega);\ \mathscr{D}(F_{1c}^{\omega}))$, $0 \leqslant \theta < 1/4$, by Proposition 7.3, the assertion of (i) is now immediate.

(Case II) The proof is somewhat simpler than in the proof of Case I, since the operator F_{2c} is m-accretive in our case. An operator similar to T_1 appears later in the third step. In order to apply this operator, however, we must introduce the operator $L_{2c} - F_2$ similar to M_{2c} in the first step.

First step (*Operator* $L_2 - E_2$). The differential equation

$$\frac{du}{dt} + F_2 u = 0, \quad u(0) = u_0 \in L^2(\Omega) \tag{7.20}$$

is well posed in $L^2(\Omega)$, and generates an analytic semigroup, e^{-tF_2}, $t > 0$. As in (6.3), by setting $x(t) = L_{2c}^{-\theta} u(t)$, $1/4 < \theta < 3/4$, the equation is transformed into

$$\begin{aligned} &\frac{dx}{dt} + (L_2 - E_2)x = 0, \quad x(0) = x_0 = L_{2c}^{-\theta} u_0, \\ \text{where} \quad &E_2 x = \sum_{k=1}^{M} \left\langle L_{2c}^{\theta} x, w_k \right\rangle L_{2c}^{1-\theta} N_{-c} g_k, \quad \mathscr{D}(L_2) \subset \mathscr{D}(E_2). \end{aligned} \tag{7.21}$$

Since θ is smaller than 3/4, the following lemma is immediate:

Lemma 7.4. *The operator* $L_2 - E_2$ *has a compact resolvent. There is an* $a > 0$ *such that* $\overline{\Sigma} - a$ *is contained in* $\rho(L_2 - E_2)$, *and that*

$$\|(\lambda - L_2 + E_2)^{-1}\| \leqslant \frac{\text{const}}{1 + |\lambda|}, \quad \lambda \in \overline{\Sigma} - a.$$

Since the problem (7.21) generates an analytic semigroup $e^{-t(L_2 - E_2)}$, $t > 0$, we see that, for $u_0 \in L^2(\Omega)$ and $\operatorname{Re}\lambda < -a$

$$\begin{aligned} (\lambda - L_2 + E_2)^{-1} x_0 &= -\int_0^{\infty} e^{\lambda t} e^{-t(L_2 - E_2)} x_0 \, dt \qquad (x_0 = L_{2c}^{-\theta} u_0) \\ &= -\int_0^{\infty} e^{\lambda t} L_{2c}^{-\theta} e^{-tF_2} u_0 \, dt = L_{2c}^{-\theta} (\lambda - F_2)^{-1} u_0, \end{aligned}$$

and thus

$$(\lambda - F_2)^{-1} = L_{2c}^{\theta} (\lambda - L_2 + E_2)^{-1} L_{2c}^{-\theta}, \quad \operatorname{Re}\lambda < -a. \tag{7.22}$$

The right-hand side of (7.22) is analytic in $\lambda \in \rho(L_2 - E_2)$. Thus, $(\lambda - F_2)^{-1}$

has an extension to an operator analytic in $\lambda \in \rho(L_2 - E_2)$. The extension is, however, nothing but the resolvent of F_2 [15, Part 2]. This shows that $\rho(L_2 - E_2)$ is contained in $\rho(F_2)$ and that (7.22) holds for $\lambda \in \rho(L_2 - E_2)$.

Second Step (*Proof of* (i)). Choose a $c > 0$ large enough so that fractional powers for F_{2c} and $L_{2c} - E_2$ are well defined. According to (7.22), we observe the following relation as

$$\begin{aligned} L_{2c}^{-\theta} F_{2c}^{-\theta} &= \frac{-1}{2\pi \mathrm{i}} \int_C \lambda^{-\theta} L_{2c}^{-\theta} (\lambda - F_{2c})^{-1} \, d\lambda \\ &= \frac{-1}{2\pi \mathrm{i}} \int_C \lambda^{-\theta} (\lambda - L_{2c} + E_2)^{-1} L_{2c}^{-\theta} \, d\lambda \\ &= (L_{2c} - E_2)^{-\theta} L_{2c}^{-\theta}, \end{aligned}$$

where C denotes a contour of a suitable right shift of $\partial \overline{\Sigma}$. Thus,

$$F_{2c}^{-\theta} = L_{2c}^{\theta} (L_{2c} - F_2)^{-\theta} L_{2c}^{-\theta}. \tag{7.23}$$

We need to characterize the domain of $(L_{2c} - E_2)^{\theta}$. The operator F_2 is subordinate to L_{2c}^{θ}. But when $\theta \geqslant \frac{1}{2}$, the m-accretiveness of $L_{2c} - E_2$ is not expected. Nevertheless, we have the following result, the proof of which is stated later in the last step.

Proposition 7.5. *The equivalence relation* $\mathscr{D}\big((L_{2c} - E_2)^{\omega}\big) = \mathscr{D}\big(L_{2c}^{\omega}\big)$ *holds for* $0 \leqslant \omega < \frac{3}{4} + \theta$ *algebraically and topologically.*

According to Proposition 7.5, we see that

$$L_{2c}^{\theta} (L_{2c} - E_2)^{\theta} L_{2c}^{-2\theta} = L_{2c}^{\theta} (L_{2c} - E_2)^{-\theta} (L_{2c} - E_2)^{2\theta} L_{2c}^{-2\theta} \in \mathscr{L}(L^2(\Omega)),$$

since 2θ is smaller than $\frac{3}{4} + \theta$. Thus the relation (7.23) implies that, for any $u \in \mathscr{D}\big(L_{2c}^{\theta}\big)$,

$$F_{2c}^{-\theta} \big(L_{2c}^{\theta} (L_{2c} - E_2)^{\theta} L_{2c}^{-\theta} u\big) = u, \quad \text{or} \quad F_{2c}^{\theta} u = L_{2c}^{\theta} (L_{2c} - E_2)^{\theta} L_{2c}^{-\theta} u,$$

which shows that $\mathscr{D}\big(L_{2c}^{\theta}\big)$ is contained in $\mathscr{D}\big(F_{2c}^{\theta}\big)$, and that

$$\left\| F_{2c}^{\theta} u \right\| \leqslant \text{const} \left\| L_{2c}^{\theta} u \right\|, \quad u \in \mathscr{D}(L_{2c}^{\theta}).$$

As for the converse relation, set $x = F_{2c}^{\theta} u$ for $u \in \mathscr{D}\big(F_{2c}^{\theta}\big)$. Then,

$$\begin{aligned} u &= L_{2c}^{\theta} (L_{2c} - E_2)^{-\theta} L_{2c}^{-\theta} x \\ &= L_{2c}^{-\theta} L_{2c}^{2\theta} (L_{2c} - E_2)^{-2\theta} (L_{2c} - E_2)^{\theta} L_{2c}^{-\theta} x \in \mathscr{D}(L_{2c}^{\theta}), \end{aligned}$$

which shows that $\mathscr{D}\big(F_{2c}^{\theta}\big)$ is contained in $\mathscr{D}\big(L_{2c}^{\theta}\big)$, and that

$$\|L_{2c}^{\theta} u\| \leqslant \text{const} \, \|F_{2c}^{\theta} u\|, \quad u \in \mathscr{D}(F_{2c}^{\theta}).$$

Therefore, we have shown that

$$\mathscr{D}\big(F_{2c}^{\theta}\big) = \mathscr{D}\big(L_{2c}^{\theta}\big), \quad \frac{1}{4} < \theta < \frac{3}{4}$$

with equivalent graph norms. We note that, since both F_{2c} and L_{2c} are m-accretive, the same is true for F_{2c}^{θ} and L_{2c}^{θ}. For a fixed θ, $1/4 < \theta < 3/4$, a generalization of Heinz's inequality [25] is applied to F_{2c}^{θ} and L_{2c}^{θ} to derive that

$$\mathscr{D}\big(F_{2c}^{\omega}\big) = \mathscr{D}\big((F_{2c}^{\theta})^{\omega/\theta}\big) = \mathscr{D}\big((L_{2c}^{\theta})^{\omega/\theta}\big) = \mathscr{D}\big(L_{2c}^{\omega}\big), \quad 0 \leqslant \omega \leqslant \theta$$

with equivalent graph norms, which proves (i) of the theorem.

Last Step. (*Proof of Proposition* 7.5). Since $\mathscr{D}(L_2) = \mathscr{D}(L_2 - F_2)$, we see that the relations

$$\mathscr{D}\big((L_{2c} - E_2)^{\beta}\big) \subset \mathscr{D}\big(L_{2c}^{\alpha}\big), \quad \text{and} \quad \mathscr{D}\big(L_{2c}^{\beta}\big) \subset \mathscr{D}\big((L_{2c} - E_2)^{\alpha}\big), \quad 0 \leqslant \alpha < \beta$$

hold algebraically and topologically [26]. So, just as in the proof of Proposition 7.2, we are able to show that

$$\mathscr{D}\big((L_{2c} - E_2)^{\omega}\big) = \mathscr{D}\big(L_{2c}^{\omega}\big), \quad 0 \leqslant \omega \leqslant 1.$$

The equivalence relation for ω, $1 < \omega < 3/4 + \theta$ is proven as follows: Take any $u \in \mathscr{D}(L_{2c}^{1+\kappa})$, $0 < \kappa < \theta - 1/4$. In view of the relation

$$(L_{2c} - E_2)u = L_{2c}u - \sum_{k=1}^{M} \left\langle L_{2c}^{\theta}u, w_k \right\rangle L_{2c}^{1-\theta} N_{-c} g_k,$$

both terms of the right-hand side belong to $\mathscr{D}\big(L_{2c}^{\kappa}\big) = \mathscr{D}\big((L_{2c} - E_2)^{\kappa}\big)$, since $1 - \theta + \kappa < 3/4$. Thus, u belongs to $\mathscr{D}((L_{2c} - E_2)^{1+\kappa})$. Moreover,

$$\left\| (L_{2c} - E_2)^{1+\kappa} u \right\| \leqslant \text{const} \left\| L_{2c}^{1+\kappa} u \right\|, \quad u \in \mathscr{D}(L_{2c}^{1+\kappa}).$$

The converse inclusion relation is similarly proven. This finishes the proof of the proposition. The proof of Theorem 7.1 is thereby complete (except for (ii) and (iii) of the two cases I and II). □

Chapter 5

Stabilization of linear systems with Riesz Bases: Dynamic feedback

5.1 Introduction

In Chapter 4, stabilization problems are successfully solved by various approaches for a general class of parabolic boundary control systems, in which *no* Riesz basis associated with them is assumed. We introduce in this chapter another type of dynamic compensators, namely, *identity* compensators, when a control system admits a Riesz basis. A merit is that a suitable sequence of finite-dimensional linear control systems reasonably approximates the original control system through the Riesz basis. In this case, identity compensators are a useful means for an explicit construction of a feedback scheme. In constructing feedback control schemes, we propose two different frameworks, which finally turn out to be algebraically connected with each other. The results of this chapter are based on those discussed in [37, 42, 44, 49].

There are various elliptic operators admitting Riesz bases. As a typical example among them, we study in this chapter self-adjoint operators with compact resolvent. Of course, the arguments here can be generalized in a straightforward manner to other such control systems with a little technical changes. Let $(\mathscr{L}, \tau)$ be a pair of differential operators defined as

$$\mathscr{L}u = -\sum_{i,j=1}^{m} \frac{\partial}{\partial x_i}\left(a_{ij}(x)\frac{\partial u}{\partial x_j}\right) + c(x)u,$$
$$\tau u = \alpha(\xi)u + (1-\alpha(\xi))\frac{\partial u}{\partial \nu}. \tag{1.1}$$

Here the assumptions on the coefficients and their regularities are the same as in the pair $(\mathscr{L}, \tau)$ in (1.1), Chapter 4, that is, $a_{ij}(x) = a_{ji}(x)$ for $1 \leqslant i,\ j \leqslant m$, $x \in \overline{\Omega}$; for some positive δ,

$$\sum_{i,j=1}^{m} a_{ij}(x)\xi_i\xi_j \geqslant \delta|\xi|^2, \quad \forall \xi = (\xi_1, \ldots, \xi_m) \in \mathbb{R}^m, \quad \forall x \in \overline{\Omega};$$

and

$$0 \leqslant \alpha(\xi) \leqslant 1, \qquad \frac{\partial u}{\partial \nu} = \sum_{i,j=1}^{m} a_{ij}(\xi)\nu_i(\xi)\left.\frac{\partial u}{\partial x_j}\right|_{\Gamma},$$

$\boldsymbol{\nu}(\xi) = (\nu_1(\xi), \ldots, \nu_m(\xi))$ being the unit outer normal at each point $\xi \in \Gamma$. Thus, the pair is the same as the one defined in (1.1), Chapter 4, except for nonexistence of the terms $\sum_{i=1}^{m} b_i(x)\dfrac{\partial u}{\partial x_i}$. All arguments are based on the $L^2(\Omega)$-framework.

As in Chapter 4, let $\hat{L}$ be the closable operator defined by

$$\hat{L}u = \mathscr{L}u, \quad u \in \mathscr{D}(\hat{L}),$$
$$\mathscr{D}(\hat{L}) = \left\{u \in C^2(\Omega) \cap C^1(\overline{\Omega});\ \mathscr{L}u \in L^2(\Omega),\ \tau u = 0\right\}.$$

The closure of $\hat{L}$ in $L^2(\Omega)$ is denoted by L. The domain $\mathscr{D}(L)$ consists of all $u \in L^2(\Omega)$ with the following properties: (i) There is a sequence $\{u_n\} \subset \mathscr{D}(\hat{L})$ such that $u_n \to u$ in $L^2(\Omega)$, and (ii) $\hat{L}u_n$ converges in $L^2(\Omega)$ as $n \to \infty$. It is well known [24] that L is a self-adjoint operator with compact resolvent $(\lambda - L)^{-1}$. By the Hilbert-Schmidt theorem, the spectrum $\sigma(L)$ consists only of real eigenvalues with finite multiplicities. Thus there is a set of eigenpairs $\{\lambda_i,\ \varphi_{ij}\}$ such that

(i) $\sigma(L) = \{\lambda_1, \lambda_2, \ldots, \lambda_i, \ldots\}, \quad \lambda_1 < \lambda_2 < \cdots < \lambda_i < \cdots \to \infty;$

(ii) $(\lambda_i - L)\varphi_{ij} = 0, \quad i \geqslant 1,\ 1 \leqslant j \leqslant m_i\,(< \infty)$; and

(iii) the system $\{\varphi_{ij}\}$ forms an *orthonormal* basis for $L^2(\Omega)$.

There arises *no* generalized eigenfunction in the present L, so that the spectral structure of L becomes rather simple. The set of eigenfunctions $\{\psi_{ij}\}$ for $L^* = L$ is simply replaced by $\{\varphi_{ij}\}$. Each function $u \in L^2(\Omega)$ has a unique Fourier series expansion,

$$u = \sum_{i=1}^{\infty}\sum_{j=1}^{m_i} u_{ij}\varphi_{ij}, \quad u_{ij} = \langle u, \varphi_{ij}\rangle. \tag{1.2$_1$}$$

The resolvent $(\lambda - L)^{-1}$ is expressed as

$$(\lambda - L)^{-1}u = \sum_{i,j} \frac{u_{ij}}{\lambda - \lambda_i} \varphi_{ij}, \quad \lambda \in \rho(L). \tag{1.2$_2$}$$

As in Chapter 4, our aim is to construct a stabilizing boundary feedback control system which will be finally described by the following system of linear differential equations for (u, v):

$$\begin{cases} \dfrac{\partial u}{\partial t} + \mathscr{L}u = 0 \quad \text{in } \mathbb{R}^1_+ \times \Omega, \\ \tau u = \displaystyle\sum_{k=1}^{M} \langle v, \rho_k \rangle_{\mathbb{R}^\ell} g_k \quad \text{on } \mathbb{R}^1_+ \times \Gamma, \\ \dfrac{dv}{dt} + B_1 v = \displaystyle\sum_{k=1}^{N} p_k(u)\xi_k \quad \text{in } \mathbb{R}^1_+, \\ u(0,\cdot) = u_0(\cdot) \quad \text{in } \Omega, \qquad v(0) = v_0. \end{cases} \tag{1.3}$$

Here, $u(t,\cdot)$ is the state of the controlled plant Σ_p, and $p_k(u)$, $1 \leqslant k \leqslant N$, outputs of Σ_p defined as

$$p_k(u) = \begin{cases} \langle u, w_k \rangle_\Gamma, & \text{in the case where } \alpha(\xi) \not\equiv 1, \quad 1 \leqslant k \leqslant N, \\ \left\langle \dfrac{\partial u}{\partial \nu}, w_k \right\rangle_\Gamma, & \text{in the case where } \alpha(\xi) \equiv 1, \quad 1 \leqslant k \leqslant N, \end{cases} \tag{1.4}$$

where $w_k \in L^2(\Gamma)$. The functions $g_k \in C^{2+\omega}(\Gamma)$, $1 \leqslant k \leqslant M$, are actuators of Σ_p. The differential equation for v in $\mathbb{R}^\ell$ means the compensator Σ_c, where B_1 is an $\ell \times \ell$ coefficient matrix; ρ_k sensors in $\mathbb{R}^\ell$; and ξ_k actuators in $\mathbb{R}^\ell$.

By assuming that

$$\lambda_1 \leqslant 0, \quad \text{and } 0 < \lambda_{\nu+1}, \tag{1.5}$$

the stabilization problem for (1.3) is the same as in Chapter 4. The difference is, however, that a conceptual compensator of infinite dimension is constructed not in a general separable Hilbert space H, but in *the same space as in the controlled plant*, that is, $L^2(\Omega)$. In the following sections, two different models of identity compensators are proposed. We do not need much preparation in describing the corrresponding differential equations in $L^2(\Omega) \times L^2(\Omega)$.

5.2 Boundary Control Systems

The purpose of this section is to formulate the boundary control system (2.1) below which is finally reduced to (1.3). As in Chapter 4, the state $u(t,\cdot)$ of the controlled plant Σ_p always stays in $L^2(\Omega)$. By the assumption (1.5), the semigroup e^{-tL} is not exponentially stable without any control. Let us consider the following control system Σ_p in $L^2(\Omega)$ (see (4.5), Chapter 4):

$$\begin{cases} \dfrac{du}{dt} + \mathscr{L}u = 0 \quad \text{in } \mathbb{R}^1_+ \times \Omega, \\ \tau u = \displaystyle\sum_{k=1}^{M} f_k(t) g_k \quad \text{on } \mathbb{R}^1_+ \times \Gamma, \\ u(0,\cdot) = u_0 \quad \text{in } \Omega. \end{cases} \tag{2.1}$$

In (2.1), the inputs $f_k(t)$, $1 \leqslant k \leqslant M$, will be finally constructed as the outputs of the compensator Σ_c in the form, $\langle v(t), \rho_k \rangle$, $v(t)$ being the state of Σ_c. The outputs of the system Σ_p are given by $p_k(u)$, $1 \leqslant k \leqslant N$, in (1.4). Let $c > 0$ be a constant such that $\lambda_1 + c > 0$, and set $L_c = L + c$. Recall that, for each k, $1 \leqslant k \leqslant M$, the function $N_{-c} g_k \in H^2(\Omega)$ means the unique solution to the boundary value problem:

$$\mathscr{L}_c N_{-c} g_k = 0 \quad \text{in } \Omega, \qquad \tau N_{-c} g_k = g_k \quad \text{on } \Gamma.$$

To transform the equation for u with boundary inputs into another equation with distributed inputs, we employ the method in Section 4, Chapter 4: Assuming for a moment that f_k are of class C^1 and setting

$$q(t) = u(t) - \sum_{k=1}^{M} f_k(t) N_{-c} g_k \in \mathscr{D}(L),$$

we obtain the equation for $q(t)$:

$$\begin{aligned} \frac{dq}{dt} + Lq &= \sum_{k=1}^{M} \left(c f_k(t) - f_k'(t) \right) N_{-c} g_k, \\ q(0) = q_0 &= u_0 - \sum_{k=1}^{M} f_k(0) N_{-c} g_k. \end{aligned} \tag{2.2}$$

On the analogy of (4.7), Chapter 4, let us consider the system of differential equations in $L^2(\Omega) \times L^2(\Omega)$:

$$\begin{cases} \dfrac{dq}{dt} + Lq = \displaystyle\sum_{k=1}^{M} \left(c f_k(t) - f_k'(t) \right) N_{-c} g_k, \quad q(0) = q_0 \in L^2(\Omega), \\ \dfrac{dv}{dt} + Bv = -Cq + \displaystyle\sum_{k=1}^{M} \left(c f_k(t) - f_k'(t) \right) N_{-c} g_k, \quad v(0) = v_0 \in L^2(\Omega), \end{cases} \tag{2.3}$$

where $C = -\sum_{k=1}^{N} p_k(\cdot) \xi_k$, and the functions $\xi_k \in L^2(\Omega)$ will be determined later. In Section 2, Chapter 4, the operator B and the actuators ξ_k are given in advance in an arbitrary separable Hilbert space H, and connected with L by Sylvester's equation: $XL - BX = C = -\sum_{k=1}^{N} p_k(\cdot)\xi_k$. Now we set $H = L^2(\Omega)$ and $X = 1$. Then,

$$B = L - C = L + \sum_{k=1}^{N} p_k(\cdot)\xi_k. \tag{2.4}$$

It is clear that $-B$ is an infinitesimal generator of an analytic semigroup, e^{-tB}, $t > 0$. Taking the difference of these two equations, we obtain

$$\frac{d}{dt}(q - v) + B(q - v) = 0, \quad (q - v)(0) = q_0 - v_0 \in L^2(\Omega), \tag{2.5}$$

and thus $q(t) - v(t) = e^{-tB}(q_0 - v_0)$, $t \geqslant 0$. The equation for v in (2.3) is a so called *identity* compensator which plays a role in estimating the state $q(t)$ by $v(t)$ asymptotically as $t \to \infty$. Let us recall the observability condition on the sensors w_k (see the first condition of (3.5), Chapter 4):

$$\text{rank}\begin{pmatrix} \hat{W}_i & \hat{W}_i\Lambda_i & \dots & \hat{W}_i\Lambda_i^{m_i-1} \end{pmatrix}^{\mathrm{T}} = m_i, \quad 1 \leqslant i \leqslant \nu,$$

$$\text{where } \hat{W}_i = \left(p_k(\varphi_{ij}); \begin{array}{ccc} j & \to & 1,\dots,m_i \\ k & \downarrow & 1,\dots,N \end{array} \right).$$

Since L does not admit any generalized eigenfunction of L, the matrices Λ_i are simply $\lambda_i I_{m_i}$. Thus, the observability condition is rewritten as

$$\text{rank}\, \hat{W}_i = m_i, \quad 1 \leqslant i \leqslant \nu. \tag{2.6}$$

As a necessary condition, the number N must be greater than or equal to $\max_{1 \leqslant i \leqslant \nu} m_i$. Assuming (2.6), we then find suitable $\xi_k \in P_\nu L^2(\Omega)$, $1 \leqslant k \leqslant N$, such that

$$\left\| e^{-tB} \right\| = \left\| e^{-t(L-C)} \right\| \leqslant \text{const}\, e^{-\lambda_{\nu+1}t}, \quad t \geqslant 0. \tag{2.7}$$

Here, the decay rate is exactly $\lambda_{\nu+1}$, since there arises no generalized eigenfunction for the eigenvalue $\lambda_{\nu+1}$. Thus, whatever the inputs $f_k(t)$ may be in (2.3), we see that

$$\|q(t) - v(t)\| \leqslant \text{const}\, e^{-\lambda_{\nu+1}t}\left(\|q_0\| + \|v_0\|\right), \quad t \geqslant 0. \tag{2.8}$$

Let F be the closure in $L^2(\Omega)$ of the operator $\hat{F}$ (see (3.14), Chapter 4):

$$\begin{gathered} \hat{F}u = \mathscr{L}u, \quad u \in \mathscr{D}(\hat{F}), \text{ where} \\ \mathscr{D}(\hat{F}) = \left\{ u \in C^2(\Omega) \cap C^1(\overline{\Omega});\ \mathscr{L}u \in L^2(\Omega),\ \tau_f u = 0 \text{ on } \Gamma \right\}, \\ \text{and } \tau_f u = \tau u - \sum_{k=1}^{M} \langle u, y_k \rangle g_k, \quad y_k \in L^2(\Omega). \end{gathered} \tag{2.9}$$

Then, the semigroup e^{-tF}, $t > 0$, is analytic. The controllability condition on the actuators g_k (see the second of (3.5), Chapter 4) is similarly rewritten as

$$\operatorname{rank} \hat{G}_i = m_i, \quad 1 \leqslant i \leqslant \nu,$$
$$\text{where } \hat{G}_i = \left(\langle g_k, \sigma\varphi_{ij} \rangle_\Gamma ; \begin{array}{ccc} j & \downarrow & 1, \ldots, m_i \\ k & \to & 1, \ldots, M \end{array} \right), \tag{2.10}$$

so that the number M must be also greater than or equal to $\max_{1\leqslant i \leqslant \nu} m_i$. Here the operator σ on Γ is defined by (3.2), Chapter 4, and takes the form,

$$\sigma\varphi_{ij} = \varphi_{ij} - \frac{\partial \varphi_{ij}}{\partial \nu}$$

in (2.10). By assuming (2.10), Proposition 3.4, (ii) of Chapter 4 ensures a set of $y_k \in P_\nu L^2(\Omega)$, $1 \leqslant k \leqslant M$, such that

$$\left\| e^{-tF} \right\| \leqslant \text{const}\, e^{-\lambda_{\nu+1} t}, \quad t \geqslant 0. \tag{2.11}$$

At this stage, we set

$$f_i(t) = \langle v(t), \zeta_i \rangle, \quad \zeta_i = \sum_{j=1}^{M} \theta_{ij} y_j \in P_\nu L^2(\Omega), \quad 1 \leqslant i \leqslant M, \tag{2.12}$$

where θ_{ij} denote the parameters to be determined later. By (2.7) and (2.8),

$$\begin{aligned} |f_i(t) - \langle q(t), \zeta_i \rangle| &= \left| f_i(t) - \langle u(t), \zeta_i \rangle + \sum_{k=1}^{M} \langle N_{-c} g_k, \zeta_i \rangle f_k(t) \right| \\ &\leqslant \text{const}\, e^{-\lambda_{\nu+1} t} (\|q_0\| + \|v_0\|), \quad t \geqslant 0, \quad 1 \leqslant i \leqslant M, \end{aligned}$$

or in vector form

$$\left| \boldsymbol{f}(t) - \overline{\Theta} \langle u(t), \boldsymbol{y} \rangle + \overline{\Theta} G_1 \boldsymbol{f}(t) \right| \leqslant \text{const}\, e^{-\lambda_{\nu+1} t} (\|q_0\| + \|v_0\|), \quad t \geqslant 0,$$

where $\boldsymbol{f}(t) = \left(f_1(t) \ldots f_M(t) \right)^{\mathrm{T}}$, and

$$\Theta = \left(\theta_{ij} ; \begin{array}{ccc} i & \downarrow & 1, \ldots, M \\ j & \to & 1, \ldots, M \end{array} \right), \quad \text{and}$$

$$G_1 = \left(\langle N_{-c} g_k, y_j \rangle ; \begin{array}{ccc} j & \downarrow & 1, \ldots, M \\ k & \to & 1, \ldots, M \end{array} \right).$$

As in the proof of Proposition 3.4, Chapter 4, we may assume that $\det(1 - G_1) \neq 0$. Setting $\overline{\Theta} = (1 - G_1)^{-1}$ or $(1 + \overline{\Theta} G_1)^{-1} \overline{\Theta} = 1$, we see that

$$|\boldsymbol{f}(t) - \langle u(t), \boldsymbol{y} \rangle| \leqslant \text{const}\, e^{-\lambda_{\nu+1} t} (\|q_0\| + \|v_0\|), \quad t \geqslant 0. \tag{2.13}$$

Similar calculations show that

$$|\boldsymbol{f}'(t) - \langle u_t(t), \boldsymbol{y}\rangle| \leqslant \begin{cases} \text{const}\, t^{-1/4-\varepsilon} e^{-\lambda_{\nu+1}t} \left(\|q_0\| + \|v_0\|\right), \quad t>0, \\ \qquad \text{in the case where } \alpha(\xi) \not\equiv 1, \\ \text{const}\, t^{-3/4-\varepsilon} e^{-\lambda_{\nu+1}t} \left(\|q_0\| + \|v_0\|\right), \quad t>0, \\ \qquad \text{in the case where } \alpha(\xi) \not\equiv 1. \end{cases} \tag{2.13'}$$

In fact, in the relation:

$$\begin{aligned} f_i'(t) - \langle q_t(t), \zeta_i\rangle &= \langle (v-q)_t, \zeta_i\rangle = -\big\langle (L-C)e^{-t(L-C)}(v_0-q_0), \zeta_i\big\rangle \\ &= -\big\langle e^{-t(L-C)}(v_0-q_0), L\zeta_i\big\rangle \\ &\quad - \sum_k p_k\big(e^{-t(L-C)}(v_0-q_0)\big)\, \langle \xi_k, \zeta_i\rangle, \end{aligned}$$

the first term has no problem. In the second term, recall that

$$p_k\big(e^{-t(L-C)}(v_0-q_0)\big) = \begin{cases} \left\langle e^{-t(L-C)}(v_0-q_0), w_k\right\rangle_\Gamma, & \text{in the case where } \alpha(\xi) \not\equiv 1, \\ \left\langle \dfrac{\partial}{\partial \nu} e^{-t(L-C)}(v_0-q_0), w_k\right\rangle_\Gamma, & \text{in the case where } \alpha(\xi) \equiv 1, \end{cases}$$

In the case where $\alpha(\xi) \not\equiv 1$, for example, note that $\mathscr{D}\big(L_c^{1/2}\big) = H_\alpha^1(\Omega) \subset H^1(\Omega)$, so that $\mathscr{D}\big(L_c^{\omega/2}\big) \subset H^\omega(\Omega)$, $0 \leqslant \omega \leqslant 1$, by a generalization of Heinz's inequality [25] (see (3.27), (3.30), Chapter 2, and (6.11), Chapter 4 for $H_\alpha^1(\Omega)$). Thus, by the trace theorem ((2.9$_1$), Chapter 2, and [32])

$$\begin{aligned} &\left|p_k\big(e^{-t(L-C)}(v_0-q_0)\big)\right| \\ &\leqslant \text{const} \left\|e^{-t(L-C)}(v_0-q_0)\right\|_{H^{1/2+2\varepsilon}(\Omega)} \\ &\leqslant \text{const} \left\|L_c^{1/4+\varepsilon} e^{-t(L-C)}(v_0-q_0)\right\| \\ &\leqslant \text{const} \left\|L_c^{1/4+\varepsilon}(L-C)^{-(1/4+\varepsilon)}(L-C)^{1/4+\varepsilon} e^{-t(L-C)}(v_0-q_0)\right\| \\ &\leqslant \text{const} \left\|(L-C)^{1/4+\varepsilon} e^{-t(L-C)}(v_0-q_0)\right\|, \quad t>0. \end{aligned}$$

A similar evaluation is made in the other case: $\alpha(\xi) \equiv 1$, by noting that $\mathscr{D}(L) = H^2(\Omega) \cap H_0^1(\Omega) \subset H^2(\Omega)$.

Remark: In the case where (i) $0 \leqslant \alpha(\xi) < 1$ or (ii) $\alpha(\xi) \equiv 1$, another approach to (2.13′) is possible. In both cases, the adjoint operator $(L-C)^*$ is given as

$$(L-C)^*\varphi = \mathscr{L}\varphi, \quad \varphi \in \mathscr{D}\big((L-C)^*\big),$$

where

$$\mathscr{D}\big((L-C)^*\big) = \begin{cases} \left\{\varphi \in H^2(\Omega);\ \tau\varphi = -\sum_{k=1}^{N} \langle \varphi, \xi_k\rangle (1-\alpha(\xi))w_k\right\}, & \text{in the case (i),} \\ \text{and} & \\ \left\{\varphi \in H^2(\Omega);\ \varphi\big|_{\Gamma} = \sum_{k=1}^{N} \langle \varphi, \xi_k\rangle w_k\right\}, & \text{in the case (ii).} \end{cases} \tag{2.14}$$

Note that $\sigma(L-C) \subset \mathbb{C}_+$. Thus fractional powers, $(L-C)^\theta$ and $(L-C)^{*\theta}$ are well defined. Then we can show that

$$\mathscr{D}\big((L-C)^{*\theta}\big) = H^{2\theta}(\Omega), \quad 0 \leqslant \theta < \frac{3}{4} \quad \text{in the case (i), and}$$
$$\mathscr{D}\big((L-C)^{*\theta}\big) = H^{2\theta}(\Omega), \quad 0 \leqslant \theta < \frac{1}{4} \quad \text{in the case (ii).}$$

When $\theta \geqslant \frac{3}{4}$ in (i) or $\theta \geqslant \frac{1}{4}$ in (ii), the boundary condition enters respective Sobolev spaces (see Theorem 7.1, Section 7, Chapter 4 for a general nonself-adjoint L). In both cases, ζ_i belong to $H^2(\Omega)$. The estimates in (2.13′) then follow from the expression:

$$\begin{aligned} &\left\langle (L-C)e^{-t(L-C)}(v_0-q_0), \zeta_i\right\rangle \\ &= \left\langle (L-C)^\theta e^{-t(L-C)}(v_0-q_0), (L-C)^{*1-\theta}\zeta_i\right\rangle, \end{aligned}$$

where $\theta = \frac{1}{4}+\varepsilon$ in the case (i), and $\theta = \frac{3}{4}+\varepsilon$ in the case (ii).

Let us go back to the decay estimate of the system. By setting

$$\varphi_i = \sum_{j=1}^{M} c_{ij} N_{-c} g_j, \quad \begin{pmatrix} c_{i1} & \ldots & c_{iM}\end{pmatrix}^{\mathrm{T}} = (1-G_1)^{-1}\boldsymbol{e}_i,$$
$$\text{where } \boldsymbol{e}_i\big|_j = \begin{cases} 1, & i=j, \\ 0, & i \neq j, \end{cases}$$

φ_i uniquely solve the boundary value problems:

$$\mathscr{L}_c\varphi_i = 0 \quad \text{in } \Omega, \quad \tau_f\varphi_i = \tau\varphi_i - \sum_{k=1}^{M} \langle \varphi_i, y_k\rangle g_k = g_i \quad \text{on } \Gamma, \quad 1 \leqslant i \leqslant M.$$

Set

$$\begin{aligned} p(t) &= u(t) - \begin{pmatrix}\varphi_1 & \ldots & \varphi_M\end{pmatrix}\big(\boldsymbol{f}(t) - \langle u(t), \boldsymbol{y}\rangle\big) \\ &= u(t) - \sum_{k=1}^{M} \big(f_k(t) - \langle u(t), y_k\rangle\big)\varphi_k. \end{aligned}$$

Then,

$$\frac{dp}{dt} + Fp = (\varphi_1 \ \dots \ \varphi_M)\left(c(\boldsymbol{f}(t) - \langle u(t), \boldsymbol{y}\rangle) - (\boldsymbol{f}'(t) - \langle u_t(t), \boldsymbol{y}\rangle)\right),$$

or

$$\begin{aligned} p(t) = e^{-tF}p(0) &+ \int_0^t e^{-(t-s)F}(\varphi_1 \ \dots \ \varphi_M)\, c\left(\boldsymbol{f}(s) - \langle u(s), \boldsymbol{y}\rangle\right) ds \\ &- \int_0^t e^{-(t-s)F}(\varphi_1 \ \dots \ \varphi_M)\left(\boldsymbol{f}'(s) - \langle u_s(s), \boldsymbol{y}\rangle\right) ds. \end{aligned}$$

In view of the estimate (2.13′), we see that

$$\|p(t)\| \leqslant \text{const}\, e^{-rt}, \quad 0 < r < \lambda_{\nu+1}, \quad t \geqslant 0.$$

This immediately gives the estimate

$$\|u(t)\| + \|v(t)\| \leqslant \text{const}\, e^{-rt}\left(\|u_0\| + \|v_0\|\right), \quad t \geqslant 0. \tag{2.15}$$

The presence of $f_k'(t) = \langle v_t(t), \zeta_k\rangle$ in (2.3) makes our control system somewhat unclear regarding its well-posedness. Let us express $cf_k(t) - f_k'(t)$ in terms of q and v including no derivative in time. In view of the equation for v in (2.3), we calculate as

$$\boldsymbol{f}'(t) + \langle (L-C)v, \boldsymbol{\zeta}\rangle = \overline{\Theta} G_2 \boldsymbol{p}(q) + \overline{\Theta} G_1 (c\boldsymbol{f}(t) - \boldsymbol{f}'(t)),$$

where

$$G_2 = \left(\langle \xi_i, y_j \rangle \,; \begin{array}{ccc} i & \rightarrow & 1, \dots, N \\ j & \downarrow & 1, \dots, M \end{array} \right), \quad \text{and} \quad \boldsymbol{p}(q) = \left(p_1(q) \quad \dots \quad p_N(q)\right)^{\mathrm{T}}.$$

Thus, $\boldsymbol{f}'(t) + \langle (L-C)v, \boldsymbol{y}\rangle = G_2\boldsymbol{p}(q) + cG_1\boldsymbol{f}(t)$, and

$$\begin{aligned} c\boldsymbol{f}(t) - \boldsymbol{f}'(t) &= \langle v, L_c\boldsymbol{y}\rangle - \langle Cv, \boldsymbol{y}\rangle - G_2\boldsymbol{p}(q) \\ &= \langle v, L_c\boldsymbol{y}\rangle - G_2\boldsymbol{p}(q - v) \\ &= \boldsymbol{g}(q, v). \end{aligned} \tag{2.16}$$

Replacing $c\boldsymbol{f}(t) - \boldsymbol{f}'(t)$ by $\boldsymbol{g}(q,v)$ in (2.3), we obtain

$$\begin{cases} \dfrac{dq}{dt} + Lq = (N_{-c}g_1 \ \dots \ N_{-c}g_M)\,\boldsymbol{g}(q,v), \quad q(0) = q_0 \in L^2(\Omega), \\ \dfrac{dv}{dt} + Bv = -Cq + (N_{-c}g_1 \ \dots \ N_{-c}g_M)\,\boldsymbol{g}(q,v), \quad v(0) = v_0 \in L^2(\Omega), \end{cases} \tag{2.17}$$

which is our basic system of differential equations. Eqn. (2.17) is well posed in $L^2(\Omega) \times L^2(\Omega)$. In (2.17), it is readily examined that $\boldsymbol{f}(t) = \langle v(t), \boldsymbol{\zeta}\rangle = \overline{\Theta}\,\langle v(t), \boldsymbol{y}\rangle$ actually satisfies the relation: $c\boldsymbol{f}(t) - \boldsymbol{f}'(t) = \boldsymbol{g}(q,v)$. Thus, the preceding arguments in this section are justified.

To reduce the compensator to a finite-dimensional equation, we add a small perturbation to (2.17). The perturbed system of equations is described by

$$\begin{cases} \dfrac{dq}{dt} + Lq = +\left(N_{-c}g_1 \ \dots \ N_{-c}g_M\right)\boldsymbol{g}_n(q,v), \\ \dfrac{dv}{dt} + (L - C_n)v = -Cq + \left(P_nN_{-c}g_1 \ \dots \ P_nN_{-c}g_M\right)\boldsymbol{g}_n(q,v), \\ q(0,\cdot) = q_0, \quad v(0,\cdot) = v_0, \end{cases} \tag{2.18}$$

where

$$C_n v = -\sum_{k=1}^{N} p_k(P_n v)\xi_k, \quad \text{and}$$

$$\boldsymbol{g}_n(q,v) = \langle v, L_c\boldsymbol{y}\rangle - G_2\boldsymbol{p}(q) + G_2\boldsymbol{p}(P_n v).$$

Note that the resolvent of the coefficient operator in (2.17) exists in a closed set consisting of the union of some sector and the half-plane, $\{\lambda;\ \mathrm{Re}\,\lambda \leqslant r\}$, and satisfies a decay estimate of order $|\lambda|^{-1}$ there. Although the perturbation contains the unbounded terms $p_k(Q_n v)$, that is, $\langle Q_n v, w_k\rangle_\Gamma$ or $\langle (Q_n v)_\nu, w_k\rangle_\Gamma$, the stability property of the perturbed system (2.18) is *little* affected when $n \geqslant \nu$ is chosen large enough. In fact, we only have to show that the resolvent of the perturbed operator exists at least in the above closed set, as long as n is large enough. Consequently the solutions (q,v) to (2.18) satisfy the decay estimate

$$\|q(t)\| + \|v(t)\| \leqslant \text{const}\, e^{-rt}\left(\|u_0\| + \|v_0\|\right), \quad t \geqslant 0.$$

In (2.18), $v(t)$ stays in $P_nL^2(\Omega)$, as long as v_0 is in $P_nL^2(\Omega)$. Thus the equation for v in (2.18) is regarded as an equation in the finite-dimensional subspace $P_nL^2(\Omega)$. Genuine regularity of q in space variables is examined just as in Theorem 2.1, Chapter 4.

As in (2.17), it is readily seen that the function $\boldsymbol{f}(t) = \langle v(t), \boldsymbol{\zeta}\rangle = \overline{\Theta}\,\langle v(t), \boldsymbol{y}\rangle$ satisfies the relation: $c\boldsymbol{f}(t) - \boldsymbol{f}'(t) = \boldsymbol{g}_n(q,v)$. Thus, by setting

$$u(t,\cdot) = q(t,\cdot) + \left(N_{-c}g_1 \ \dots \ N_{-c}g_M\right)\langle v(t), \boldsymbol{\zeta}\rangle,$$

the system of differential equations for $(u(t,\cdot), v(t)) \in L^2(\Omega) \times P_nL^2(\Omega)$, which is equivalent to (2.18), is described by

$$\begin{cases} \dfrac{du}{dt} + \mathscr{L}u = 0, \qquad \tau u = \displaystyle\sum_{k=1}^{M} \langle v, \zeta_k\rangle g_k, \\ \dfrac{dv}{dt} + (L - C_n)v = \displaystyle\sum_{k=1}^{N} p_k(u)\xi_k + \sum_{k=1}^{M} \langle v, \zeta_k\rangle CN_{-c}g_k \\ \qquad\qquad + \left(P_nN_{-c}g_1 \ \dots \ P_nN_{-c}g_M\right)\tilde{\boldsymbol{g}}_n(u,v), \end{cases} \tag{2.19}$$

where $\tilde{g}_n(u,v) = g_n\big(u - (N_{-c}g_1 \ \dots \ N_{-c}g_M)\langle v, \zeta\rangle, \ v\big)$, and the solutions (u,v) to (2.19) satisfy the decay estimate:

$$\|u(t,\cdot)\| + \|v(t)\|_{P_nL^2(\Omega)} \leqslant \text{const}\, e^{-rt}\left(\|u_0\| + \|v_0\|_{P_nL^2(\Omega)}\right), \quad t \geqslant 0. \quad (2.20)$$

Equation (2.19) is the desired control system, and interpreted as (1.3), by setting $\ell = m_1 + \cdots + m_n$. The result is summerized as the following theorem:

Theorem 2.1. *Suppose that the sensors w_k and the actuators g_k on the boundary Γ satisfy the conditions*

$$\text{rank}\, \hat{W}_i = m_i, \quad \textit{and} \quad \text{rank}\, \hat{G}_i = m_i, \quad 1 \leqslant i \leqslant \nu, \tag{2.21}$$

respectively, where

$$\hat{W}_i = \left(p_k(\varphi_{ij}); \ \begin{matrix} j & \rightarrow & 1,\dots,m_i \\ k & \downarrow & 1,\dots,N \end{matrix} \right), \quad \textit{and}$$

$$\hat{G}_i = \left(\langle g_k, \sigma\varphi_{ij}\rangle_\Gamma; \ \begin{matrix} j & \downarrow & 1,\dots,m_i \\ k & \rightarrow & 1,\dots,M \end{matrix} \right).$$

Then, we find an integer ℓ; an $\ell \times \ell$ matrix B_1; $\rho_k \in \mathbb{R}^\ell$, $1 \leqslant k \leqslant M$; and $\xi_k \in \mathbb{R}^\ell$, $1 \leqslant k \leqslant N$, such that every solution $(u(t,\cdot), v(t)) \in L^2(\Omega) \times \mathbb{R}^\ell$ to (1.3) satisfies the decay estimate,

$$\|u(t,\cdot)\| + |v(t)|_\ell \leqslant \text{const}\, e^{-rt}\left(\|u_0\| + |v_0|_\ell\right), \quad t \geqslant 0. \tag{2.22}$$

Remark: In the approach through identity compensators, we have to add a small perturbation to the stabilized system (2.17) to obtain a finite-dimensional stabilizing compensator. The situation is the same as in the following section and in other settings with distributed inputs. This procedure comes from the structure setting of the coupling system with state (q, v) or (u,v), and is different from the approaches developed in Chapter 4.

An application to a class of second order equations:

We have so far assumed that L is a self-adjoint operator. As remarked in Section 1, however, self-adjointness of L is not an essential assumption in this chapter. The algebraic approach developed in this and the next sections is equally applied, with slight technical changes, to a class of linear boundary control systems of second order in time. Let us consider the following linear differential equation with state $(u(t,\cdot), u_t(t,\cdot))$ in the interval $I = (0, 1)$:

$$\begin{cases} u_{tt} - 2\alpha u_{txx} + u_{xxxx} = 0, \\ u_x(t,0) = f(t), \quad u(t,1) = 0, \qquad u_{xxx}(t,0) = u_{xx}(t,1) = 0, \\ u(0,\cdot) = u_0(\cdot), \qquad u_t(0,\cdot) = u_1(\cdot). \end{cases} \tag{2.23}$$

Here, α, $0 < \alpha < 1$, denotes a constant, and $f(t)$ the boundary input. We first consider a static feedback control scheme and then proceed to a dynamic feedback scheme containing an identity compensator.

A static feedback control scheme:

A single output is assumed. Given a real-valued $w \in L^2(I)$, the output is given by $\langle u, w \rangle = \int_0^1 u(t,x)w(x)\,dx$, and set $f(t) = \langle u, w \rangle$. Setting

$$Au = \mathscr{A} u = -\frac{d^2u}{dx^2}, \quad u \in \mathscr{D}(A),$$
$$\text{where} \quad \mathscr{D}(A) = \left\{u \in H^2(0,1);\ u'(0) = u(1) = 0\right\},$$

and $u_1 = u$, $u_2 = u_t$, we have

$$\frac{d}{dt}\begin{pmatrix} u_1 \\ u_2 \end{pmatrix} + \begin{pmatrix} 0 & -1 \\ \mathscr{A}^2 & 2\alpha\mathscr{A} \end{pmatrix}\begin{pmatrix} u_1 \\ u_2 \end{pmatrix} = \begin{pmatrix} 0 \\ 0 \end{pmatrix}.$$

Let T be an operator defined by $Tu = u - \langle u, w \rangle \varphi$ in $L^2(I)$, where $\varphi(x) = x - 1$, and set

$$\boldsymbol{z} = \begin{pmatrix} z_1 \\ z_2 \end{pmatrix}, \quad \text{where} \quad z_1 = Tu_1, \quad z_2 = Tu_2.$$

When $\langle \varphi, w \rangle \neq 1$, the bounded inverse T^{-1} exists. Let $H = \mathscr{D}(A) \times L^2(I)$ be a Hilbert space with inner product $\langle \cdot, \cdot \rangle_H$ and norm $\|\cdot\|_H$ [1]. By assuming that w belongs to $\mathscr{D}(A)$, then $\boldsymbol{z}$ satisfies the equation in H:

$$\frac{d\boldsymbol{z}}{dt} + L\boldsymbol{z} = \langle \boldsymbol{z}, \hat{\boldsymbol{w}} \rangle_H \boldsymbol{\varphi}, \quad \boldsymbol{z}(0,\cdot) = \boldsymbol{z}_0, \tag{2.24}$$

where

$$L = \begin{pmatrix} 0 & -1 \\ A^2 & 2\alpha A \end{pmatrix}, \quad \mathscr{D}(L) = \mathscr{D}(A^2) \times \mathscr{D}(A),$$
$$\boldsymbol{\varphi} = \begin{pmatrix} 0 \\ \varphi \end{pmatrix}, \quad \text{and } \hat{\boldsymbol{w}} = \begin{pmatrix} w \\ 2\alpha Aw \end{pmatrix} \in H.$$

It is clear that

(i) $\sigma(L) = \{\mu_n \omega^{\pm}\}_{n \geqslant 0}$, where $\mu_n = \left(n + \frac{1}{2}\right)^2 \pi^2$, $\omega^{\pm} = \alpha \pm \mathrm{i}\sqrt{1 - \alpha^2}$;

(ii) $(\mu_n \omega^{\pm} - L)\eta_n^{\pm} = 0$, $n \geqslant 0$, where $\eta_n^{\pm} = \frac{1}{\sqrt{2}\mu_n}\begin{pmatrix} \psi_n \\ -\mu_n \omega^{\pm} \psi_n \end{pmatrix}$, $\psi_n = \sqrt{2}\cos\left(n + \frac{1}{2}\right)\pi x$; and

(iii) the set $\{\eta_n^{\pm}\}_{n \geqslant 0}$ forms a normalized Riesz basis for H.

---

[1] $\langle \boldsymbol{z}, \boldsymbol{q} \rangle_H = \langle Az_1, Aq_1 \rangle + \langle z_2, q_2 \rangle$, $\|\boldsymbol{z}\|_H = \langle \boldsymbol{z}, \boldsymbol{z} \rangle_H^{1/2}$ for $\boldsymbol{z} = (z_1\ z_2)^{\mathrm{T}}$ and $\boldsymbol{q} = (q_1\ q_2)^{\mathrm{T}} \in H$.

Thus the semigroup e^{-tL} satisfies the decay estimate

$$\left\|e^{-tL}\right\|_{\mathscr{L}(H)} \leqslant \text{const}\, e^{-\alpha\mu_0 t}, \quad t \geqslant 0. \tag{2.25}$$

The set $\{\psi_n\}_{n\geqslant 0}$ forms an orthonormal basis for $L^2(I)$. Let P_n, $n \geqslant 0$, be the projector in $L^2(I)$ corresponding to the eigenvalues μ_i of A, $i \leqslant n$, and let P_n^H be the projector in H corresponding to the eigenvalues $\mu_i\omega^{\pm}$ of L, $i \leqslant n$:

$$P_n^H \boldsymbol{z} = \sum_{i=0}^{n} \left(z_i^+ \eta_i^+ + z_i^- \eta_i^-\right) = \begin{pmatrix} P_n z_1 \\ P_n z_2 \end{pmatrix}$$

for $\boldsymbol{z} = \begin{pmatrix} z_1 & z_2 \end{pmatrix}^{\mathrm{T}} = \sum_{i=0}^{\infty} \left(z_i^+ \eta_i^+ + z_i^- \eta_i^-\right)$. When $n = 0$, we have

$$P_0^H \boldsymbol{z} = z_0^+ \eta_0^+ + z_0^- \eta_0^-, \quad \text{and} \quad P_0^H \boldsymbol{\varphi} = \frac{\mathrm{i}\,\varphi_0}{\sqrt{2}\sqrt{1-\alpha^2}} \left(\eta_0^+ - \eta_0^-\right),$$

where $\varphi_0 = \langle \varphi, \psi_0 \rangle = \sqrt{2}\left(\frac{2}{\pi}\right)^2$. Let us construct a w simply as a scalar multiple of ψ_0. Then,

$$\langle \boldsymbol{z}, \hat{\boldsymbol{w}} \rangle_H = \frac{\mu_0 w_0}{\sqrt{2}} \begin{pmatrix} 1 - 2\alpha\omega^+ & 1 - 2\alpha\omega^- \end{pmatrix} \begin{pmatrix} z_0^+ \\ z_0^- \end{pmatrix}, \quad w_0 = \langle w, \psi_0 \rangle.$$

The equation for $(z_0^+ \; z_0^-)^{\mathrm{T}}$ is written as

$$\begin{aligned} &\frac{d}{dt} \begin{pmatrix} z_0^+ \\ z_0^- \end{pmatrix} + \left(\mu_0 \begin{pmatrix} \omega^+ & 0 \\ 0 & \omega^- \end{pmatrix} \right. \\ &\qquad \left. + \frac{\mu_0 \varphi_0 \mathrm{i}}{2\sqrt{1-\alpha^2}} w_0 \begin{pmatrix} -1 \\ 1 \end{pmatrix} \begin{pmatrix} 1 - 2\alpha\omega^+ & 1 - 2\alpha\omega^- \end{pmatrix} \right) \begin{pmatrix} z_0^+ \\ z_0^- \end{pmatrix} = \begin{pmatrix} 0 \\ 0 \end{pmatrix}. \end{aligned}$$

We can choose a $w_0 = \langle w, \psi_0 \rangle$ such that the minimum κ of the real part of the spectrum of the above coefficient matrix is greater than $\alpha\mu_0$. For such a w_0, we have the estimate

$$\left\| \begin{pmatrix} z_0^+(t) \\ z_0^-(t) \end{pmatrix} \right\|_H \leqslant \text{const}\, e^{-\kappa t} \left\| \begin{pmatrix} z_0^+(0) \\ z_0^-(0) \end{pmatrix} \right\|_H, \quad t \geqslant 0,$$

which immediately leads to the estimate

$$\|\boldsymbol{z}(t)\|_H \leqslant \text{const}\, e^{-\min(\kappa, \alpha\mu_1)t} \|\boldsymbol{z}_0\|_H, \quad t \geqslant 0$$

for solutions $\boldsymbol{z}(t)$ to (2.24). In other words,

$$\left\| \exp\left(-t\left(L - \langle \cdot, \hat{\boldsymbol{w}} \rangle_H \boldsymbol{\varphi}\right)\right) \right\|_{\mathscr{L}(H)} \leqslant \text{const}\, e^{-\min(\kappa, \alpha\mu_1)t}, \quad t \geqslant 0, \quad \kappa > \alpha\mu_0.$$

Thus we obtain an improvement of the stability estimate (2.25):

$$\begin{aligned} &\|u(t,\cdot)\|_{H^2(0,1)} + \|u_t(t,\cdot)\| \\ &\leqslant \text{const}\, e^{-\min(\kappa, \alpha\mu_1)t} \left(\|u_0\|_{H^2(0,1)} + \|u_1\| \right), \quad t \geqslant 0. \end{aligned} \tag{2.26}$$

A dynamic feedback control scheme:
Instead of the implausible output $\langle u, w\rangle$, the output here is assumed to be $u(t,0)$ and $u_t(t,0)$, $t \geqslant 0$. We construct a dynamic compensator for enhancing the stability of the whole control system. In (2.23) set

$$\boldsymbol{q}(t) = \begin{pmatrix} q_1(t,\cdot) \\ q_2(t,\cdot) \end{pmatrix} = \begin{pmatrix} u(t,\cdot) \\ u_t(t,\cdot) \end{pmatrix} - \begin{pmatrix} f(t) \\ f'(t) \end{pmatrix} \varphi. \tag{2.27}$$

Assuming that $f(t)$ is of class C^2, the equation for $\boldsymbol{q}$ is described in H by

$$\frac{d\boldsymbol{q}}{dt} + L\boldsymbol{q} + f''(t)\boldsymbol{\varphi} = \mathbf{0}. \tag{2.28}$$

Given a $\boldsymbol{\xi} \in H$, let C be the bounded operator defined by $C\boldsymbol{q} = -q_1(0)\boldsymbol{\xi}$ for $\boldsymbol{q}(\cdot) \in H$. Our compensator with state $\boldsymbol{v}(t) = \begin{pmatrix} v_1(t) \\ v_2(t) \end{pmatrix} \in H$ is formally given by

$$\frac{d\boldsymbol{v}}{dt} + (L-C)\boldsymbol{v} = -C\boldsymbol{q} - f''(t)\boldsymbol{\varphi}, \quad \boldsymbol{v}(0) = \boldsymbol{v}_0. \tag{2.29}$$

As before, we see that $\boldsymbol{q}(t) - \boldsymbol{v}(t) = e^{-t(L-C)}(\boldsymbol{q}_0 - \boldsymbol{v}_0)$, $t \geqslant 0$. Let us find a $\boldsymbol{\xi} = \begin{pmatrix} \xi_1 & \xi_2 \end{pmatrix}^{\mathrm{T}} \in P_0^H H$ so that $\min \operatorname{Re} \sigma(L-C)$ is greater than $\alpha\mu_0$. It is enough to investigate the structure of the restriction $P_0^H(L-C)P_0^H$. By setting $\boldsymbol{\xi} = \xi_0^+ \eta_0^+ + \xi_0^- \eta_0^-$, the operator $P_0^H(L-C)P_0^H$ is equivalent to the matrix

$$\mu_0 \begin{pmatrix} \omega^+ & 0 \\ 0 & \omega^- \end{pmatrix} + \frac{1}{\mu_0} \begin{pmatrix} \xi_0^+ \\ \xi_0^- \end{pmatrix} \begin{pmatrix} 1 & 1 \end{pmatrix}.$$

Since $\left(\begin{pmatrix} 1 & 1 \end{pmatrix}, \mu_0 \begin{pmatrix} \omega^+ & 0 \\ 0 & \omega^- \end{pmatrix} \right)$ is an observable pair, there is a vector $\begin{pmatrix} \xi_0^+ & \xi_0^- \end{pmatrix}^{\mathrm{T}}$ such that the spectrum $\sigma\left(P_0^H(L-C)P_0^H\right)$ is freely assigned (see, e.g., Proposition 2.2, Chapter 1). Thus we can choose a $\boldsymbol{\xi} \in P_0^H H$ such that $\min \operatorname{Re} \sigma(L-C) = \kappa$, where $\alpha\mu_0 < \kappa \leqslant \alpha\mu_1$. With this choice of $\boldsymbol{\xi}$ we have the decay estimate

$$\left\|\boldsymbol{q}(t) - \boldsymbol{v}(t)\right\|_H \leqslant \text{const}\, e^{-\kappa t} \left\|\boldsymbol{q}_0 - \boldsymbol{v}_0\right\|_H, \quad t \geqslant 0. \tag{2.30}$$

At this stage we define $f(t)$ as

$$f(t) = \langle \boldsymbol{v}(t), \boldsymbol{\rho} \rangle_H, \qquad \boldsymbol{\rho} = \theta \begin{pmatrix} A^{-2}w \\ 0 \end{pmatrix}, \quad \theta \in \mathbb{R}^1,$$

where $w \in P_0 L^2(I)$ is the function stated in (2.24). We may assume with no loss of generality that $\langle \varphi, w\rangle \neq 1$. Set $\theta = (1 - \langle \varphi, w\rangle)^{-1}$. As in (2.13), we obtain the decay estimate

$$\left| f(t) - \langle u(t,\cdot), w\rangle \right| \leqslant \text{const}\, e^{-\kappa t}, \quad t \geqslant 0. \tag{2.31}$$

As in (2.16), let us express $f''(t) = \frac{d^2}{dt^2}\langle \boldsymbol{v}(t), \boldsymbol{\rho}\rangle_H$ in terms of $\boldsymbol{q}(t)$ and $\boldsymbol{v}(t)$ including no derivative in time. Looking at (2.29) and noting that $\boldsymbol{\rho}$ is in $\mathscr{D}\left(L^{*2}\right)$, we see that

$$0 = f'(t) + \langle \boldsymbol{v}(t), L^*\boldsymbol{\rho}\rangle_H + (v_1(t,0) - q_1(t,0))\langle \boldsymbol{\xi}, \boldsymbol{\rho}\rangle_H,$$
$$\text{where } L^*\boldsymbol{\rho} = -\theta\begin{pmatrix}0\\ w\end{pmatrix}, \quad \text{and } L^{*2}\boldsymbol{\rho} = -\theta\hat{\boldsymbol{w}}.$$

Differentiating the both sides in t, we calculate as

$$\begin{aligned}0 =& f''(t) + \langle \boldsymbol{v}_t(t), L^*\boldsymbol{\rho}\rangle_H + ((v_1)_t(t,0) - u_t(t,0) - f'(t))\langle \boldsymbol{\xi}, \boldsymbol{\rho}\rangle_H\\ =& f''(t) + \theta\langle \boldsymbol{v}(t), \hat{\boldsymbol{w}}\rangle_H - (v_1(t,0) - u(t,0) - f(t))\langle \boldsymbol{\xi}, L^*\boldsymbol{\rho}\rangle_H\\ &+ f''(t)\theta\langle \varphi, w\rangle + ((v_1)_t(t,0) - u_t(t,0))\langle \boldsymbol{\xi}, \boldsymbol{\rho}\rangle_H\\ &+ \left(\langle \boldsymbol{v}(t), L^*\boldsymbol{\rho}\rangle_H + (v_1(t,0) - u(t,0) - f(t))\langle \boldsymbol{\xi}, \boldsymbol{\rho}\rangle_H\right)\langle \boldsymbol{\xi}, \boldsymbol{\rho}\rangle_H,\end{aligned}$$

$$\begin{aligned}\theta f''(t) =& -\theta\langle \boldsymbol{v}(t), \hat{\boldsymbol{w}}\rangle_H - \langle \boldsymbol{\xi}, \boldsymbol{\rho}\rangle_H\langle \boldsymbol{v}(t), L^*\boldsymbol{\rho}\rangle_H\\ &+ \left(\langle \boldsymbol{\xi}, L^*\boldsymbol{\rho}\rangle_H - \langle \boldsymbol{\xi}, \boldsymbol{\rho}\rangle_H^2\right)(v_1(t,0) - u(t,0) - f(t))\\ &- ((v_1)_t(t,0) - u_t(t,0))\langle \boldsymbol{\xi}, \boldsymbol{\rho}\rangle_H\\ =& -\theta\langle \boldsymbol{v}(t), \hat{\boldsymbol{w}}\rangle_H + (v_1(t,0) - q_1(t,0))\langle \boldsymbol{\xi}, L^*\boldsymbol{\rho}\rangle_H\\ &- \left(v_2(t,0) - q_2(t,0) - (v_1(t,0) - q_1(t,0))\xi_1(0)\right)\langle \boldsymbol{\xi}, \boldsymbol{\rho}\rangle_H.\end{aligned} \tag{2.32}$$

The last term of the right-hand side of (2.32) is denoted by $\Xi(t)$. The terms $|v_2(t,0)|$ and $|q_2(t,0)|$ in $\Xi(t)$ are bounded from above, respectively, by $\left\|\sqrt{L}\boldsymbol{v}(t)\right\|_H$ and $\left\|\sqrt{L}\boldsymbol{q}(t)\right\|_H$. Replacing $f''(t)$ by $\theta^{-1}\Xi(t)$ in (2.28) and (2.29), we obtain the basic system of equations:

$$\begin{cases}\dfrac{d\boldsymbol{q}}{dt} + L\boldsymbol{q} + \theta^{-1}\Xi(t)\boldsymbol{\varphi} = \boldsymbol{0},\\[2mm] \dfrac{d\boldsymbol{v}}{dt} + (L - C)\boldsymbol{v} + C\boldsymbol{q} + \theta^{-1}\Xi(t)\boldsymbol{\varphi} = \boldsymbol{0},\end{cases} \tag{2.33}$$

which is well posed in $H \times H$ and generates an analytic semigroup. In (2.33), it is not hard to verify that the second derivative $f''(t) = \frac{d^2}{dt^2}\langle \boldsymbol{v}(t), \boldsymbol{\rho}\rangle_H$ is actually equal to $\theta^{-1}\Xi(t)$. Thus we can go back to (2.28) and (2.29). The first equation of (2.33) is rewritten as

$$\frac{d\boldsymbol{q}}{dt} + L\boldsymbol{q} - \langle \boldsymbol{q}, \hat{\boldsymbol{w}}\rangle_H\boldsymbol{\varphi} = \varepsilon(t)\boldsymbol{\varphi}, \quad t > 0, \quad \varepsilon(t) = -f''(t) - \langle \boldsymbol{q}(t), \hat{\boldsymbol{w}}\rangle_H.$$

The error term $\varepsilon(t)$ is evaluated as

$$\begin{aligned}|\varepsilon(t)| =& \left|\langle \boldsymbol{q}(t) - \boldsymbol{v}(t), \hat{\boldsymbol{w}}\rangle_H + \theta^{-1}(v_1(t,0) - q_1(t,0))\langle \boldsymbol{\xi}, L^*\boldsymbol{\rho}\rangle_H\right.\\ &\left.- \theta^{-1}(v_2(t,0) - q_2(t,0) - (v_1(t,0) - q_1(t,0))\xi_1(0))\langle \boldsymbol{\xi}, \boldsymbol{\rho}\rangle_H\right|\\ \leqslant& \text{ const}\,\frac{e^{-\kappa t}}{\sqrt{t}}\left(\|\boldsymbol{q}_0\|_H + \|\boldsymbol{v}_0\|_H\right), \qquad t > 0.\end{aligned}$$

In view of the decay estimate of the semigroup $\exp\left(-t\left(L - \langle \cdot, \hat{\boldsymbol{w}} \rangle_H \boldsymbol{\varphi}\right)\right)$ and $\varepsilon(t)$, we obtain

$$\|\boldsymbol{q}(t)\|_H \leqslant \text{const}\, e^{-\kappa t} \left(\|\boldsymbol{q}_0\|_H + \|\boldsymbol{v}_0\|_H \right), \quad t \geqslant 0,$$

and a similar decay estimate for $\boldsymbol{v}(t)$. We have thus estabished the stability enhancement.

To reduce the compensator to a finite-dimensional equation, we add a small perturbation to (2.33). The perturbed equation is described as

$$\begin{cases} \dfrac{d\boldsymbol{q}}{dt} + L\boldsymbol{q} + \theta^{-1} \Xi_n(t) \boldsymbol{\varphi} = \boldsymbol{0}, \\ \dfrac{d\boldsymbol{v}}{dt} + L\boldsymbol{v} + (P_n v_1)(t,0)\boldsymbol{\xi} + C\boldsymbol{q} + \theta^{-1} \Xi_n(t) P_n^H \boldsymbol{\varphi} = \boldsymbol{0}, \end{cases} \tag{2.34}$$

where

$$\begin{aligned} \Xi_n(t) = &- \theta \langle \boldsymbol{v}(t), \hat{\boldsymbol{w}} \rangle_H + ((P_n v_1)(t,0) - q_1(t,0)) \langle \boldsymbol{\xi}, L^* \boldsymbol{\rho} \rangle_H \\ &- ((P_n v_2)(t,0) - q_2(t,0) - ((P_n v_1)(t,0) - q_1(t,0))\xi_1(0)) \langle \boldsymbol{\xi}, \boldsymbol{\rho} \rangle_H . \end{aligned}$$

When n is chosen large enough, the stability of (2.34) is little affected. Thus the estimate

$$\|q(t)\|_H + \|v(t)\|_H \leqslant \text{const}\, e^{-\kappa t} \left(\|q_0\|_H + \|v_0\|_H \right), \quad t \geqslant 0 \tag{2.35}$$

holds for the solutions to (2.34). In (2.34), $\boldsymbol{v}(t)$ stays in $P_n^H H$, as long as $\boldsymbol{v}_0$ is in $P_n^H H$. Just as before, we see in (2.34) that $f''(t) = \frac{d^2}{dt^2} \langle \boldsymbol{v}(t), \boldsymbol{\rho} \rangle_H$ is nothing but $\theta^{-1} \Xi_n(t)$.

Setting $u(t,\cdot) = q_1(t,\cdot) + \langle \boldsymbol{v}(t), \boldsymbol{\rho} \rangle_H \varphi$, we go back to the original state $(u, u_t)^{\mathrm{T}}$. By recalling that $\boldsymbol{q}(t)$ belongs to $\mathscr{D}(A^2) \times \mathscr{D}(A)$, $(u(t,\cdot), \boldsymbol{v}(t))$ satisfies the equation

$$\begin{cases} u_{tt} + 2\alpha \mathscr{A} u_t + \mathscr{A}^2 u = 0, \\ u_x(t,0) = f(t), \quad u(t,1) = 0, \qquad u_{xxx}(t,0) = u_{xx}(t,1) = 0, \\ \dfrac{d\boldsymbol{v}}{dt} + L\boldsymbol{v} + ((P_n v_1)(t,0) - u(t,0) - f(t))\boldsymbol{\xi} + \theta^{-1} \Xi_n(t) P_n^H \boldsymbol{\varphi} = 0, \\ u(0,\cdot) = u_0(\cdot), \quad u_t(0,\cdot) = u_1(\cdot), \quad \boldsymbol{v}(0) = \boldsymbol{v}_0 \in P_n^H H, \end{cases} \tag{2.36}$$

where $f(t) = \langle \boldsymbol{v}(t), \boldsymbol{\rho} \rangle_H$, and

$$\begin{aligned} \Xi_n(t) = &- \theta \langle v(t), \hat{\boldsymbol{w}} \rangle_H \\ &+ ((P_n v_1)(t,0) - u(t,0) - f(t)) \left(\langle \boldsymbol{\xi}, L^* \boldsymbol{\rho} \rangle_H + \xi_1(0) \langle \boldsymbol{\xi}, \boldsymbol{\rho} \rangle_H - \langle \boldsymbol{\xi}, \boldsymbol{\rho} \rangle_H^2 \right) \\ &- \left((P_n v_2)(t,0) - u_t(t,0) + \langle \boldsymbol{v}; L^* \boldsymbol{\rho} \rangle_H \right) \langle \boldsymbol{\xi}, \boldsymbol{\rho} \rangle_H . \end{aligned}$$

5.3 Another Model of Identity Compensators

In Section 2, the stabilizing identity compensator is constructed in (2.3), which is based on the transformed equation of the controlled plant Σ_p with distributed inputs. In this section, we propose another setting of stabilizing compensators. The assumptions on the controlled plant Σ_p is the same as in Section 2, such as the pair $(\mathscr{L}, \tau)$ of differential operators leading to the self-adjoint operator L in $L^2(\Omega)$.

Let us introduce an auxiliary feedback control system, and first establish stabilization for the system. The differential equation describing a dynamic compensator has state w in $L^2(\Omega)$. Then, the auxiliary system containing a compensator is described as

$$\begin{cases} \dfrac{\partial u}{\partial t} + \mathscr{L}u = 0 \quad \text{in } \mathbb{R}^1_+ \times \Omega, \\ \tau u = \displaystyle\sum_{k=1}^{M} \langle w, y_k \rangle g_k(\xi) \quad \text{on } \mathbb{R}^1_+ \times \Gamma, \\ \dfrac{\partial w}{\partial t} + \mathscr{B}w = \displaystyle\sum_{k=1}^{N} p_k(u)\xi_k \quad \text{in } \mathbb{R}^1_+ \times \Omega, \\ \tau w = \displaystyle\sum_{k=1}^{M} \langle w, y_k \rangle g_k \quad \text{on } \mathbb{R}^1_+ \times \Gamma, \\ u(0,x) = u_0(x) \in L^2(\Omega), \quad w(0,x) = w_0(x) \in L^2(\Omega), \end{cases} \tag{3.1}$$

where $p_k(u)$, $1 \leqslant k \leqslant N$, denote the output of the controlled plant characterized by (1.4), and $\mathscr{B}$ a formal differential operator defined by

$$\mathscr{B} = \mathscr{L} - C = \mathscr{L} + \sum_{k=1}^{N} p_k(\cdot)\xi_k.$$

The closure of the operator $\mathscr{B}|_{\ker \tau}$ in $L^2(\Omega)$ is denoted as B, that is, the operator already given by (2.4):

$$B = L - C = L + \sum_{k=1}^{N} p_k(\cdot)\xi_k, \quad \mathscr{D}(B) = \mathscr{D}(L). \tag{3.2}$$

In the setting of eqn. (3.1), note that the boundary conditions on u and w are the same. Similarly, set

$$E = L - \sum_{k=1}^{M} \langle \cdot, L_c y_k \rangle N_{-c} g_k, \quad \mathscr{D}(E) = \mathscr{D}(L). \tag{3.3}$$

The readers may remember that the operator E is derived from F, where F is the closure of $\hat{F}$ in (2.9) (see also (3.14), Chapter 4): Let us determine the vectors y_k in (3.1) and (3.3). According to Green's formula, note that

$$\langle N_{-c}g_k, L_c\varphi_{ij}\rangle = \left\langle g_k, \varphi_{ij} - \frac{\partial \varphi_{ij}}{\partial \nu}\right\rangle_\Gamma, \quad \text{or } \langle N_{-c}g_k, \varphi_{ij}\rangle = \frac{1}{\lambda_i + c}\langle g_k, \sigma\varphi_{ij}\rangle_\Gamma .$$

If rank $\hat{G}_i = m_i$, $1 \leqslant i \leqslant \nu$ (see (2.21)), then Corollary 2.2, Chapter 3 ensures a set of $L_c y_k$ or equivalently $y_k \in P_\nu L^2(\Omega)$, $1 \leqslant k \leqslant M$, such that

$$\left\| e^{-tE} \right\| \leqslant \text{const}\, e^{-\lambda_{\nu+1}t}, \quad t \geqslant 0. \tag{3.4}$$

Let $T_{-c} = 1 - \sum_{k=1}^{M} \langle \cdot, y_k\rangle N_{-c}g_k \in \mathscr{L}(L^2(\Omega))$. As mentioned in the proof of Proposition 3.4, Chapter 4, we may assume with no loss of generality that the bounded inverse T_{-c}^{-1} exists, and

$$e^{-tF} = T_{-c}^{-1} e^{-tE} T_{-c}, \quad t \geqslant 0, \tag{3.5}$$

where

$$\begin{aligned} T_{-c}^{-1} &= 1 + (N_{-c}g_1 \dots N_{-c}g_M)(1 - \Phi_{-c})^{-1}\langle \cdot, \mathbf{y}\rangle, \\ \Phi_{-c} &= \left(\langle N_{-c}g_k, y_j\rangle;\ \begin{matrix} j & \downarrow & 1, \dots, M \\ k & \rightarrow & 1, \dots, M \end{matrix} \right). \end{aligned} \tag{3.6}$$

By assuming the existence and suitable regularity of solutions to (3.1), $u - w$ belongs to $\mathscr{D}(B)$ and is subject to the equation:

$$\frac{d}{dt}(u - w) + B(u - w) = 0, \quad (u - w)(0) = u_0 - w_0,$$

so that $u(t) - w(t) = e^{-tB}(u_0 - w_0)$, $t \geqslant 0$. If rank $\hat{W}_i = m_i$, $1 \leqslant i \leqslant \nu$ (see (2.21)), Theorem 2.1, Chapter 3 ensures a set of $\xi_k \in P_\nu L^2(\Omega)$, $1 \leqslant k \leqslant N$, such that

$$\|e^{-tB}\| \leqslant \text{const}\, e^{-\lambda_{\nu+1}t}, \quad t \geqslant 0. \tag{3.7}$$

Thus,

$$\|u(t) - w(t)\| \leqslant \text{const}\, e^{-\lambda_{\nu+1}t}\left(\|u_0\| + \|w_0\|\right), \quad t \geqslant 0.$$

Setting $y = u - w$ and $v = T_{-c}w$, or

$$\begin{pmatrix} y \\ v \end{pmatrix} = \begin{pmatrix} 1 & -1 \\ 0 & T_{-c} \end{pmatrix} \begin{pmatrix} u \\ w \end{pmatrix}, \tag{3.8}$$

we derive the equation for $(y\ \ v)$:

$$\begin{aligned} &\frac{d}{dt}\begin{pmatrix} y \\ v \end{pmatrix} + \Lambda \begin{pmatrix} y \\ v \end{pmatrix} = \begin{pmatrix} 0 \\ 0 \end{pmatrix}, \quad \begin{pmatrix} y(0) \\ v(0) \end{pmatrix} = \begin{pmatrix} y_0 \\ v_0 \end{pmatrix}, \\ &\text{where } \Lambda = \begin{pmatrix} L - C & 0 \\ T_{-c}C & L - \sum_{k=1}^{M}\langle \cdot, L_c\rho_k\rangle N_{-c}g_k \end{pmatrix} = \begin{pmatrix} B & 0 \\ T_{-c}C & E \end{pmatrix}. \end{aligned} \tag{3.9}$$

The operator $-\Lambda$ is the infinitesimal generator of an analytic semigroup $e^{-t\Lambda}$, $t > 0$. In view of the decays (3.4) and (3.7), we see that

$$\left\| e^{-t\Lambda} \right\|_{\mathscr{L}(L^2(\Omega)\times L^2(\Omega))} \leqslant \text{const}\, e^{-\lambda_{\nu+1}t}, \quad t \geqslant 0. \tag{3.10}$$

Thus, (3.1) is stabilized, since $\begin{pmatrix} u \\ w \end{pmatrix} = \begin{pmatrix} 1 & T_{-c}^{-1} \\ 0 & T_{-c}^{-1} \end{pmatrix} \begin{pmatrix} y \\ v \end{pmatrix}$. By the decay (3.10), the resolvent $(\lambda - \Lambda)^{-1}$ exists in a closed set Σ^* consisting of the union of some sector and the half-plane, $\{\lambda;\ \mathrm{Re}\,\lambda \leqslant r\}$, $0 < \forall r < \lambda_{\nu+1}$.

We add a small perturbation Θ_n to Λ, so that the perturbed operator $\Lambda + \Theta_n$ is described as

$$\Lambda + \Theta_n = \begin{pmatrix} L + \sum_{k=1}^{N} p_k(\cdot) T_{-c}^{-1} P_n T_{-c}\xi_k & -\sum_{k=1}^{M} \langle \cdot, L_c y_k\rangle T_{-c}^{-1} Q_n N_{-c} g_k \\ -\sum_{k=1}^{N} p_k(\cdot) P_n T_{-c}\xi_k & L - \sum_{k=1}^{M} \langle \cdot, L_c y_k\rangle P_n N_{-c} g_k \end{pmatrix},$$

$$\text{where } \Theta_n = \begin{pmatrix} -\sum_{k=1}^{N} p_k(\cdot) T_{-c}^{-1} Q_n T_{-c}\xi_k & -\sum_{k=1}^{M} \langle \cdot, L_c y_k\rangle T_{-c}^{-1} Q_n N_{-c} g_k \\ \sum_{k=1}^{N} p_k(\cdot) Q_n T_{-c}\xi_k & \sum_{k=1}^{M} \langle \cdot, L_c y_k\rangle Q_n N_{-c} g_k \end{pmatrix}. \tag{3.11}$$

As indicated in the remark after Theorem 2.1, adding such a perturbation seems indispensable to obtain a finite-dimensional stabilizing feedback scheme, as long as an identity compensator is employed. In the perturbed equation,

$$\frac{d}{dt}\begin{pmatrix} y \\ v \end{pmatrix} + (\Lambda + \Theta_n)\begin{pmatrix} y \\ v \end{pmatrix} = \begin{pmatrix} 0 \\ 0 \end{pmatrix}, \quad \begin{pmatrix} y(0) \\ v(0) \end{pmatrix} = \begin{pmatrix} y_0 \\ v_0 \end{pmatrix}, \tag{3.12}$$

it is easily seen that the solutions $(y(t), v(t))$ stay in $L^2(\Omega) \times P_n L^2(\Omega)$, as long as v_0 belong to $P_n L^2(\Omega)$. Thus, (3.12) is regarded as the equation in $L^2(\Omega) \times P_n L^2(\Omega)$ which is well posed. The operator Θ_n in (3.11) containing unbounded terms $p_k(y)$ is subordinate to Λ. However, as long as n is large enough, the resolvent $(\lambda - (\Lambda + \Theta_n))^{-1}$ exists in the above closed set Σ^*, due to the presence of $Q_n T_{-c}\xi_k$ and $Q_n N_{-c} g_k$. Thus, every solution $(y, v) \in L^2(\Omega) \times P_n L^2(\Omega)$ to (3.12) satisfies the decay estimate,

$$\|y(t)\| + \|v(t)\|_{P_n L^2(\Omega)} \leqslant \text{const}\, e^{-rt}\left(\|u_0\| + \|v_0\|_{P_n L^2(\Omega)}\right), \quad t \geqslant 0. \tag{3.13}$$

The equation for (u, v) will be derived from (3.12). Eqn. (3.12) is concretely written as

$$\frac{dy}{dt} + Ly + \sum_{k=1}^{N} p_k(y) T_{-c}^{-1} P_n T_{-c} \xi_k - \sum_{k=1}^{M} \langle v, L_c y_k \rangle T_{-c}^{-1} Q_n N_{-c} g_k = 0,$$

$$\frac{dv}{dt} - \sum_{k=1}^{N} p_k(y) P_n T_{-c} \xi_k + Lv - \sum_{k=1}^{M} \langle v, L_c y_k \rangle P_n N_{-c} g_k = 0.$$

Since $y = u - w = u - T_{-c}^{-1} v$, we calculate as

$$\begin{aligned} \boldsymbol{p}(y) = \left(p_1(y) \ \dots \ p_N(y)\right)^{\mathrm{T}} &= \boldsymbol{p}\left(u - T_{-c}^{-1} v\right) \\ &= \boldsymbol{p}(u) - \boldsymbol{p}(v) - V(1 - \Phi_{-c})^{-1} \langle v, \boldsymbol{y} \rangle, \end{aligned}$$

where

$$V = \left(p_j(N_{-c} g_k); \ \begin{matrix} j & \downarrow & 1, \dots, N \\ k & \rightarrow & 1, \dots, M \end{matrix} \right).$$

The equation for v is rewritten as

$$\begin{aligned} &\frac{dv}{dt} - (P_n T_{-c} \xi_1 \ \dots \ P_n T_{-c} \xi_N) \left(\boldsymbol{p}(u) - V(1 - \Phi_{-c})^{-1} \langle v, \boldsymbol{y} \rangle\right) \\ &+ (P_n T_{-c} \xi_1 \ \dots \ P_n T_{-c} \xi_N) \boldsymbol{p}(v) + Lv - \sum_{k=1}^{M} \langle v, L_c y_k \rangle P_n N_{-c} g_k = 0, \end{aligned}$$

or

$$\begin{aligned} &\frac{dv}{dt} + (L - C)v - (P_n N_{-c} g_1 \ \dots \ P_n N_{-c} g_M) \left(G_2 \boldsymbol{p}(v) + \langle v, L_c \boldsymbol{y} \rangle\right) \\ &\quad - \left((\xi_1 \dots \xi_N) - (P_n N_{-c} g_1 \ \dots \ P_n N_{-c} g_M) G_2\right) \\ &\qquad \times \left(\boldsymbol{p}(u) - V(1 - \Phi_{-c})^{-1} \langle v, \boldsymbol{y} \rangle\right) = 0, \end{aligned}$$

where

$$G_2 = \left(\langle \xi_k, y_j \rangle; \ \begin{matrix} j & \downarrow & 1, \dots, M \\ k & \rightarrow & 1, \dots, N \end{matrix} \right).$$

The equation for $u = y + T_{-c}^{-1} v$ is obtained in the following manner: Substituting the equation for v into the equation for y in (3.12), we calculate as

$$\begin{aligned} &\frac{du}{dt} - T_{-c}^{-1} \left(\sum_{k=1}^{N} p_k(y) P_n T_{-c} \xi_k - Lv + \sum_{k=1}^{M} \langle v, L_c y_k \rangle P_n N_{-c} g_k \right) + \mathscr{L}_c \left(u - T_{-c}^{-1} v\right) \\ &+ \sum_{k=1}^{N} p_k(y) T_{-c}^{-1} P_n T_{-c} \xi_k - \sum_{k=1}^{M} \langle v, L_c y_k \rangle T_{-c}^{-1} Q_n N_{-c} g_k = c\left(u - T_{-c}^{-1} v\right). \end{aligned}$$

Thus,

$$\frac{du}{dt} + T_{-c}^{-1} \left(L_c v - \sum_{k=1}^{M} \langle L_c v, y_k \rangle N_{-c} g_k \right) + \mathscr{L} u - L_c v = 0,$$

$$\text{or} \quad \frac{du}{dt} + \mathscr{L} u = 0, \quad \tau u = (g_1 \ \dots \ g_M)(1 - \Phi_{-c})^{-1} \langle v, \boldsymbol{y} \rangle.$$

Finally, (u, v) is subject to the equation in $L^2(\Omega) \times P_n L^2(\Omega)$:

$$\begin{cases} \dfrac{du}{dt} + \mathscr{L}u = 0, \qquad \tau u = \left(g_1 \ \dots \ g_M\right)(1 - \Phi_{-c})^{-1} \langle v, \mathbf{y} \rangle, \\ \dfrac{dv}{dt} + (L - C)v = (P_n N_{-c} g_1 \ \dots \ P_n N_{-c} g_M)\left(\langle v, L_c \mathbf{y} \rangle + G_2 \mathbf{p}(v)\right) \\ \qquad + \left((\xi_1 \ \dots \ \xi_N) - (P_n N_{-c} g_1 \ \dots \ P_n N_{-c} g_M) G_2\right) \\ \qquad \times \left(\mathbf{p}(u) - V(1 - \Phi_{-c})^{-1} \langle v, \mathbf{y} \rangle\right). \end{cases} \tag{3.14}$$

Note that both Cv and $\mathbf{p}(v)$ in (3.14) are bounded, since v belongs to $P_n L^2(\Omega)$. By assuming the same rank conditions as in (2.21), the decay estimate,

$$\|u(t,\cdot)\| + \|v(t)\|_{P_n L^2(\Omega)} \leqslant \text{const}\, e^{-rt} \left(\|u_0\| + \|v_0\|_{P_n L^2(\Omega)}\right), \quad t \geqslant 0$$

is ensured for every solution (u, v) to (3.14).

Algebraic equivalence of the two mathematical settings

We close this section with the following observation: The control system (3.1) looks fairly different from (2.17). However, there is a close algebraic relationship between the two control systems. Let us clarify this in the following.

Let (u, w) be a solution to (3.1), and set

$$\begin{pmatrix} q \\ v \end{pmatrix} = \begin{pmatrix} 1 & T_{-c} - 1 \\ 0 & T_{-c} \end{pmatrix} \begin{pmatrix} u \\ w \end{pmatrix}, \quad T_{-c} = 1 - \sum_{k=1}^{N} \langle \cdot, y_k \rangle N_{-c} g_k. \tag{3.15}$$

It is clear that $\tau v = 0$ and $\mathscr{L}_c v = L_c v = \mathscr{L}_c w$. Applying T_{-c} to the equation for w, we obtain

$$\frac{dv}{dt} + Lv - \sum_{k=1}^{M} \langle v, L_c y_k \rangle N_{-c} g_k = \sum_{k=1}^{N} p_k(u - w) T_{-c} \xi_k.$$

By (3.15), the above right-hand side is

$$\begin{aligned} \sum_{k=1}^{N} p_k(u - w) T_{-c} \xi_k &= \left((\xi_1 \ \dots \ \xi_N) - (N_{-c} g_1 \ \dots \ N_{-c} g_M) G_2\right) \mathbf{p}(u - w) \\ &= \left((\xi_1 \ \dots \ \xi_N) - (N_{-c} g_1 \ \dots \ N_{-c} g_M) G_2\right) \mathbf{p}(q - v). \end{aligned}$$

Thus,

$$\begin{aligned} \frac{dv}{dt} + Lv + (\xi_1 \ \dots \ \xi_N)\mathbf{p}(v) = (N_{-c} g_1 \ \dots \ N_{-c} g_M) \langle v, L_c \mathbf{y} \rangle + (\xi_1 \ \dots \ \xi_N) \mathbf{p}(q) \\ - (N_{-c} g_1 \ \dots \ N_{-c} g_M) G_2 \mathbf{p}(q - v), \end{aligned}$$

or

$$\begin{aligned} \frac{dv}{dt} + (L - C)v &= -Cq + (N_{-c} g_1 \ \dots \ N_{-c} g_M)\left(\langle v, L_c \mathbf{y} \rangle - G_2 \mathbf{p}(q - v)\right) \\ &= -Cq + (N_{-c} g_1 \ \dots \ N_{-c} g_M)\, \mathbf{g}(q, v). \end{aligned}$$

As for the equation for q, note that

$$0 = \tau(u - w) = \tau(q - v) = \tau q - \tau v = \tau q,$$

since the boundary conditions for u and w are the same. Thus, q is in $\mathscr{D}(L)$. Since $u = q - v + w$,

$$\frac{d}{dt}(q - v + w) + \mathscr{L}(q - v + w) = 0,$$
$$\frac{dq}{dt} - \frac{dv}{dt} + \frac{dw}{dt} + Lq - Lv + \mathscr{L}w = 0.$$

In (3.1), the equation for w is rewritten as

$$\begin{aligned}\frac{dw}{dt} + \mathscr{L}w &= \sum_{k=1}^{N} p_k(u - w)\xi_k \\ &= (\xi_1 \ \dots \ \xi_N)\boldsymbol{p}(q - v).\end{aligned}$$

Thus,

$$\begin{aligned}\frac{dq}{dt} + Lq &= \frac{dv}{dt} + Lv - \left(\frac{dw}{dt} + \mathscr{L}w\right) \\ &= -C(q - v) + (N_{-c}g_1 \ \dots \ N_{-c}g_M)\boldsymbol{g}(q, v) - (\xi_1 \ \dots \ \xi_N)\boldsymbol{p}(q - v) \\ &= (N_{-c}g_1 \ \dots \ N_{-c}g_M)\boldsymbol{g}(q, v).\end{aligned}$$

The equation for (q, v) is summerized as follows:

$$\begin{aligned}&\frac{dq}{dt} + Lq = (N_{-c}g_1 \ \dots \ N_{-c}g_M)\boldsymbol{g}(q, v), \\ &\frac{dv}{dt} + (L - C)v = -Cq + (N_{-c}g_1 \ \dots \ N_{-c}g_M)\boldsymbol{g}(q, v).\end{aligned}$$

The equation is, however, nothing but (2.17). Thus, we have shown

Theorem 3.1. *In the control system* (3.1) *with state* (u, w), *we introduce the new state* (q, v) *by* (3.15). *Then the state* (q, v) *satisfies* (2.17). *In other words, the system* (3.1) *is algebraically connected with the system* (2.17) *by the boundedly invertible transformation* (3.15).

Chapter 6

Output stabilization : lack of the observability and/or the controllability conditions

6.1 Introduction

In the preceding chapters on stabilization problems, we have constructed suitable feedback laws such that the state of the system decays with the designated decay rate as $t \to \infty$. The outputs are (bounded or unbounded) linear functionals of the state. In this chapter, we focus our attention on stability enhancement of outputs and related linear functional for a class of linear parabolic systems by means of feedback control, and present sufficient conditions which turn out to be fairly different from those for regular state stabilization. Stability enhancement of outputs is regarded as "stabilization" of outputs to strengthen the stability property of outputs. Regular feedback laws are such that the state of the system strongly converge to zero as $t \to \infty$ in respective topology of function spaces. Let H be a separable Hilbert space equipped with inner product $\langle \cdot, \cdot \rangle_H$ and norm $\|\cdot\|_H$. Let us consider the following control system with state $u(\cdot)$ in H:

$$\frac{du}{dt} + Lu = \sum_{k=1}^{M} f_k(t) g_k, \quad u(0) = u_0. \tag{1.1}$$

The output of the system is N nontrivial linear functionals defined as

$$\langle u, w_k \rangle_H, \quad 1 \leqslant k \leqslant N, \tag{1.2}$$

for solutions $u = u(t)$ to (1.1), where $f_k(t)$, $1 \leqslant k \leqslant M$, denote inputs; $g_k \in H$ actuators; and $w_k \in H$ weighting vectors. The setting of the system in this form is just for simplicity of the following arguments. Later, it is naturally generalized to a problem arising from boundary control systems.

In regular state stabilization problems, we have seen that the vectors w_k and g_k satisfying, respectively, the finite-dimensional observability and controllability conditions ensure a stabilizing feedback scheme of the system. In a simple system such that there arises no generalized eigenspace of L, for example, these conditions mean

$$\begin{aligned} &\text{rank}\, W_i = m_i \quad \text{(observability condition), and} \\ &\text{rank}\, G_i = m_i \quad \text{(controllability condition),} \end{aligned}$$

for $1 \leqslant i \leqslant \nu$, where W_i and G_i are, respectively, $N \times m_i$ and $m_i \times M$ matrices defined in (2.1) in the next section (m_i is the albebraic multiplicity of the eigenvalue λ_i of L). In the case of the scheme of boundary observation/boundary control, W_i and G_i are replaced by $\hat{W}_i$ and $\hat{G}_i$, respectively (see (3.1), Chapter 4). Thus we have to choose both M and N greater than or equal to $\max m_i$. We then ask the following question: "*what control theoretic results do we expect with smaller numbers* M, $N < \max m_i$*?*" Stability enhancement of output or output stabilization answers this question, which is the main contribution of the chapter. Stabilization of the output does not necessarily guarantee stabilization of the state when the system without any control is unstable. In many physical systems equipped with control schemes, however, particular nonlinearities will protect actual systems from destruction even though the states in *linearized* mathematical models grow or blow up as $t \to \infty$. In such systems, we may restrict our attention to the output only. Therefore the study of output stabilization for such systems is justified. Some readers might be, however, concerned about stability of state itself. In order to guarantee the state stability of *linear* systems under consideration, we thus limit ourselves to eqn. (1.1) which is originally stable (see the assumption (iii) below). The results of this chapter [1] are based on those discussed in [43, 47].

There appears an essential distinction between the state and the output stabilization: In fact, in the latter problem w_k and g_k must be determined, dependent on each other (see the rank conditions (2.3) in the next section). To avoid the possibility of generalized eigenspaces, it is assumed in Sections 1 through 3 that

[1] reprinted with permission of Cambridge University Press.

(i) the operator L is self-adjoint with dense domain $\mathscr{D}(L)$;
(ii) L has a compact resolvent; and
(iii) the spectrum $\sigma(L)$ is contained in $\mathbb{C}_+$.

The last assumption (iii) is of a technical nature: It is set for ensuring the stability of the state u while the output (1.2) is stabilized. In order to avoid non-essential technical difficulty, the output (1.2) is a bounded linear functional regarding u and L is self-adjoint. It is possible, however, to weaken these assumptions. The output (1.2) may admit some unboundedness derived from, e.g., boundary observation, and L may be replaced by spectral operators [15, Part 3] which often appear in flexible structures (see also [13]). In fact, the only difference is that there appear complex eigenvalues in the case of spectral operators, whereas the corresponding Riesz basis is available in each class of operators. Due to assumption (iii), our problem is to enhance stability of the output (1.2) when the observability and controllability conditions are lost.

In the static feedback scheme such that $f_k(t) = \langle u, w_k \rangle_H$, $1 \leqslant k \leqslant N(= M)$, the output stabilization consists of constructing g_k under the rank conditions on the matrices W_i. However, a severe restriction on g_k must be imposed [38], which is not satisfied, e.g., in boundary control systems. Generally speaking, both w_k and g_k cannot be freely designed, and have "spillovers." In other words, what we can construct is limited to a finite number of parameters corresponding to the Fourier series expansions of these vectors (we are lucky, however, since the Fourier series expansions are available in our problem setting). The other parameters belonging to the infinite-dimensional subspace remain as uncontrollable residues or spillovers. Thus we introduce a compensator in $\mathbb{C}^m$:

$$\frac{dv}{dt} + B_1 v = \sum_{k=1}^{N} \langle u, w_k \rangle_H \xi_k + \sum_{k=1}^{M} f_k(t) \alpha_k, \quad v(0) = v_0 \qquad (1.3)$$
$$\text{with } f_k(t) = \langle v, \rho_k \rangle_{\mathbb{C}^m},$$

and construct a feedback control scheme in (1.1) and (1.3) so that the output (1.2) decays exponentially as $t \to \infty$ for an arbitrary initial value. This control scheme contains several parameters. Given w_k and g_k, parameters to be determined are the dimension m; vectors ξ_k; α_k; and ρ_k in (1.3).

There are a few papers in the literature such as [8, 62] for the control of the output. Their problems are briefly stated as follows: Given a set of specific scalar functions generated by a class of finite-dimensional oscillations, construct feedback control schemes such that the output of the infinite-dimensional systems asymptotically track these functions, while the state stability of the infinite-dimensional systems are guaranteed. To achieve this, the above observability and controllability conditions are indispensable in their works. On the contrary, our setting of the problem is entirely different from theirs: We are seeking *some* control theoretic result and assertion when the observability and controllability conditions are *lost*.

In Section 2, the same principle is applied with no essential change to boundary control systems, where the output and the actuators admit some unboundedness. The same problem is studied again in Section 3 via a different approach. Throughout the chapter, it is assumed that L admits *no* generalized eigenvector associated with the first ν eigenvalues of L. In fact, for a class of L admitting generalized eigenvectors, it is shown in Section 4 that the output stabilization scheme implies state stabilization, too. In Section 5, we construct a specific feedback scheme, under the observability condition, such that some nontrivial functional decay *faster* than the state: In constructing the control scheme, we solve a new problem of an arbitrary allocation of the eigenvalues of some coefficient matrix which is *subject to constraint*, and give the necessary and sufficient condition in terms of controllability and observabililty assumptions. This is an essential extension of the celebrated theory of pole allocation [70] by W. M. Wonham.

6.2 Output Stabilization

According to the property (ii) in Section 1, the spectrum $\sigma(L)$ consists only of eigenvalues, λ_i, $i \geqslant 1$. According to Hilbert-Schmidt theory, in addition, there is a set of eigenpairs $\{\lambda_i, \varphi_{ij}\}$ satisfying the conditions:

(i) $\sigma(L) = \{\lambda_i\}_{i\geqslant 1}, \quad 0 < \lambda_1 < \cdots < \lambda_i < \cdots \to \infty$;

(ii) $(\lambda_i - L)\varphi_{ij} = 0, \quad i \geqslant 1, \ 1 \leqslant j \leqslant m_i (< \infty)$; and

(iii) the set $\{\varphi_{ij}\}$ forms an *orthonormal* basis for H.

Thus, the assumption on L is the same as in Chapter 5. The condition: $\lambda_1 > 0$ is derived from our (technical) assumption (iii) in Section 1. Let P_n be the projector corresponding to the first n eigenvalues, $\lambda_1, \ldots, \lambda_n$, and set $Q_n = 1 - P_n$. Then, for $n \geqslant 1$,

$$u = P_n u + Q_n u = \sum_{i,j\,(i\leqslant n)} u_{ij}\varphi_{ij} + Q_n u, \qquad u_{ij} = \langle u, \varphi_{ij}\rangle_H .$$

After similar Fourier expansions of w_k and g_k, let W_i and G_i, $1 \leqslant i \leqslant n$, be the matrices defined, respectively, as

$$\begin{aligned} W_i &= \left(w_{ij}^k ; \begin{array}{ccc} j & \to & 1, \ldots, m_i \\ k & \downarrow & 1, \ldots, N \end{array} \right), \quad w_{ij}^k = \langle \varphi_{ij}, w_k \rangle_H, \quad \text{and} \\ G_i &= \left(g_{ij}^k ; \begin{array}{ccc} j & \downarrow & 1, \ldots, m_i \\ k & \to & 1, \ldots, M \end{array} \right), \quad g_{ij}^k = \langle g_k, \varphi_{ij} \rangle_H . \end{aligned} \tag{2.1}$$

As stated in Section 1, we are trying to find what we could obtain with smaller integers M and N. Thus we assume that

$$M, N < \max_{1\leqslant i\leqslant \nu} m_i. \tag{2.2}$$

The following result ensures a stabilizing compensator for the output. The result will be later applied to a class of boundary control systems in Section 3, where the matrices W_i and G_i take somewhat different forms.

Theorem 2.1. *Let* μ, $\lambda_\nu < \mu < \lambda_{\nu+1}$, *be an arbitrary number, and set* $M = N$. *Let*

$$\mathbb{N}_1 = \{i \in \mathbb{N};\ m_i > N,\ 1 \leqslant i \leqslant \nu\} \neq \varnothing,$$
$$\mathbb{N}_2 = \{i \in \mathbb{N};\ m_i \leqslant N,\ 1 \leqslant i \leqslant \nu\}.$$

Suppose that

$$\begin{cases} \text{rank}\ W_i G_i = N, & i \in \mathbb{N}_1, \\ \text{rank}\ W_i = \text{rank}\ G_i = m_i, & i \in \mathbb{N}_2. \end{cases} \tag{2.3}$$

Then we can find the compensator (1.3) *of suitable dimension m such that*

$$|\langle u(t), w_k \rangle_H| + |v(t)|_{\mathbb{C}^m} \leqslant \text{const}\, e^{-\mu t} \left(\|u_0\|_H + |v_0|_{\mathbb{C}^m} \right), \quad t \geqslant 0 \tag{2.4}$$

for arbitrary $(u_0, v_0) \in H \times \mathbb{C}^m$, $1 \leqslant k \leqslant N$. *In addition, the state* $\|u(t)\|_H$ *stays bounded as* $t \to \infty$. *In fact,* $\|u(t)\|_H$ *has an upper bound,* $\text{const}\, e^{-\lambda_1 t}$, $t \geqslant 0$.

Proof of Theorem 2.1.

The idea of the proof is to derive a differential equation for the output and to reduce the problem to the state stabilization problem of this equation.

(*Equation for the output*). Changing the order of λ_i if necessary, we may assume, with no loss of generality, that $\mathbb{N}_1 = \{i \in \mathbb{N};\ 1 \leqslant i \leqslant \nu_1\}$ and $\mathbb{N}_2 = \{i \in \mathbb{N};\ \nu_1 < i \leqslant \nu\}$. Set

$$\hat{u}_i = \left(u_{i1} \ \dots \ u_{im_i} \right)^{\mathrm{T}}, \quad \text{and } \hat{w}_i^k = \left(\langle \varphi_{i1}, w_k \rangle_H \ \cdots \ \langle \varphi_{im_i}, w_k \rangle_H \right),$$

and let $\hat{y}$ and u_2 be defined, respectively, by

$$\hat{y} = \left(\hat{y}_1 \ \dots \ \hat{y}_N \right)^{\mathrm{T}}, \quad \hat{y}_k = \left(\hat{w}_1^k \hat{u}_1 \ \dots \ \hat{w}_{\nu_1}^k \hat{u}_{\nu_1} \right)^{\mathrm{T}}, \quad 1 \leqslant k \leqslant N,$$
$$\text{and } u_2 = (1 - P_{\nu_1}) u = Q_{\nu_1} u.$$

The output is then rewritten as

$$\langle u, w_k \rangle_H = (\underbrace{1 \dots 1}_{\nu_1\ 1\text{s}}) \hat{y}_k + \langle u_2, w_k \rangle_H, \quad 1 \leqslant k \leqslant N,$$

or in vector form

$$\langle u, \mathbf{w} \rangle_H = \Phi \hat{y} + \langle u_2, \mathbf{w} \rangle_H = \begin{pmatrix} \Phi & \langle \cdot, \mathbf{w} \rangle_H \end{pmatrix} \begin{pmatrix} \hat{y} \\ u_2 \end{pmatrix},$$

where Φ denotes the $N \times \nu_1 N$ matrix defined by

$$\Phi = \begin{pmatrix} 1 \dots 1 & 0 \dots 0 & \dots & 0 \dots 0 \\ 0 \dots 0 & 1 \dots 1 & \dots & 0 \dots 0 \\ \vdots & \vdots & \ddots & \vdots \\ 0 \dots 0 & 0 \dots 0 & \dots & 1 \dots 1 \end{pmatrix}.$$

Setting

$$x = \begin{pmatrix} \hat{y} \\ u_2 \end{pmatrix} \in \mathbb{C}^{\nu_1 N} \times Q_{\nu_1} H,$$

we obtain

$$\frac{dx}{dt} + \boldsymbol{L}x = \begin{pmatrix} WG \\ Q_{\nu_1} g_1 \dots Q_{\nu_1} g_N \end{pmatrix} \begin{pmatrix} f_1 \\ \vdots \\ f_N \end{pmatrix} = K\boldsymbol{f}, \tag{2.5}$$

where

(i) $\boldsymbol{L} = \begin{pmatrix} \Lambda_N & 0 \\ 0 & L_2 \end{pmatrix}$, $\Lambda_N = \text{diag}\,(\underbrace{\Lambda\,\Lambda\,\dots\,\Lambda}_{N\,\Lambda\text{s}})$, and $\Lambda = \text{diag}\,(\lambda_1\ \lambda_2\ \dots\ \lambda_{\nu_1})$;

(ii) the operator L_2 denotes the restriction: $L|_{Q_{\nu_1} H}$ with $\mathscr{D}(L_2) = \mathscr{D}(L) \cap Q_{\nu_1} H$; and

(iii)

$$W = \begin{pmatrix} \tilde{W}_1 \\ \vdots \\ \tilde{W}_N \end{pmatrix}, \quad \tilde{W}_k = \begin{pmatrix} \hat{w}_1^k & 0 & \dots & 0 \\ 0 & \hat{w}_2^k & \dots & 0 \\ \vdots & \vdots & \ddots & \vdots \\ 0 & 0 & \dots & \hat{w}_{\nu_1}^k \end{pmatrix},$$

$$\text{and } G = \begin{pmatrix} G_1 \\ \vdots \\ G_{\nu_1} \end{pmatrix} = \left(g_{ij}^k;\ \begin{matrix} (i,j) & \downarrow & (1,1), \dots, (\nu_1, m_{\nu_1}) \\ k & \rightarrow & 1, \dots, N \end{matrix} \right).$$

It is apparent that $\mathscr{D}(\boldsymbol{L}) = \mathbb{C}^{\nu_1 N} \times \mathscr{D}(L_2)$ and $\sigma(\boldsymbol{L}) = \sigma(L)$ with different multiplicities.

(*Compensator design*). The compensator for (1.1) is here an *identity compensator* employed in Chapter 5, and is first given as a differential equation in the space $\mathscr{H} = \mathbb{C}^{\nu_1 N} \times Q_{\nu_1} H$:

$$\begin{aligned} \frac{dv}{dt} + Bv &= \sum_{k=1}^{N} \langle u, w_k \rangle_H\, \xi_k + \sum_{k=1}^{N} f_k(t) \alpha_k \\ &= (\xi_1 \dots \xi_N)\big(\Phi\ \langle \cdot, \boldsymbol{w} \rangle_H\big)x + \big(\alpha_1\ \dots\ \alpha_N\big)\boldsymbol{f} \\ &= -Cx + \big(\alpha_1\ \dots\ \alpha_N\big)\boldsymbol{f}. \end{aligned} \tag{2.6}$$

Equation (2.6) will be finally reduced to a finite-dimensional equation equivalent to (1.3). Set

$$B = \boldsymbol{L} - C = \boldsymbol{L} + \big(\xi_1\ \dots\ \xi_N\big)\big(\Phi \quad \langle \cdot, \boldsymbol{w} \rangle_H\big). \tag{2.7}$$

Here, ξ_k are the parameters to be constructed. The operator $-B$ is the generator of an analytic semigroup. Let $\mathscr{P}_n$, $n > \nu_1$, be the projector in $\mathscr{H}$, corresponding to the eigenvalues $\lambda_1, \dots, \lambda_n$ of $\boldsymbol{L}$. Then,

$$\mathscr{P}_\nu B \mathscr{P}_\nu \iff \begin{pmatrix} \Lambda_N & 0 \\ 0 & \Sigma \end{pmatrix} + (\xi_1 \ \dots \ \xi_N) \begin{pmatrix} \Phi & W_{\nu_1+1} & \dots & W_\nu \end{pmatrix},$$

where $\Sigma = \operatorname{diag}\begin{pmatrix} \lambda_{\nu_1+1} I_{m_{\nu_1+1}} & \dots & \lambda_\nu I_{m_\nu} \end{pmatrix}$. Let us choose ξ_k in the subspace $\mathscr{P}_\nu \mathscr{H} = \mathbb{C}^{\nu_1 N+J} \times \{0\}$, $J = \sum_{i=\nu_1+1}^{\nu} m_i$. It is clear that the stability of e^{-tB} is determined by $\sigma(\mathscr{P}_\nu B \mathscr{P}_\nu)$. Let Π_1 be the $\nu_1 N \times \nu_1 N$ matrix defined by

$$\Pi_1 = \begin{pmatrix} \overbrace{10\dots0}^{N} & \overbrace{}^{N} & \cdots & \overbrace{00\dots0}^{N} \\ & 10\dots0 & \cdots & \\ \vdots & \vdots & \ddots & \vdots \\ 00\dots0 & & \cdots & 10\dots0 \\ \cdots\cdots & \cdots\cdots & \cdots\cdots & \cdots\cdots \\ \vdots & \vdots & \vdots & \vdots \\ \cdots\cdots & \cdots\cdots & \cdots\cdots & \cdots\cdots \\ 0\dots01 & & \cdots & 00\dots0 \\ & 0\dots01 & & \\ \vdots & \vdots & \ddots & \vdots \\ 00\dots0 & & \cdots & 0\dots01 \end{pmatrix} \begin{matrix} \left.\vphantom{\begin{matrix}1\\1\\1\\1\end{matrix}}\right\} \nu_1 \\ \\ \\ \\ \left.\vphantom{\begin{matrix}1\\1\\1\\1\end{matrix}}\right\} \nu_1 \end{matrix}.$$

It is easily seen that

$$\begin{aligned} &\begin{pmatrix} \Pi_1 & 0 \\ 0 & I_J \end{pmatrix}^{-1} \left(\begin{pmatrix} \Lambda_N & 0 \\ 0 & \Sigma \end{pmatrix} + (\xi_1 \ \dots \ \xi_N)(\Phi \ W_{\nu_1+1} \ \dots \ W_\nu) \right) \begin{pmatrix} \Pi_1 & 0 \\ 0 & I_J \end{pmatrix} \\ &= \operatorname{diag}\begin{pmatrix} \lambda_1 I_N & \dots & \lambda_{\nu_1} I_N & \Sigma \end{pmatrix} \\ &\quad + \begin{pmatrix} \Pi_1 & 0 \\ 0 & I_J \end{pmatrix}^{-1} (\xi_1 \ \dots \ \xi_N) \cdot \begin{pmatrix} I_N & \dots & I_N & W_{\nu_1+1} & \dots & W_\nu \end{pmatrix}. \end{aligned}$$

Since rank $I_N = N$, $1 \leqslant i \leqslant \nu_1$, and rank $W_i = m_i$, $\nu_1 < i \leqslant \nu$, the observability condition is satisfied. Thus we can find suitable $\xi_1, \dots, \xi_N \in \mathbb{C}^{\nu_1 N+J}$ such that $\min \operatorname{Re} \sigma(\mathscr{P}_\nu B \mathscr{P}_\nu) > \lambda_{\nu+1}$ (see Chapter 1). This immediately gives the estimate:

$$\left\| e^{-tB} \right\|_{\mathscr{L}(\mathscr{H})} \leqslant \text{const}\, e^{-\lambda_{\nu+1} t}, \quad t \geqslant 0. \tag{2.8}$$

Setting

$$(\alpha_1 \ \dots \ \alpha_N) = \begin{pmatrix} WG \\ Q_{\nu_1} g_1 \dots Q_{\nu_1} g_N \end{pmatrix} \in \mathscr{H}^N,$$

we see that $x(t) - v(t) = e^{-tB}(x_0 - v_0)$, $t \geqslant 0$. At this stage, set $f_k(t) = \langle v, \rho_k \rangle_{\mathscr{H}}$, $1 \leqslant k \leqslant N$, or $\boldsymbol{f}(t) = \langle v(t), \boldsymbol{\rho} \rangle_{\mathscr{H}}$. Then,

$$\frac{dx}{dt} + (\boldsymbol{L} - K\langle \cdot, \boldsymbol{\rho} \rangle_{\mathscr{H}})x = K\langle e^{-tB}(v_0 - x_0), \boldsymbol{\rho} \rangle_{\mathscr{H}}. \tag{2.9}$$

By assuming that the ρ_k are constructed in $\mathscr{P}_\nu \mathscr{H} = \mathbb{C}^{\nu_1 N+J} \times \{0\}$, the set $\{\rho_1, \ldots, \rho_N\}$ is identified with an $N \times (\nu_1 N + J)$ matrix, say, Ψ. Then, we see that

$$\mathscr{P}_\nu (\boldsymbol{L} - K\langle \cdot, \boldsymbol{\rho} \rangle_{\mathscr{H}}) \mathscr{P}_\nu \iff \begin{pmatrix} \Lambda_N & 0 \\ 0 & \Sigma \end{pmatrix} - \begin{pmatrix} WG \\ G_{\nu_1+1} \\ \vdots \\ G_\nu \end{pmatrix} \Psi. \tag{2.10}$$

Let Π_2 be the $\nu_1 N \times n_1 N$ matrix defined by

$$\Pi_2 = \begin{pmatrix} \overbrace{10\ldots0}^{\nu_1} & \overbrace{}^{\nu_1} & \cdots & \overbrace{00\ldots0}^{\nu_1} \\ & 10\ldots0 & \cdots & \\ \vdots & \vdots & \ddots & \vdots \\ 00\ldots0 & & \cdots & 10\ldots0 \\ \cdots\cdots & \cdots\cdots & \cdots\cdots & \cdots\cdots \\ \vdots & \vdots & \vdots & \vdots \\ \cdots\cdots & \cdots\cdots & \cdots\cdots & \cdots\cdots \\ 0\ldots01 & & \cdots & 00\ldots0 \\ & 0\ldots01 & & \\ \vdots & \vdots & \ddots & \vdots \\ 00\ldots0 & & \cdots & 0\ldots01 \end{pmatrix} \begin{matrix} \left.\vphantom{\begin{matrix}a\\a\\a\\a\end{matrix}}\right\} N \\ \\ \\ \\ \left.\vphantom{\begin{matrix}a\\a\\a\\a\end{matrix}}\right\} N \end{matrix} .$$

Then, it is clear that

$$\Pi_2 = \Pi_1^{-1}.$$

The pole assignment problem of the right-hand side of (2.10) is equivalent to the problem of

$$\begin{aligned} &\begin{pmatrix} \Pi_1 & 0 \\ 0 & I_J \end{pmatrix}^{-1} \left\{ \begin{pmatrix} \Lambda_N & 0 \\ 0 & \Sigma \end{pmatrix} - \begin{pmatrix} WG \\ G_{\nu_1+1} \\ \vdots \\ G_\nu \end{pmatrix} \Psi \right\} \begin{pmatrix} \Pi_1 & 0 \\ 0 & I_J \end{pmatrix} \\ &= \operatorname{diag}\left(\lambda_1 I_N \quad \ldots \quad \lambda_{\nu_1} I_N \quad \Sigma\right) - \begin{pmatrix} \Pi_2 WG \\ G_{\nu_1+1} \\ \vdots \\ G_\nu \end{pmatrix} \Psi \begin{pmatrix} \Pi_1 & 0 \\ 0 & I_J \end{pmatrix}. \end{aligned} \tag{2.10$'$}$$

In $(2.10')$ note that

$$\Pi_2 WG = \begin{pmatrix} W_1 G_1 \\ \vdots \\ W_{\nu_1} G_{\nu_1} \end{pmatrix}.$$

Since rank $W_i G_i = N$, $1 \leqslant i \leqslant \nu_1$, and rank $G_i = m_i$, $\nu_1 < i \leqslant \nu$, the controllability condition is satisfied. Thus, there is a Ψ such that the spectrum of the matrix $(2.10')$ is freely assigned. More precisely, for an arbitrarily preassigned set of $\nu_1 N + J$ complex numbers, we can find a Ψ such that the spectrum of the matrix coincides with this set. Choosing a Ψ so that the spectrum lies in the right half-plane: $\mathrm{Re}\,\lambda > \lambda_{\nu+1}$, i.e.,

$$\min \mathrm{Re}\, \sigma\left(\mathscr{P}_\nu\left(\boldsymbol{L} - K\langle\cdot, \boldsymbol{\rho}\rangle_{\mathscr{H}}\right)\mathscr{P}_\nu\right) > \lambda_{\nu+1},$$

we see that

$$\left\|\exp\left(t\left(-\boldsymbol{L} + K\langle\cdot, \boldsymbol{\rho}\rangle_{\mathscr{H}}\right)\right)\right\|_{\mathscr{L}(\mathscr{H})} \leqslant \mathrm{const}\, e^{-\lambda_{\nu+1} t}, \quad t \geqslant 0. \tag{2.11}$$

We find that, for an arbitrary μ', $\mu < \mu' < \lambda_{\nu+1}$,

$$\|x(t)\|_{\mathscr{H}} + \|v(t)\|_{\mathscr{H}} \leqslant \mathrm{const}\, e^{-\mu' t}\left(\|x(0)\|_{\mathscr{H}} + \|v(0)\|_{\mathscr{H}}\right), \quad t \geqslant 0. \tag{2.12}$$

Since $\{\varphi_{ij}\}$ forms an orthonormal basis for H, the following lemma is immediate.

Lemma 2.2. *The operator $\mathscr{P}_n B - B\mathscr{P}_n$ has a unique extension to an element in $\mathscr{L}(\mathscr{H})$, and satisfies the estimate*

$$\|(\mathscr{P}_n B - B\mathscr{P}_n)v\|_{\mathscr{H}} \leqslant \varepsilon_n \|v\|_{\mathscr{H}}, \quad v \in \mathscr{D}(B), \quad \varepsilon_n \to 0 \text{ as } n \to \infty.$$

Let us add a small perturbation $\mathscr{P}_n B - B\mathscr{P}_n$ on the right-hand side of (2.6), and denote the perturbed equation as $(2.6')$. If n (greater than ν) is large enough, the solutions (x, v) to eqns. (2.5) and $(2.6')$ satisfy the estimate

$$\|x(t)\|_{\mathscr{H}} + \|v(t)\|_{\mathscr{H}} \leqslant \mathrm{const}\, e^{-\mu t}\left(\|x(0)\|_{\mathscr{H}} + \|v(0)\|_{\mathscr{H}}\right), \quad t \geqslant 0.$$

By setting $v_1 = \mathscr{P}_n v$ for such an n, (x, v_1) satisfies the equation:

$$\begin{aligned} &\frac{dx}{dt} + \boldsymbol{L}x = K\langle v_1, \boldsymbol{\rho}\rangle_{\mathscr{P}_n\mathscr{H}}, \\ &\frac{dv_1}{dt} + \mathscr{P}_n B\mathscr{P}_n v_1 = -\mathscr{P}_n C x + \mathscr{P}_n\left(\alpha_1 \ \dots \ \alpha_N\right)\langle v_1, \boldsymbol{\rho}\rangle_{\mathscr{P}_n\mathscr{H}}. \end{aligned} \tag{2.13}$$

Here, $\mathscr{P}_n B\mathscr{P}_n$ is bounded in the finite-dimensional subspace $\mathscr{P}_n\mathscr{H}$. The second equation of (2.13) means a finite-dimensional compensator. The solutions to (2.13) satisfy the estimate

$$\|x(t)\|_{\mathscr{H}} + \|v_1(t)\|_{\mathscr{P}_n\mathscr{H}} \leqslant \mathrm{const}\, e^{-\mu t}\left(\|x(0)\|_{\mathscr{H}} + \|v_1(0)\|_{\mathscr{P}_n\mathscr{H}}\right), \quad t \geqslant 0. \tag{2.14}$$

The resultant feedback control system corresponding to (2.13) is then described as

$$\begin{aligned} &\frac{du}{dt} + Lu = \left(g_1 \ \dots \ g_N\right)\langle v_1, \boldsymbol{\rho}\rangle_{\mathscr{P}_n\mathscr{H}}, \\ &\frac{dv_1}{dt} + \mathscr{P}_n B\mathscr{P}_n v_1 = \left(\xi_1 \ \dots \ \xi_N\right)\langle u, \boldsymbol{w}\rangle_H + \mathscr{P}_n\left(\alpha_1 \ \dots \ \alpha_N\right)\langle v_1, \boldsymbol{\rho}\rangle_{\mathscr{P}_n\mathscr{H}}. \end{aligned}$$

The last assertion of the theorem, that is, the state stability of the system is clear by looking at (1.1) with $f_k(t) = \langle v_1, \rho_k \rangle_{\mathbb{C}^m}$. The proof of Theorem 2.1 is hereby completed. □

Remark: Recall that the parameters ξ_k, ρ_k, and α_k of the compensator are constructed in the following manner:

(i) $\min \mathrm{Re}\, \sigma\left(\mathscr{P}_\nu B \mathscr{P}_\nu\right) > \lambda_{\nu+1}$;

(ii) $\min \mathrm{Re}\, \sigma\left(\mathscr{P}_\nu \left(\boldsymbol{L} - K\langle \cdot, \boldsymbol{\rho}\rangle_{\mathscr{H}}\right) \mathscr{P}_\nu\right) > \lambda_{\nu+1}$; and

(iii) $\mathscr{P}_n(\alpha_1 \ \dots \ \alpha_N) = \mathscr{P}_n \begin{pmatrix} WG \\ Q_{\nu_1} g_1 \ \dots \ Q_{\nu_1} g_N \end{pmatrix} \in (\mathscr{P}_n \mathscr{H})^N, \ n > \nu.$

6.3 Application to Boundary Control Systems

In boundary control systems, some weak unboundedness of observation/control appears. We show in this section that the principle is equally applied to a class of boundary feedback control systems with no essential difficulty. The operator L is derived from a pair of differential operators $(\mathscr{L}, \tau)$ which is similar to the one discussed in Chapter 5.

Let $\Omega \subset \mathbb{R}^m$ be a bounded domain with the boundary Γ which consists of a finite number of smooth components of $(m-1)$-dimension. Let $(\mathscr{L}, \tau)$ be a pair of differential operators in $\overline{\Omega}$ defined as

$$\mathscr{L}z = -\sum_{i,j=1}^{m} \frac{\partial}{\partial x_i}\left(a_{ij}(x)\frac{\partial z}{\partial x_j}\right) + c(x)z,$$

$$\tau z = \frac{\partial z}{\partial \nu} + \sigma(\xi)z = \sum_{i,j=1}^{m} a_{ij}(\xi)\nu_i(\xi)\frac{\partial z}{\partial x_j} + \sigma(\xi)z,$$

where $a_{ij}(x) = a_{ji}(x)$ for $1 \leqslant i, j \leqslant m$, $x \in \overline{\Omega}$, and $\mathscr{L}$ is uniformly elliptic. In the boundary operator τ, $\nu(\xi) = (\nu_1(\xi), \dots, \nu_m(\xi))$ denotes the unit outer normal at $\xi \in \Gamma$, and $\sigma(\xi)$ bounded measurable. Regularity of the other coefficients in $(\mathscr{L}, \tau)$ is assumed tacitly. The boundary operator in Chapters 4 and 5 is given in the form: $\tau z = \alpha(\xi)z + (1 - \alpha(\xi))\dfrac{\partial z}{\partial \nu}$. The present τ thus corresponds to the case where $0 \leqslant \alpha(\xi) < 1$. Consider the boundary control system described by

$$\begin{cases} \dfrac{\partial z}{\partial t} + \mathscr{L}z = 0, \quad \text{in } \mathbb{R}^1_+ \times \Omega, \\ \tau z = \displaystyle\sum_{k=1}^{N} f_k(t) g_k, \quad \text{on } \mathbb{R}^1_+ \times \Gamma, \\ z(0, \cdot) = z_0, \quad \text{in } \Omega \end{cases} \tag{3.1}$$

with a finite number of observations on Γ:

$$\langle z, w_k \rangle_\Gamma, \quad 1 \leqslant k \leqslant N, \tag{3.2}$$

$\langle \cdot, \cdot \rangle_\Gamma$ being the inner product in $L^2(\Gamma)$. It is assumed that the actuators g_k, $1 \leqslant k \leqslant N$, belong to $H^{1/2}(\Gamma)$ and the sensors w_k to $L^2(\Gamma)$. Let us briefly show how a decay estimate for the output (3.2) is obtained. We first show that (3.1) is reduced to (1.1) within the $L^2(\Omega)$-framework. As usual, set $Lz = \mathscr{L}z$, $z \in \mathscr{D}(L) = \{z \in H^2(\Omega);\ \tau z = 0 \text{ on } \Gamma\}$, and set $H = L^2(\Omega)$. As we have seen in Chapter 2, properties (i) and (ii) in Section 1 are satisfied for this L. Assume that $\inf_{\overline{\Omega}} c(x) > 0$. Then, $\sigma(L) \subset \mathbb{C}_+$, and the fractional powers of L are well defined. Following Section 6, Chapter 4, set

$$u(t) = L^{-1/4-\varepsilon} z(t, \cdot), \quad t > 0$$

for a fixed ε, $0 < \varepsilon < 1/4$. Assuming the solution $z(t)$ in $H^2(\Omega)$ holds, we see that $u(t)$ belongs to $\mathscr{D}(L)$ and satisfies the differential equation in $H = L^2(\Omega)$:

$$\frac{du}{dt} + Lu = \sum_{k=1}^{N} f_k(t) L^{3/4-\varepsilon} \psi_k, \quad u(0) = L^{-1/4-\varepsilon} z_0, \tag{3.3}$$

which is of the form same as (1.1)[2]. Here the well known relation: $\mathscr{D}(L^\theta) = H^{2\theta}(\Omega)$, $0 \leqslant \theta < 3/4$ (see (3.32), Chapter 2) has been applied, and the $\psi_k \in H^2(\Omega)$, $1 \leqslant k \leqslant N$, denote unique solutions to the boundary value problems:

$$\mathscr{L}\psi_k = 0 \quad \text{in } \Omega, \quad \tau\psi_k = g_k \quad \text{on } \Gamma, \quad 1 \leqslant k \leqslant N.$$

The matrices W_i and G_i in Theorem 2.1 are replaced by

$$\begin{aligned} W_i &= \left(\langle \varphi_{ij}, w_k \rangle_\Gamma \, ; \begin{array}{ccc} j & \rightarrow & 1, \dots, m_i \\ k & \downarrow & 1, \dots, N \end{array} \right), \quad \text{and} \\ G_i &= \left(\langle g_k, \varphi_{ij} \rangle_\Gamma \, ; \begin{array}{ccc} j & \downarrow & 1, \dots, m_i \\ k & \rightarrow & 1, \dots, N \end{array} \right), \end{aligned} \tag{3.4}$$

respectively. As in Section 2, define the vectors $\hat{y}_k$ and $\hat{y}$ with $\hat{w}_i^k$ replaced by

$$\hat{w}_i^k = \lambda_i^{1/4+\varepsilon} \left(w_{i1}^k \ \dots \ w_{im_i}^k \right), \quad w_{ij}^k = \langle \varphi_{ij}, w_k \rangle_\Gamma \, .$$

[2] If w_k belong to $H^{1/2}(\Gamma)$ in addition, then Green's formula implies that

$$\langle z, w_k \rangle_\Gamma = \left\langle L^{1/2+2\varepsilon} u(t), L^{3/4-\varepsilon} \varphi_k \right\rangle, \quad 1 \leqslant k \leqslant N,$$

$\langle \cdot, \cdot \rangle$ being the inner product in $L^2(\Omega)$. Here, the functions $\varphi_k \in H^2(\Omega)$ denote unique solutions to the boundary value problems:

$$\mathscr{L}\varphi_k = 0 \quad \text{in } \Omega, \quad \tau\varphi_k = w_k \quad \text{on } \Gamma, \quad 1 \leqslant k \leqslant N.$$

Then the output is rewritten in vector form as

$$\langle z, \boldsymbol{w}\rangle_\Gamma = \Phi\hat{y} + \left\langle L^{1/4+\varepsilon}u_2, \boldsymbol{w}\right\rangle_\Gamma = \begin{pmatrix}\Phi & \langle L^{1/4+\varepsilon}\cdot, \boldsymbol{w}\rangle_\Gamma\end{pmatrix}\begin{pmatrix}\hat{y}\\ u_2\end{pmatrix}.$$

The equation for $x = (\hat{y}\ u_2)^{\mathrm{T}}$ is described by (2.5) with g_k replaced by $L^{3/4-\varepsilon}\psi_k$. The equation for v is described by (2.6) with C replaced by

$$Cx = -\begin{pmatrix}\xi_1 & \dots & \xi_N\end{pmatrix}\begin{pmatrix}\Phi & \langle L^{1/4+\varepsilon}\cdot, \boldsymbol{w}\rangle_\Gamma\end{pmatrix}x, \quad x = (\hat{y}\ u_2)^{\mathrm{T}}.$$

Here ξ_k, $1 \leqslant k \leqslant N$, are chosen in $\mathscr{P}_\nu\mathscr{H}$ so that $\min \operatorname{Re}\sigma(\mathscr{P}_\nu B\mathscr{P}_\nu) > \lambda_{\nu+1}$, where $B = \boldsymbol{L} - C$, and α_k, $1 \leqslant k \leqslant N$, are given as

$$\begin{pmatrix}\alpha_1 & \dots & \alpha_N\end{pmatrix} = \begin{pmatrix} WG \\ Q_{\nu_1}L^{3/4-\varepsilon}\psi_1 \ \dots \ Q_{\nu_1}L^{3/4-\varepsilon}\psi_N\end{pmatrix}.$$

The vectors ρ_k, $1 \leqslant k \leqslant N$, are chosen in $\mathscr{P}_\nu\mathscr{H} = \mathbb{C}^{\nu_1 N+J} \times \{0\}$ so that

$$\min \operatorname{Re}\sigma\left(\mathscr{P}_\nu(\boldsymbol{L} - K\langle\cdot, \boldsymbol{\rho}\rangle_{\mathscr{H}})\mathscr{P}_\nu\right) > \lambda_{\nu+1},$$

where $K = \begin{pmatrix}\alpha_1 & \dots & \alpha_N\end{pmatrix} \in \mathscr{L}(\mathbb{C}^N; \mathbb{C}^{\nu_1 N} \times Q_{\nu_1}H)$.

Our goal is a decay estimate for (3.2). In $\mathscr{H} = \mathbb{C}^{\nu_1 N} \times Q_{\nu_1}H$, set

$$\boldsymbol{S} = \begin{pmatrix}1 & 0\\ 0 & L\end{pmatrix}, \quad \text{and } \boldsymbol{M} = \boldsymbol{L} - K\langle\cdot, \boldsymbol{\rho}\rangle_{\mathscr{H}},$$

where $\mathscr{D}(\boldsymbol{S}) = \mathscr{D}(\boldsymbol{M}) = \mathbb{C}^{\nu_1 N} \times \mathscr{D}(L_2)$. Both $\boldsymbol{S}$ and $\boldsymbol{M}$ are m-accretive with $\sigma(\boldsymbol{S})$ and $\sigma(\boldsymbol{M})$ lying in the right half-plane $\mathbb{C}_+$. It is clear that both $\boldsymbol{M}\boldsymbol{S}^{-1}$ and $\boldsymbol{S}\boldsymbol{M}^{-1}$ are bounded, or $\mathscr{D}(\boldsymbol{S}) = \mathscr{D}(\boldsymbol{M})$ with equivalence of the graph norms. This implies that

$$\mathscr{D}(\boldsymbol{S}^\omega) = \mathscr{D}(\boldsymbol{M}^\omega), \quad \boldsymbol{S}^\omega = \begin{pmatrix}1 & 0\\ 0 & L^\omega\end{pmatrix}, \quad 0 \leqslant \omega \leqslant 1$$

with equivalence of the graph norms [25]. Let us go back to the expression for solutions derived from (2.9). Note that x_0 is in $\mathscr{D}(\boldsymbol{S}^{\beta/2})$, since $u_0 = L^{-\beta/2}z_0$ is in $\mathscr{D}(L^{\beta/2})$, where $\beta = \frac{1}{2} + 2\varepsilon < 1$. Then,

$$\begin{aligned}\boldsymbol{S}^\beta x(t) &= \boldsymbol{S}^\beta\boldsymbol{M}^{-\beta}\boldsymbol{M}^{\beta/2}e^{-t\boldsymbol{M}}\boldsymbol{M}^{\beta/2}\boldsymbol{S}^{-\beta/2}\boldsymbol{S}^{\beta/2}x_0\\ &\quad + \int_0^t \boldsymbol{S}^\beta\boldsymbol{M}^{-\beta}\boldsymbol{M}^\beta e^{-(t-s)\boldsymbol{M}}K\langle e^{-sB}(v_0 - x_0), \boldsymbol{\rho}\rangle_{\mathscr{H}}\,ds.\end{aligned}$$

Thus the above expression immediately implies that

$$\left\|\boldsymbol{S}^\beta x(t)\right\|_{\mathscr{H}} \leqslant \begin{cases}\text{const}\, t^{-\beta/2}e^{-\mu t}\left(\left\|\boldsymbol{S}^{\beta/2}x_0\right\|_{\mathscr{H}} + \|v_0\|_{\mathscr{H}}\right), & t > 0, \quad \text{or}\\ \text{const}\, e^{-\mu t}\left(\left\|\boldsymbol{S}^\beta x_0\right\|_{\mathscr{H}} + \|v_0\|_{\mathscr{H}}\right), & t \geqslant 0.\end{cases}$$

The reduction procedure to a finite-dimensional compensator is the same as before. We finally obtain the estimate

$$\begin{aligned}
&\left|\langle z(t), \boldsymbol{w}\rangle_{\Gamma}\right| + |v(t)|_{\mathbb{C}^m} \\
&\leqslant \begin{cases} \text{const}\, t^{-\beta/2} e^{-\mu t} \left(\|z_0\| + |v_0|_{\mathbb{C}^m}\right), & t > 0, \quad \text{or} \\ \text{const}\, e^{-\mu t} \left(\|z_0\|_{H^{1/2+2\varepsilon}(\Omega)} + |v_0|_{\mathbb{C}^m}\right), & t \geqslant 0. \end{cases}
\end{aligned} \tag{3.5}$$

6.3.1 Algebraic approach to boundary control systems

Let us go back to the boundary control system (3.1) with the output (3.2) on Γ. Since the boundary condition consists simply of the third kind (the Robin boundary), we could employ the (integral) transform: $u(t) = L^{-1/4-\varepsilon} z(t, \cdot)$ in order to cancel the effect of the feedback term on Γ. However, this approach is no longer available, when the boundary operator τ is replaced by a more complicated operator such as the one in Chapters 4 and 5, where the Dirichlet boundary is locally *continuously* connected with the Neumann boundary.

In (3.1), let τ be replaced by

$$\tau z = \alpha(\xi) + (1 - \alpha(\xi)) \frac{\partial z}{\partial \nu}, \quad 0 \leqslant \alpha(\xi) \leqslant 1, \quad \alpha(\xi) \not\equiv 1,$$

where $\alpha(\xi)$ is smooth on Γ. The output (2.16) is unchanged. Following Section 3, Chapter 5, let us just outline the approach briefly. First note that

$$\frac{dz}{dt} + L\left(z - \sum_{k=1}^{N} f_k(t)\psi_k\right) = 0. \tag{3.6}$$

Set

$$z = P_\nu z + Q_\nu z = \sum_{i,j(i\leqslant\nu)} z_{ij}\varphi_{ij} + z_2, \quad z_{ij} = \langle z, \varphi_{ij}\rangle, \quad \hat{z}_i = \left(z_{i1} \ \dots \ z_{im_i}\right)^{\mathrm{T}},$$

$$\hat{y} = \left(\hat{y}_1 \ \dots \ \hat{y}_N\right)^{\mathrm{T}}, \quad \hat{y}_k = \left(\hat{w}_1^k \hat{z}_1 \ \dots \ \hat{w}_{\nu_1}^k \hat{z}_{\nu_1}\right)^{\mathrm{T}}, \quad 1 \leqslant k \leqslant N,$$

$$x = \begin{pmatrix} \hat{y} \\ \hat{z} \\ z_2 \end{pmatrix} \in \mathbb{C}^{\nu_1 N} \times \mathbb{C}^J \times Q_\nu L^2(\Omega), \quad \hat{z} = \left(\hat{z}_{\nu_1+1} \ \dots \ \hat{z}_\nu\right)^{\mathrm{T}}, \quad \text{and}$$

$$\boldsymbol{L} = \operatorname{diag}\left(\Lambda_N \quad \Sigma \quad L_2\right), \quad L_2 = L|_{Q_\nu L^2(\Omega)}, \quad \mathscr{D}(\boldsymbol{L}) = \mathbb{C}^{\nu_1 N} \times \mathbb{C}^J \times \mathscr{D}(L_2).$$

Then we have the equation for x:

$$\frac{dx}{dt} + \boldsymbol{L}\left(x - \sum_{k=1}^{N} f_k(t) \begin{pmatrix} 0 \\ 0 \\ Q_\nu \psi_k \end{pmatrix}\right) - \begin{pmatrix} \Lambda_N W G \\ \Sigma G' \\ 0 \end{pmatrix} \boldsymbol{f}(t) = 0. \tag{3.7}$$

Here,

$$G = \begin{pmatrix} G_1 \\ \vdots \\ G_{\nu_1} \end{pmatrix}, \quad G_i = \left(\langle \psi_k, \varphi_{ij} \rangle; \begin{array}{ccc} j & \downarrow & 1, \ldots, m_i \\ k & \rightarrow & 1, \ldots, N \end{array} \right)$$
$$= \frac{1}{\lambda_i} \left(\langle g_k, \varphi_{ij} - (\varphi_{ij})_\nu \rangle_\Gamma; \begin{array}{ccc} j & \downarrow & 1, \ldots, m_i \\ k & \rightarrow & 1, \ldots, N \end{array} \right). \tag{3.8}$$

The last expression is derived from Green's formula, and $G^r = (G_{\nu_1+1} \ \ldots \ G_\nu)^{\mathrm{T}}$. The output is written as

$$\langle z, \boldsymbol{w} \rangle_\Gamma = \Phi \hat{y} + \sum_{\nu_1 < i \leqslant \nu} W_i \hat{z}_i + \langle z_2, \boldsymbol{w} \rangle_\Gamma$$
$$= \left(\Phi \quad W_{\nu_1+1} \quad \ldots \quad W_\nu \quad \langle \cdot, \boldsymbol{w} \rangle_\Gamma \right) x. \tag{3.9}$$

The operator $\boldsymbol{L}$ admits a Riesz basis, so that it is possible to design an identity compensator for the equation (3.7). We take here, however, a more general compensator design, following Chapter 4. Let $\hat{H}$ be an arbitrary separable Hilbert space equipped with the orthonormal basis $\left\{ \eta_{ij}^{\pm}; \ i \geqslant 1, \ 1 \leqslant j \leqslant n_i \right\}$, with $n_i < \infty$ for each i. Every vector $v \in \hat{H}$ is expressed as a Fourier series in terms of $\{\eta_{ij}^{\pm}\}$ as

$$v = \sum_{i,j} \left(v_{ij}^+ \eta_{ij}^+ + v_{ij}^- \eta_{ij}^- \right), \quad v_{ij}^{\pm} = \left\langle v, \eta_{ij}^{\pm} \right\rangle_{\hat{H}},$$

$\langle \cdot, \cdot \rangle_{\hat{H}}$ being the inner product in $\hat{H}$. As in Section 2, Chapter 4, let $\{\mu_i\}$ be a sequence of increasing positive numbers: $0 < \mu_1 < \mu_2 < \cdots \to \infty$, and define the closed operator B as

$$Bv = \sum_{i,j} \left(\mu_i \omega^+ v_{ij}^+ \eta_{ij}^+ + \mu_i \omega^- v_{ij}^- \eta_{ij}^- \right), \quad v \in \mathscr{D}(B), \tag{3.10}$$

where $\omega^{\pm} = a \pm \mathrm{i}\sqrt{1-a^2}$, $0 < a < 1$; and $\mathscr{D}(B) = \left\{ v \in \hat{H}; \ \sum_{i,j} |v_{ij}^{\pm} \mu_i|^2 < \infty \right\}$. Then, we recall that (i) $\sigma(B) = \{\mu_i \omega^{\pm}; \ i \geqslant 1\}$; (ii) $(\mu_i \omega^{\pm} - B)\eta_{ij}^{\pm} = 0$ for $i \geqslant 1$, $1 \leqslant j \leqslant n_i$; and (iii) $-B$ is the generator of an analytic semigroup e^{-tB}, $t > 0$:

$$e^{-tB} v = \sum_{i,j} \left(e^{-\mu_i \omega^+ t} v_{ij}^+ \eta_{ij}^+ + e^{-\mu_i \omega^- t} v_{ij}^- \eta_{ij}^- \right),$$

which satisfies the decay estimate, $\left\| e^{-tB} \right\|_{\mathscr{L}(\hat{H})} \leqslant e^{-a\mu_1 t}$ for $t \geqslant 0$.

Note that $\sigma(\boldsymbol{L}) = \sigma(L) \, (\subset \mathbb{R}^1_+)$ with different multiplicities. Thus, $\sigma(\boldsymbol{L}) \cap \sigma(B) = \varnothing$. Given vectors $\xi_k \in \hat{H}$, $1 \leqslant k \leqslant N$, let us consider Sylvester's equation in $\mathscr{D}(\boldsymbol{L})$:

$$X\boldsymbol{L} - BX = C,$$
$$C = -\left(\xi_1 \ \ldots \ \xi_N \right) \left(\Phi \quad W_{\nu_1+1} \quad \ldots \quad W_\nu \quad \langle \cdot, \boldsymbol{w} \rangle_\Gamma \right). \tag{3.11}$$

The following propositions are proven in exactly the same way as in Propositions 3.2 and 3.3, Chapter 4:

Proposition 3.1. *The operator equation* (3.11) *admits a unique solution* $X \in \mathscr{L}(\mathbb{C}^{\nu_1 N} \times \mathbb{C}^J \times Q_\nu L^2(\Omega); \hat{H})$. *By setting* $\xi_k = \sum_{i,j}\left(\xi_{ij}^k \eta_{ij}^+ + \overline{\xi_{ij}^k}\eta_{ij}^-\right)$, *the solution* X *is expressed as*

$$Xx = \sum_{i,j}\sum_k f_k(\mu_i\omega^+; x)\xi_{ij}^k \eta_{ij}^+ + \sum_{i,j}\sum_k f_k(\mu_i\omega^-; x)\overline{\xi_{ij}^k}\eta_{ij}^-, \tag{3.12$_1$}$$

where

$$\begin{aligned} f_k(\lambda; x) &= (\underbrace{1 \ldots 1}_{\nu_1 \text{ 1s}})(\lambda - \Lambda)^{-1}\hat{y}_k + \sum_{\nu_1 < i \leqslant \nu} \frac{1}{\lambda - \lambda_i} W_i \hat{z}_i\big|_k \\ &\quad + \left\langle (\lambda - L_2)^{-1} z_2, w_k \right\rangle_\Gamma, \quad x = \begin{pmatrix} \hat{y} \\ \hat{z} \\ z_2 \end{pmatrix}, \quad \lambda \in \rho(\boldsymbol{L}). \end{aligned} \tag{3.12$_2$}$$

Assume in addition that $\sum_{i,j}\left|\xi_{ij}^k \mu_i^{1/4+\varepsilon}\right|^2 < \infty$, $1 \leqslant k \leqslant N$. *Then the range of* X *is contained in* $\mathscr{D}(B)$.

Proposition 3.2. *Assume that*

(i) $\mu_i \leqslant \text{const}\, i^\gamma$, $i \geqslant 1$ *for* $0 < \exists\gamma < 2$;

(ii) $\text{rank}\, \Xi_i = N$, $i \geqslant 1$, *where*

$$\Xi_i = \left(\xi_{ij}^k; \begin{array}{ccc} j & \downarrow & 1, \ldots, n_i \\ k & \rightarrow & 1, \ldots, N \end{array}\right), \quad i \geqslant 1; \quad \textit{and}$$

(iii) $\text{rank}\, W_i = m_i$, $\nu_1 + 1 \leqslant i \leqslant \nu$.

Then we have the inclusion relation:

$$\overline{X^* \hat{H}} \supset \mathbb{C}^{\nu_1 N + J} \times \{0\}. \tag{3.13}$$

Remark: The inclusion relation (3.13) is derived from the relation, $f_k(\lambda; x) \equiv 0$, $1 \leqslant k \leqslant N$.

Through the solution X in (3.12$_1$), (3.12$_2$), the compensator with state $v(t) \in \hat{H}$, $t \geqslant 0$, is defined as

$$\begin{aligned} &\frac{dv}{dt} + Bv = -Cx + F(t), \quad \text{where} \\ &F(t) = (BX + C)\begin{pmatrix} 0 \\ 0 \\ Q_\nu \boldsymbol{\psi} \end{pmatrix} \boldsymbol{f}(t) + X \begin{pmatrix} \Lambda_N W G \\ \Sigma G^r \\ 0 \end{pmatrix} \boldsymbol{f}(t). \end{aligned} \tag{3.14}$$

Then, we see that $Xx(t) - v(t) = e^{-tB}(Xx_0 - v_0),\ t \geqslant 0$. Set $\boldsymbol{f}(t) = \langle v(t), \boldsymbol{\rho} \rangle_{\hat{H}}$. According to Proposition 3.2, choose vectors $\boldsymbol{\rho}$ so that $X^*\boldsymbol{\rho}$ is arbitrarily close to the prescribed vectors $\boldsymbol{\zeta} \in (\mathbb{C}^{\nu_1 N+J})^N$. These $\boldsymbol{\zeta}$ will be constructed below to determine the spectral property of the matrix A_{11} in (3.18). Since

$$\langle v, \boldsymbol{\rho} \rangle_{\hat{H}} - \langle x, X^*\boldsymbol{\rho} \rangle_{\mathbb{C}^{\nu_1 N+J} \times Q_\nu L^2(\Omega)} \to 0 \quad \text{as } t \to \infty,$$

we have to examine the stability of the equation:

$$\begin{aligned} &\frac{dx}{dt} + \boldsymbol{L}\left(x - \begin{pmatrix} 0 \\ 0 \\ Q_\nu \boldsymbol{\psi} \end{pmatrix} \langle x, \boldsymbol{\zeta} \rangle_{\mathbb{C}^{\nu_1 N+J} \times Q_\nu H} \right) \\ &\qquad - \begin{pmatrix} \Lambda_N W G \\ \Sigma G^r \\ 0 \end{pmatrix} \langle x, \boldsymbol{\zeta} \rangle_{\mathbb{C}^{\nu_1 N+J} \times Q_\nu L^2(\Omega)} = 0, \end{aligned} \tag{3.15}$$

Let $T \in \mathscr{L}\left(\mathbb{C}^{\nu_1 N+J} \times Q_\nu L^2(\Omega)\right)$ be defined by

$$\varphi = Tx = x - \begin{pmatrix} 0 \\ 0 \\ Q_\nu \boldsymbol{\psi} \end{pmatrix} \langle x, \boldsymbol{\zeta} \rangle_{\mathbb{C}^{\nu_1 N+J} \times Q_\nu L^2(\Omega)}. \tag{3.16}$$

It is clear that the bounded inverse T^{-1} exists and is given by

$$x = T^{-1}\varphi = \varphi + \begin{pmatrix} 0 \\ 0 \\ Q_\nu \boldsymbol{\psi} \end{pmatrix} \langle \varphi, \boldsymbol{\zeta} \rangle_{\mathbb{C}^{\nu_1 N+J} \times Q_\nu L^2(\Omega)}.$$

Thus eqn. (3.15) is equivalent to the equation for φ:

$$\frac{d\varphi}{dt} + \boldsymbol{L}\varphi - \begin{pmatrix} 0 \\ 0 \\ Q_\nu \boldsymbol{\psi} \end{pmatrix} \langle \varphi, \boldsymbol{L}\boldsymbol{\zeta} \rangle - T \begin{pmatrix} \Lambda_N W G \\ \Sigma G^r \\ 0 \end{pmatrix} \langle \varphi, \boldsymbol{\zeta} \rangle = 0, \tag{3.17a}$$

or simply

$$\frac{d\varphi}{dt} + A\varphi = 0. \tag{3.17b}$$

Let $\hat{P}_n$ be the projector from $\mathbb{C}^{\nu_1 N+J} \times Q_\nu L^2(\Omega)$ onto $\mathbb{C}^{\nu_1 N+J} \times \{0\}$ and set $\hat{Q}_\nu = 1 - \hat{P}_\nu$. Setting $\varphi = (\varphi_1\ \varphi_2)^{\mathrm{T}},\ \varphi_1 = \hat{P}_\nu \varphi,\ \varphi_2 = \hat{Q}_\nu \varphi$, we have

$$\frac{d\varphi_1}{dt} + A_{11}\varphi_1 = 0, \quad \frac{d\varphi_2}{dt} + A_{21}\varphi_1 + A_{22}\varphi_2 = 0, \tag{3.18}$$

where

$$A_{11} = \hat{P}_\nu A \hat{P}_\nu = \begin{pmatrix} \Lambda_N & 0 \\ 0 & \Sigma \end{pmatrix} - \begin{pmatrix} \Lambda_N W G \\ \Sigma G^r \end{pmatrix} \langle \cdot, \boldsymbol{\zeta} \rangle,$$

$$A_{21} = \hat{Q}_\nu A \hat{P}_\nu = -Q_\nu \boldsymbol{\psi} \langle \cdot, \boldsymbol{L\zeta} \rangle - \hat{Q}_\nu T \begin{pmatrix} \Lambda_N W G \\ \Sigma G^r \\ 0 \end{pmatrix} \langle \cdot, \boldsymbol{\zeta} \rangle, \quad \text{and}$$

$$A_{22} = \hat{Q}_\nu A \hat{Q}_\nu = L_2.$$

Theorem 2.1 holds with the matrices W_i and G_i replaced at this time by

$$W_i = \left(\langle \varphi_{ij}, w_k \rangle_\Gamma \, ; \begin{array}{ccc} j & \rightarrow & 1, \dots, m_i \\ k & \downarrow & 1, \dots, N \end{array} \right), \quad \text{and}$$

$$G_i = \left(\langle g_k, \varphi_{ij} - (\varphi_{ij})_\nu \rangle_\Gamma \, ; \begin{array}{ccc} j & \downarrow & 1, \dots, m_i \\ k & \rightarrow & 1, \dots, N \end{array} \right),$$

respectively. The expression of the above G_i is somewhat different from those in (3.4), merely due to the form of the boundary operator τ.

According to our assumption (2.3), it is not hard to find vectors $\boldsymbol{\zeta} \in (\mathbb{C}^{\nu_1 N + J})^N$ such that $\min \operatorname{Re} \sigma(A_{11}) > \lambda_{\nu+1}$. In fact, we only have to apply the similarity transformation, $\operatorname{diag}\begin{pmatrix} \Pi_1 & I_J \end{pmatrix}$ to A_{11}, so that the matrix for the actuator, $\begin{pmatrix} \Lambda_N W G & \Sigma G^r \end{pmatrix}^{\mathrm{T}}$ is changed to

$$\begin{pmatrix} \lambda_1 W_1 G_1 & \dots & \lambda_{\nu_1} W_{\nu_1} G_{\nu_1} & \lambda_{\nu_1+1} G_{\nu_1+1} & \dots & \lambda_\nu G_\nu \end{pmatrix}^{\mathrm{T}}$$

(see the proof of Theorem 2.1). Then, the rest of the arguments to derive the stability property of (3.17) or (3.15) with a designated decay rate and reduce to the system containing a finite-dimensional compensator is the same as before.

6.3.2 Some generalization

We have so far assumed that L is a self-adjoint or a spectral operator. The class of L is generalized to some extent. The basic equation is (1.1) in H. Instead of the assumptions (i), (ii) in Section 1, we assume that

(i) the operator L with dense domain $\mathscr{D}(L)$ is closed.

(ii) L has a compact resolvent (thus, according to the Riesz-Schauder theory, the spectrum $\sigma(L)$ consists only of eigenvalues).

(iii) The resolvent satisfies the decay estimate

$$\|(\lambda - L)^{-1}\|_{\mathscr{L}(H)} \leqslant \frac{\text{const}}{1+|\lambda|}, \quad \lambda \in \overline{\Sigma},$$

where $\overline{\Sigma}$ denotes some sector described by $\overline{\Sigma} = \{\lambda - b;\ \theta_0 \leqslant |\arg\lambda| \leqslant \pi\}$, $0 < \theta_0 < \pi/2$, $b \in \mathbb{R}^1$.

(iv) $\sigma(L) \subset \mathbb{C}_+$.

(v) L admits *no* generalized eigenvector associated with the first ν eigenvalues under consideration, but may admit generalized eigenvectors for the other eigenvalues.

Items (i) – (iv) are general assumptions, whereas (v) looks somewhat restrictive. The eigenvalues, denoted by λ_i, $i \geqslant 1$, are distinct from each other and labelled according to increasing Re λ_i as:

$$0 < \mathrm{Re}\,\lambda_1 \leqslant \mathrm{Re}\,\lambda_2 \leqslant \cdots \leqslant \mathrm{Re}\,\lambda_\nu < \mathrm{Re}\,\lambda_{\nu+1} \leqslant \cdots \to \infty.$$

Associated with each λ_i, $1 \leqslant i \leqslant \nu$, is the eigenspace spanned by the basis $\{\varphi_{i1}, \ldots, \varphi_{im_i}\}$, where $m_i < \infty$. Note that there is no assumption on the existence of a Riesz basis associated with L. Thus, Lemma 2.2 is no longer available.

For each n, let P_n be the projector corresponding to the first n eigenvalues, $\lambda_1, \ldots, \lambda_n$, and set $Q_n = 1 - P_n$. After the expansion:

$$g_k = \sum_{i,j\,(i \leqslant \nu)} g_{ij}^k \varphi_{ij} + Q_\nu g_k,$$

set

$$W_i = \left(\langle \varphi_{ij}, w_k \rangle_H ;\ \begin{matrix} j & \rightarrow & 1, \ldots, m_i \\ k & \downarrow & 1, \ldots, N \end{matrix} \right), \quad \text{and}$$

$$G_i = \left(g_{ij}^k ;\ \begin{matrix} j & \downarrow & 1, \ldots, m_i \\ k & \rightarrow & 1, \ldots, N \end{matrix} \right),$$

where W_i are the same as in (2.1). Then, for an arbitrary μ, Re $\lambda_\nu < \mu <$ Re $\lambda_{\nu+1}$, Theorem 2.1 holds. In this case, the compensator (2.6) is considered in an arbitrary separable Hilbert space $\hat{H}$ (see (3.14)). In order to ensure Propositions 3.1 and 3.2, the separation condition of the operator B in (3.10) is somewhat different from the previous one: We adjust the parameters a and μ_1 so that

$$\mu < a\mu_1; \quad \arg\omega^+ > \theta_0; \quad \sigma(\boldsymbol{L}) \cap \sigma(B) = \varnothing; \quad \text{and}$$

$$\mu_i \leqslant \text{const}\, i^\gamma, \quad i \geqslant 1 \quad \text{for } 0 < \exists\gamma < 2 - \frac{2}{\pi}\theta_0.$$

The difference seems technical. Thus the detailed examination of the proof is left for the readers.

6.4 Operator L Admitting Generalized Eigenvectors

The operator L studied so far has not admitted any *generalized* eigenvector for the eigenvalues in question. Generally, the concept of output stabilization is the one weaker than that of state stabilization. In order to see the influence of generalized eigenvectors, we extract in this section a finite-dimensional L such that output stabilization is equivalent to state stabilization, or the implication: *"output stabilization ⇒ state stabilization"* holds. The existence of such a class of L is one reason why the eigenvalues in question are assumed to admit no generalized eigenvector in the preceding sections.

Let L be the $n \times n$ Jordan matrix given by

$$L = N_\lambda(n) = \begin{pmatrix} \lambda & 1 & \dots & 0 \\ 0 & \lambda & \ddots & \vdots \\ \vdots & & \ddots & 1 \\ 0 & \dots & \dots & \lambda \end{pmatrix} = \lambda + N, \quad \lambda \in \mathbb{R}^1,$$

where N is nilpotent. Thus L cannot be diagonalized and has a generalized eigenvector. Consider the differential equation in $H = \mathbb{C}^n$

$$\begin{aligned} &\frac{du}{dt} + Lu = gwu, \quad u(0) = u_0, \quad \text{where} \\ &w = \left(w_1\, w_2\, \dots\, w_n\right), \quad g = \left(g_1\, g_2\, \dots\, g_n\right)^{\mathrm{T}}. \end{aligned} \tag{4.1}$$

Thus we have only one output, which is given by

$$\varphi(t) = wu(t). \tag{4.2}$$

Assume: $w_1 \neq 0$ for nontriviality of the output. Then,

Proposition 4.1. *Suppose that the output $\varphi(t)$ decays exponentially with decay rate $-a$, $a > 0$, for any u_0. Then we have the state stabilization of* (4.1), *that is,*

$$\left\| e^{t(-L+gw)} \right\|_{\mathscr{L}(\mathbb{C}^n)} \leqslant \text{const}\, e^{-a't}, \quad t \geqslant 0, \quad 0 < \exists a' < a, \tag{4.3}$$

where a' can be chosen arbitrarily close to a.

Remark: Proposition 4.1 is generalized to the case of a more complicated L such as $L = \text{diag}\left(N_{\lambda_1}(n_1)\, N_{\lambda_2}(n_2)\, N_{\lambda_2}(n_2)\right)$ with $n_1,\, n_2,\, n_3 \geqslant 2$.

Proof. We assume with no loss of generality that $\lambda = 0$. Thus, $L = N_0(n)$. Let us seek the equation for φ. Since $\varphi^{(i)}(t) = w(gw - L)^i u(t)$, $0 \leqslant i \leqslant n-1$, it is easily seen that

$$\psi = \left(\varphi\; \dot{\varphi}\; \dots\; \varphi^{(n-1)}\right)^{\mathrm{T}} = A\left(w\; wL\; \dots\; wL^{n-1}\right)^{\mathrm{T}} u = AWu.$$

Here A denotes the $n \times n$ nonsingular lower triangular matrix, the diagonal elements of which are $+1$ or -1. Since $w_1 \neq 0$, it is clear that $\det W \neq 0$. By setting $C = AW$, ψ satisfies the differential equation

$$\frac{d\psi}{dt} = C(gw - L)C^{-1}\psi, \quad \psi(0) = Cu_0. \tag{4.4}$$

Let $\{\mu, \psi_0\}$ be an eigenpair of $C(gw - L)C^{-1}$. For the solution u to (4.1) with $u_0 = C^{-1}\psi_0$, the function $\tilde{\psi} = \left(\varphi\ \dot{\varphi}\ \dots\ \varphi^{(n-1)}\right)^{\mathrm{T}}$ satisfies (4.4) with $\tilde{\psi}(0) = \psi_0$. Thus, $\tilde{\psi}(t) = e^{\mu t}\psi_0$, or $\varphi^{(i-1)}(t) = e^{\mu t}\psi_{0i}$ for $1 \leqslant i \leqslant n$, where $\psi_0 = \left(\psi_{01}\ \dots\ \psi_{0n}\right)^{\mathrm{T}}$. It is clear that $\psi_{01} \neq 0$: Otherwise we must have the relation: $\psi_0 = 0$. By the assumption we have the estimate

$$|\varphi(t)| = e^{(\mathrm{Re}\,\mu)t}|\psi_{01}| \leqslant \mathrm{const}\, e^{-at}, \quad t \geqslant 0,$$

or $\mathrm{Re}\,\mu \leqslant -a$ for any eigenvalue of $C(gw - L)C^{-1}$. Since $C(gw - L)C^{-1}$ is similar to $gw - L$, we see that

$$\max \mathrm{Re}\,\sigma(gw - L) \leqslant -a.$$

This turns out to be the estimate in the proposition with any a', $0 < a' < a$ which is arbitrarily close to a. □

Remark: In this proposition, the observability condition:

$$\mathrm{rank}\left(w\ wL\ \dots\ wL^{n-1}\right)^{\mathrm{T}} = n$$

is satisfied (in fact, the determinant of $(w\ wL\ \dots\ wL^{n-1})$ is equal to ${w_1}^n \neq 0$). Thus there is an actuator g which stabilizes the state u of (4.1). The argument of this section is in this sense a story different from the one in the preceding sections. It is, however, worthwhile to see how the generalized eigenvectors influence the relationship between the state stability and the output stability.

6.5 Some Functionals

In the preceding sections, we have achieved *at least* output stabilization even when the observability and controllability conditions are lost. It is then natural to expect more results when the observability or controllability conditions are satisfied. Let us consider again the control system (1.1) with M inputs $f_k(t)$ and N outputs $\langle u, w_k\rangle_H$ (see (1.2)). In regular stabilization studies, the inputs $f_k(t)$ are designed as a suitable feedback of the outputs $\langle u, w_k\rangle_H$ –via dynamic compensators. We are satisfied with obtaining a decay estimate of $\|u(t)\|_H$ as $t \to \infty$. Every linear functional of u then decays *at least* with the same decay rate. This is also true in the case where the functional is unbounded and subordinate to L. We then raise a question: Can we find a nontrivial linear functional which decays faster than $\|u(t)\|_H$? In this section, we construct a specific feedback control system such that $\|u(t)\|_H$ decays exponentially with the designated decay rate, and that some nontrivial linear functionals of $u(t)$ decay *faster than* $\|u(t)\|_H$.

As in Section 2, it is assumed that L is self-adjoint with dense domain $\mathscr{D}(L)$, and has compact resolvent. Thus there is a set of eigenpairs $\{\lambda_i, \varphi_{ij}\}$ such that

(i) $\sigma(L) = \{\lambda_i\}_{i\geqslant 1}$, $\lambda_1 < \cdots < \lambda_i \cdots \to \infty$. Here it is assumed that $\min \sigma(L) = \lambda_1 < 0$. Thus, (1.1) is unstable without input.

(ii) $(\lambda_i - L)\varphi_{ij} = 0$ for $i \geqslant 1$ and $1 \leqslant j \leqslant m_i$ $(< \infty)$.

(iii) The set $\{\varphi_{ij}\}$ forms an orthonormal basis for H. Any $u \in H$ is expressed as a Fourier series: $u = \sum_{i,j} u_{ij}\varphi_{ij}$, where $u_{ij} = \langle u, \varphi_{ij}\rangle_H$.

(iv) Let $\lambda_{\nu+1} > 0$, and assume that $m_i = 1$, $1 \leqslant i \leqslant \nu$.

The assumption (iv) seems somewhat restrictive. A typical example satisfying (iv), however, appears in one-dimensional heat conduction equations in a bounded interval, since all eigenvalues are simple [11]. The projector associated with the eigenvalue λ_i is denoted as P_{λ_i}, or $P_{\lambda_i}u = \sum_{j=1}^{m_i} u_{ij}\varphi_{ij}$. Set $P_n = \sum_{i=1}^{n} P_{\lambda_i}$. However, the projector P_ν, $1 \leqslant i \leqslant \nu$, is simply written as $P_\nu u = \sum_{i=1}^{\nu} u_i\varphi_i$, where $u_i = \langle u, \varphi_i\rangle_H$ and $\varphi_i = \varphi_{i1}$. The same convension is hereafter employed.

Choose a positive α such that $0 < \alpha < \lambda_{\nu+1}$. Our control system has state $(u(t), v(t))$. In view of the above (iv), we consider a single output, $\langle u, w\rangle_H$ of the controlled system. The system is then described as the differential equation in $H \times H$:

$$\begin{cases} \dfrac{du}{dt} + Lu = -\langle v, (\alpha - L)q\rangle_H \eta + \langle v, \rho\rangle_H \gamma, \\ \dfrac{dv}{dt} + Bv = -\langle v, (\alpha - L)q\rangle_H \eta + \langle v, \rho\rangle_H \gamma + \langle u, w\rangle_H g. \end{cases} \tag{5.1}$$

Here, η, γ, and g are the actuators in H to be designed; and $q \in \mathscr{D}(L)$ the weight producing the functional $\langle u(t), q\rangle_H$. The equation (5.1) is clearly well posed in $H \times H$. The equation for v denotes the compensator. It is hoped that $\langle u(t), q\rangle_H$ would decay with the designated decay rate $-\alpha$. In constructing regular stabilization schemes, the term $\langle v, (\alpha - L)q\rangle_H \eta$ is unnecessary: It is introduced for our specific purpose that $\langle u(t), q\rangle_H$ would decay faster than $\|u(t)\|_H$ as $t \to \infty$.

We employ a so-called *identity* compensator in (5.1), and set

$$B = L + \langle \cdot, w\rangle_H g. \tag{5.2}$$

The operators $-L$ and $-B$ generate analytic semigroups e^{-tL} and e^{-tB}, $t > 0$, respectively. It is easily seen that $u - v$ satisfies the equation:

$$\frac{d}{dt}(u - v) + B(u - v) = 0, \quad t > 0, \quad u(0) - v(0) = u_0 - v_0.$$

Thus, $u(t) - v(t) = e^{-tB}(u_0 - v_0)$, $t \geqslant 0$, whatever the feedback terms may be. In (5.2), by assuming the observability conditions: $w_i = \langle w, \varphi_i \rangle \neq 0$, $1 \leqslant i \leqslant \nu$, there is a $g \in P_\nu H$ such that the decay estimate

$$\left\| e^{-tB} \right\|_{\mathscr{L}(H)} \leqslant \text{const}\, e^{-\lambda_{\nu+1} t}, \quad t \geqslant 0 \tag{5.3}$$

holds (see Theorem 2.1, Chapter 3). Thus we see that

$$\|u(t) - v(t)\|_H \leqslant \text{const}\, e^{-\lambda_{\nu+1} t} \|u_0 - v_0\|_H, \quad t \geqslant 0.$$

Going back to the equation for u in (5.1), we will derive the equation for the functional $\langle u, q \rangle_H$. We calculate as

$$\frac{d}{dt} \langle u, q \rangle_H + \langle u, Lq \rangle_H = - \langle v, (\alpha - L) q \rangle_H \langle \eta, q \rangle_H + \langle v, \rho \rangle_H \langle \gamma, q \rangle_H .$$

For an integer μ, $1 \leqslant \mu \leqslant \nu$, we choose q in $P_\mu H$, and assume that

$$\langle \eta, q \rangle_H = \sum_{i=1}^{\mu} \overline{q_i} \eta_i = 1, \tag{5.4}$$

where $P_\mu q = \sum_{i=1}^{\mu} q_i \varphi_i$ and $P_\mu \eta = \sum_{i=1}^{\mu} \eta_i \varphi_i$. Then, we have

Theorem 5.1. (i) *Let $0 < \beta < \alpha$. Suppose that*

$$\begin{cases} w_i \neq 0, & 1 \leqslant i \leqslant \nu, \\ \eta_i \neq 0, & 1 \leqslant i \leqslant \mu, \\ \gamma_i = 0, & 1 \leqslant i \leqslant \mu, \\ \gamma_i \neq 0, & \mu < i \leqslant \nu. \end{cases} \tag{5.5}$$

Then, we can find a vector q satisfying (5.4), $g \in P_\nu H$, and $\rho \in P_\nu H$ such that the estimate

$$\|u(t)\|_H + \|v(t)\|_H \leqslant \text{const}\, e^{-\beta t} \left(\|u_0\|_H + \|v_0\|_H \right), \quad t \geqslant 0 \tag{5.6}$$

and the decay estimate

$$\left| \langle u(t), q \rangle_H \right| \leqslant \text{const}\, e^{-\alpha t}, \quad t \geqslant 0 \tag{5.7}$$

hold for every solution $(u(t), v(t))$ to (5.1). The estimate (5.6) is no longer improved.

(ii) *Suppose, in addition, that there is an integer $n \geqslant \nu$ such that*

$$\left\langle P_{\lambda_i} \eta, P_{\lambda_i} w \right\rangle_H = \left\langle P_{\lambda_i} \gamma, P_{\lambda_i} w \right\rangle_H = 0, \quad i > n. \tag{5.8}$$

Then the compensator in (5.1) *is reduced to the equation in* $\mathbb{C}^{S_n}$, $S_n = \sum_{1 \leqslant i \leqslant n} m_i$, *with state* $v_1(t) = P_n v(t)$. *The equation for* $(u(t), v_1(t)) \in H \times P_n H$ *is described by*

$$\begin{aligned} &\frac{du}{dt} + Lu = -\langle v, (\alpha - L)q \rangle_H \eta + \langle v, \rho \rangle_H \gamma, \\ &\frac{dv_1}{dt} + B_1 v = -\langle v_1, (\alpha - L)q \rangle_H P_n \eta + \langle v_1, \rho \rangle_H P_n \gamma + \langle u, w \rangle_H g, \end{aligned} \tag{5.9}$$

where B_1 *denotes the restriction of* B *onto the* S_n*-dimensional subspace* $P_n H$: $B_1 = B|_{P_n H} = L|_{P_n H} + \langle \cdot, P_n w \rangle_H g$. *The estimate*

$$\|u(t)\|_H + \|v_1(t)\|_H \leqslant \text{const}\, e^{-\beta t} \left(\|u_0\|_H + \|v_{10}\|_H \right), \quad t \geqslant 0 \tag{5.10}$$

and the decay estimate (5.7) *hold for every solution* $(u(t), v_1(t))$ *to* (5.9). *The estimate* (5.10) *is no longer improved. In fact, there is a solution such that* $\|u(t)\|_H = \text{const}\, e^{-\beta t}$ *and* $\|v_1(t)\|_H = \text{const}\, e^{-\beta t}$, $t \geqslant 0$.

Remark: When $\mu = \nu$ in the theorem, the vectors ρ and γ do not appear in (5.1). Thus the assumption on γ is removed.

A part of the proof of Theorem 5.1 consists of a finite-dimensional pole assignment argument with *constraint*. The following result, Theorem 5.2, is different from the well known pole assignment theory [70], since it is subject to some constraint. The proof will be stated at the end of this section.

Theorem 5.2. *Let*

$$\begin{aligned} \Lambda &= \operatorname{diag}\left(\lambda_1 \quad \lambda_2 \quad \dots \quad \lambda_m\right), \\ \boldsymbol{q} &= \left(q_1 \quad q_2 \quad \dots \quad q_m\right)^{\mathrm{T}}, \\ \boldsymbol{\eta} &= \left(\eta_1 \quad \eta_2 \quad \dots \quad \eta_m\right), \end{aligned}$$

and set $\Xi = \Lambda + (\alpha - \Lambda)\boldsymbol{q}\boldsymbol{\eta}$. *Consider its spectrum* $\sigma(\Xi)$, *subject to the constraint:*

$$\sum_{i=1}^{m} \overline{q_i}\, \eta_i = 1. \tag{5.11}$$

Then, α *belongs to* $\sigma(\Xi)$. *For an arbitrary set* $\{\mu_i;\ 1 \leqslant i \leqslant m-1\}$ *of complex numbers, there is a vector* $\boldsymbol{q}$ *which is subject to* (5.11), *such that* $\sigma(\Xi) = \{\alpha, \mu_1, \mu_2, \dots, \mu_{m-1}\}$, *if and only if*

$$\eta_i \neq 0, \quad 1 \leqslant i \leqslant m. \tag{5.12}$$

Proof of Theorem 5.1.

(i) We begin with the operator A:

$$A = L + \langle \cdot, (\alpha - L)q \rangle_H \eta, \quad \mathscr{D}(A) = \mathscr{D}(L). \tag{5.13}$$

The adjoint operator A^* is given by

$$A^* = L + \langle \cdot, \eta \rangle_H (\alpha - L)q, \quad \mathscr{D}(A^*) = \mathscr{D}(L).$$

Let A_1^* be the restriction of A^* onto the subspace $P_\nu H$: $A_1^* = A^*|_{P_\nu H}$. According to the basis, $\{\varphi_i\}_{1 \leqslant i \leqslant \mu}$, $L|_{P_\mu H}$ is identified with the $\mu \times \mu$ matrix $\Lambda = \operatorname{diag}\left(\lambda_1\ \lambda_2\ \dots\ \lambda_\mu\right)$. Let us consider Theorem 5.2 for $m = \mu$, and choose $q \in P_\mu H$ satisfying (5.11) such that $\min \sigma(\Xi) \geqslant \beta$, and that the eigenvalues are different from each other: One of the elements of $\sigma(\Xi)$ is, of course, α. The operator A_1^* is identified with the matrix:

$$\widehat{A_1^*} = \begin{pmatrix} \Xi & (\alpha - \Lambda)\boldsymbol{q}\tilde{\boldsymbol{\eta}} \\ 0 & \Lambda_2 \end{pmatrix}, \quad \text{where} \quad \tilde{\boldsymbol{\eta}} = \left(\eta_{\mu+1} \quad \eta_{\mu+2} \quad \dots \quad \eta_\nu\right),$$
$$\text{and} \quad \Lambda_2 = \operatorname{diag}\left(\lambda_{\mu+1} \quad \lambda_{\mu+2} \quad \dots \quad \lambda_\nu\right). \tag{5.14}$$

Let Π be the nonsingular matrix such that

$$\Pi^{-1} \Xi \Pi = \operatorname{diag}\left(\alpha_1 \quad \alpha_2 \quad \dots \quad \alpha_\mu\right) = \mathscr{A}, \quad \alpha_1 = \alpha, \quad \alpha_i \geqslant \beta.$$

The other eigenvalues of $\widehat{A_1^*}$ are λ_i, $\mu + 1 \leqslant i \leqslant \nu$: for each λ_i, the vector $(\psi_1\ \psi_2)^{\mathrm{T}}$ with

$$\psi_1 = (\lambda_i - \Xi)^{-1}(\alpha - \Lambda)\boldsymbol{q}\tilde{\boldsymbol{\eta}}\,\psi_2,$$
$$\psi_2 = \big(\overset{(\mu+1)\text{th}}{0} \quad \dots \overset{i\text{th}}{1} \dots \overset{\nu\text{th}}{0}\big)^{\mathrm{T}} = e_i, \quad \mu + 1 \leqslant i \leqslant \nu$$

is an eigenvector. Set $D = (\alpha - \Lambda)\boldsymbol{q}\tilde{\boldsymbol{\eta}}$ for simplicity. Then the nonsingular matrix

$$\Psi =$$
$$\begin{pmatrix} \Pi & (\lambda_{\mu+1} - \Xi)^{-1} D e_{\mu+1} & (\lambda_{\mu+2} - \Xi)^{-1} D e_{\mu+2} & \dots & (\lambda_\nu - \Xi)^{-1} D e_\nu \\ 0 \dots 0 & 1 & 0 & \dots & 0 \\ 0 \dots 0 & 0 & 1 & \dots & 0 \\ \vdots & \vdots & \vdots & \ddots & \vdots \\ 0 \dots 0 & 0 & 0 & \dots & 1 \end{pmatrix}$$

diagonalizes $\widehat{A_1^*}$, or

$$\Psi^{-1} \widehat{A_1^*} \Psi = \begin{pmatrix} \mathscr{A} & 0 \\ 0 & \Lambda_2 \end{pmatrix}.$$

Let us turn to the operator $K = A - \langle \cdot, \rho \rangle_H \gamma$. The restriction of K^* onto the subspace $P_\nu H$: $K^*|_{P_\nu H} = A_1^* - \langle \cdot, P_\nu \gamma \rangle_H \rho$ is then identidfied with the matrix

$$\widehat{A_1^*} - R^* \Gamma^* \iff \Psi^{-1}(\widehat{A_1^*} - R^* \Gamma^*)\Psi = \begin{pmatrix} \mathscr{A} & 0 \\ 0 & \Lambda_2 \end{pmatrix} - \Psi^{-1} R^* \, \Gamma^* \Psi,$$
$$\Gamma^* = \left(\gamma_1 \quad \gamma_2 \quad \dots \quad \gamma_\nu\right), \quad R^* = \left(\rho_1 \quad \rho_2 \quad \dots \quad \rho_\nu\right)^{\mathrm{T}}.$$

By the assumption: $\gamma_i = 0,\ 1 \leqslant i \leqslant \mu$, it is clear that

$$\Gamma^*\Psi = \begin{pmatrix} 0 & \dots & 0 & \gamma_{\mu+1} & \dots & \gamma_\nu \end{pmatrix} = \Gamma^* \quad \text{and} \quad \langle \gamma, q \rangle_H = 0.$$

Decomposing $\Psi^{-1}R^*$ as $(\tau_1 \ \ \tau_2)^{\mathrm{T}}$, τ_1: $\mu \times 1$, and τ_2: $(\nu - \mu) \times 1$, we see that

$$\begin{pmatrix} \mathscr{A} & 0 \\ 0 & \Lambda_2 \end{pmatrix} - \Psi^{-1}R^*\Gamma^*\Psi = \begin{pmatrix} \mathscr{A} & -\tau_1(\gamma_{\mu+1} \ \dots \ \gamma_\nu) \\ 0 & \Lambda_2 - \tau_2(\gamma_{\mu+1} \ \dots \ \gamma_\nu) \end{pmatrix}.$$

Thus,

$$\begin{aligned} &\sigma\left(A_1^* - \langle \cdot, P_\nu \gamma \rangle \rho\right) \\ &= \{\alpha_1, \alpha_2, \dots, \alpha_\mu\} \cup \sigma\left(\Lambda_2 - \tau_2(\gamma_{\mu+1} \ \dots \ \gamma_\nu)\right), \quad \alpha_1 = \alpha. \end{aligned}$$

By the assumption: $\gamma_i \neq 0,\ \mu + 1 \leqslant i \leqslant \nu$, we find a suitable vector τ_2 such that

$$\min \sigma\big(\Lambda_2 - \tau_2(\gamma_{\mu+1} \ \dots \ \gamma_\nu)\big) = \beta.$$

Thus we see that $\min \sigma\left(K^*|_{P_\nu H}\right) = \beta$. Note that β actually belongs to $\sigma(K^*)$, or $\beta\ (= \overline{\beta})$ is in $\sigma(K)$, and that

$$\left\| e^{-tK} \right\|_{\mathscr{L}(H)} = \left\| e^{-tK^*} \right\|_{\mathscr{L}(H)} \leqslant \text{const}\, e^{-\beta t}, \quad t \geqslant 0. \tag{5.15}$$

The equation for u in (5.1) is rewritten as

$$\begin{aligned} &\frac{du}{dt} + Ku = \langle u - v, (\alpha - L)q \rangle_H \eta - \langle u - v, \rho \rangle_H \gamma, \\ \text{or} \quad &u(t) = e^{-tK}u_0 + \int_0^t e^{-(t-s)K} \langle u(s) - v(s), (\alpha - L)q \rangle_H \eta \, ds \\ &\qquad - \int_0^t e^{-(t-s)K} \langle u(s) - v(s), \rho \rangle_H \gamma ds, \end{aligned}$$

from which we obtain the estimate:

$$\|u(t)\|_H \leqslant \text{const}\, e^{-\beta t}, \quad t \geqslant 0. \tag{5.16}$$

This is the best possible estimate we could expect. Actually, let ξ be an eigenvector of K for β: $(\beta - K)\xi = 0$. By setting $u_0 = v_0 = \xi$, the pair $(u(t), v(t)) = (e^{-\beta t}\xi, e^{-\beta t}\xi)$ is in fact a solution to (5.1), and thus the decay estimate (5.6) is no longer improved.

(ii) We begin with the following proposition. The proof is to be given later in this section.

Proposition 5.3. *Let p and q be vectors in H, and let $p = \sum_{i,j} p_{ij}\varphi_{ij}$ and $q = \sum_{i,j} q_{ij}\varphi_{ij}$. The function*

$$\left\langle e^{-tL}Q_n p, Q_n q \right\rangle_H, \quad t \geqslant 0$$

is identically equal to 0, *if and only if*

$$\sum_{l=1}^{m_i} p_{il}\overline{q_{il}} = 0, \quad or \quad \langle P_{\lambda_i}p, P_{\lambda_i}q\rangle_H = 0, \quad i > n. \tag{5.17}$$

We go back to the equation for v in (5.1):

$$\frac{dv}{dt} + Lv + \langle v, w\rangle_H g = -\langle v, (\alpha - L)q\rangle_H \eta + \langle v, \rho\rangle_H \gamma + \langle u, w\rangle_H g.$$

Recalling that $q \in P_\mu H$, ρ and $g \in P_\nu H$, we divide v into the direct sum: $v = v_1 + v_2$, where $v_1 \in P_n H$ and $v_2 \in Q_n H$, $n \geqslant \nu$. The differential equation for v is then written as a coupling system of equations for v_1 and v_2:

$$\begin{aligned}
&\frac{dv_1}{dt} + Lv_1 + \langle v_1 + v_2, w\rangle_H g \\
&\qquad = -\langle v_1, (\alpha - L)q\rangle_H P_n\eta + \langle v_1, \rho\rangle_H P_n\gamma + \langle u, w\rangle_H g, \\
&\frac{dv_2}{dt} + Lv_2 = -\langle v_1, (\alpha - L)q\rangle_H Q_n\eta + \langle v_1, \rho\rangle_H Q_n\gamma.
\end{aligned} \tag{5.18}$$

In (5.18), the state v_2 actually affects the dynamics of v_1. By the second equation,

$$\begin{aligned}
\langle v_2(t), Q_n w\rangle_H &= \left\langle e^{-tL}Q_n v_0, Q_n w\right\rangle_H \\
&\quad - \int_0^t \left\langle e^{-(t-s)L}Q_n\eta, Q_n w\right\rangle_H \langle v_1(s), (\alpha - L)q\rangle_H\, ds \\
&\quad + \int_0^t \left\langle e^{-(t-s)L}Q_n\gamma, Q_n w\right\rangle_H \langle v_1(s), \rho\rangle_H\, ds.
\end{aligned} \tag{5.19}$$

By assumption (5.8) and Proposition 5.3 with $p = \eta$ or $= \gamma$ and $q = w$, the second and the third terms of (5.19) disappear. Thus, we see that

$$\langle v_2(t), Q_n w\rangle_H = \left\langle e^{-tL}Q_n v_0, Q_n w\right\rangle_H, \quad t \geqslant 0.$$

We *choose* the initial data v_0 such that $\left\langle P_{\lambda_i}v_0, P_{\lambda_i}w\right\rangle_H = 0$ for $i > n$. Then, by Proposition 5.3 again, we see that $\langle v_2(t), Q_n w\rangle_H = 0$, $t \geqslant 0$. In the equation for v_1 in (5.18), the term $\langle v_2(t), w\rangle_H$ then does not appear.

We have come to the conclusion: As long as v_0 satisfies $\left\langle P_{\lambda_i}v_0, P_{\lambda_i}w\right\rangle_H = 0$, $i > n$, the new state $(u(t), v_1(t))$ satisfies the differential equation in $H \times P_n H$:

$$\begin{aligned}
&\frac{du}{dt} + Lu = -\langle v_1, (\alpha - L)q\rangle_H \eta + \langle v_1, \rho\rangle_H \gamma, \\
&\frac{dv_1}{dt} + Lv_1 + \langle v_1, P_n w\rangle_H g \\
&\qquad = -\langle v_1, (\alpha - L)q\rangle_H P_n\eta + \langle v_1, \rho\rangle_H P_n\gamma + \langle u, w\rangle_H g.
\end{aligned} \tag{5.20}$$

Equation (5.20) is clearly *well posed* in $H \times P_n H$, and the decay estimate (5.10) holds.

To show that (5.10) is the best possible estimate, we reconsider the eigenvector ξ for the eigenvalue β of K. Settimg $\xi = \xi_1 + \xi_2$ with $\xi_1 \in P_n H$ and $\xi_2 \in Q_n H$, we obtain

$$\beta \xi_1 = L_{\langle 1\rangle} \xi_1 + \langle \xi_1, (\alpha - L)q\rangle_H P_n \eta - \langle \xi_1, \rho\rangle_H P_n \gamma,$$
$$\beta \xi_2 = L_{\langle 2\rangle} \xi_2 + \langle \xi_1, (\alpha - L)q\rangle_H Q_n \eta - \langle \xi_1, \rho\rangle_H Q_n \gamma,$$

where $L_{\langle 1\rangle} = L|_{P_n H}$ and $L_{\langle 2\rangle} = L|_{Q_n H \cap \mathscr{D}(L)}$. Since $\sigma\left(L_{\langle 2\rangle}\right) = \{\lambda_{n+1}, \lambda_{n+2}, \ldots\}$, we see that

$$\xi_2 = \langle \xi_1, (\alpha - L)q\rangle_H \left(\beta - L_{\langle 2\rangle}\right)^{-1} Q_n \eta - \langle \xi_1, \rho\rangle_H \left(\beta - L_{\langle 2\rangle}\right)^{-1} Q_n \gamma.$$

Assumption (5.8) yields

$$\left\langle (\beta - L_{\langle 2\rangle})^{-1} Q_n \eta, P_{\lambda_i} w\right\rangle_H = \left\langle \sum_{i,l\,(i>n)} \frac{\eta_{il}}{\beta - \lambda_i} \varphi_{il}, P_{\lambda_i} w \right\rangle_H$$
$$= \frac{1}{\beta - \lambda_i} \sum_{l=1}^{m_i} \eta_{il} \overline{w_{il}} = 0, \quad i > n,$$

and similarly $\left\langle (\beta - L_{\langle 2\rangle})^{-1} Q_n \gamma, P_{\lambda_i} w\right\rangle_H = 0$ for $i > n$. This means that $\left\langle P_{\lambda_i} \xi, P_{\lambda_i} w\right\rangle_H = 0$ for $\forall i > n$. As we have seen, the pair, $(u(t), v(t)) = (e^{-\beta t}\xi, e^{-\beta t}\xi)$ is a solution to (5.1), and $v(0) = \xi$. Thus, $(u(t), v_1(t)) = (e^{-\beta t}\xi, e^{-\beta t}\xi_1)$ is a solution to (5.20), and (5.10) is the best possible decay estimate.

Proof of Proposition 5.3.
Using the necessary condition

$$\left\langle e^{-tL} Q_n p, Q_n q\right\rangle_H = \sum_{i,l\,(i>n)} e^{-\lambda_i t} p_{il} \overline{q_{il}} = 0, \quad t \geqslant 0,$$

we calculate, for λ, $\mathrm{Re}\,\lambda < \lambda_{n+1}$, as

$$0 = \int_0^\infty e^{\lambda t} \sum_{i,l\,(i>n)} e^{-\lambda_i t} p_{il} \overline{q_{il}}\, dt = \sum_{i,l\,(i>n)} \int_0^\infty e^{(\lambda - \lambda_i)t}\, dt\; p_{il} \overline{q_{il}}$$
$$= -\sum_{i>n} \frac{1}{\lambda - \lambda_i} \sum_{l=1}^{m_i} p_{il} \overline{q_{il}}.$$

The last term is an analytic function in λ ($\neq \lambda_{n+1}, \lambda_{n+2}, \ldots$). By analytic continuation, the above relation holds for arbitrary λ ($\neq \lambda_{n+1}, \lambda_{n+2}, \ldots$). Calculating the residue at each λ_i, $i \geqslant n+1$, we obtain (5.17).

Conversely, assuming (5.17), we see that

$$\left\langle e^{-tL} Q_n p, Q_n q\right\rangle_H = \sum_{i,l\,(i>n)} e^{-\lambda_i t} p_{il} \overline{q_{il}}$$
$$= \sum_{i>n} e^{-\lambda_i t} \sum_{l=1}^{m_i} p_{il} \overline{q_{il}} = 0, \quad t \geqslant 0.$$

This finishes the proof of Proposition 5.3. □

To complete the proof of Theorem 5.1, we turn to Theorem 5.2.

Proof of Theorem 5.2.

To calculate the spectrum $\sigma(\Xi)$, we consider the algebraic equation on λ of order m:

$$\det(\lambda - \Xi) = \det\big(\lambda - \Lambda - (\alpha - \Lambda)\boldsymbol{q}\boldsymbol{\eta}\big) = 0.$$

Necessity of (5.12) is easy. In fact, if one of the $\eta_1, \ldots, \eta_m$ is equal to 0, say $\eta_i = 0$, it is clear that the above equation has the solution $\lambda = \lambda_i$ for any choice of $\boldsymbol{q}$. Thus (5.12) is a necessary condition.

Now suppose conversely that (5.12) holds. By assuming that $(\lambda - \Lambda)^{-1}$ exists for a moment, the equation may be rewritten as

$$\begin{aligned}
\det(\lambda - \Xi) &= \det\big(1 - (\alpha - \Lambda)\boldsymbol{q}\boldsymbol{\eta}(\lambda - \Lambda)^{-1}\big) \cdot \det(\lambda - \Lambda) \\
&= \left(1 - \sum_{i=1}^{m}(\alpha - \lambda_i)q_i \frac{\eta_i}{\lambda - \lambda_i}\right) \cdot \det(\lambda - \Lambda) \\
&= \prod_{i=1}^{m}(\lambda - \lambda_i) - \sum_{i=1}^{m}(\alpha - \lambda_i)q_i\eta_i \prod_{j=1,\, j\neq i}^{m}(\lambda - \lambda_j) \\
&= \prod_{i=1}^{m}(\lambda - \lambda_i) - \sum_{i=1}^{m}(\alpha - \lambda + \lambda - \lambda_i)q_i\eta_i \prod_{j=1,\, j\neq i}^{m}(\lambda - \lambda_j) \\
&= \left(1 - \sum_{i=1}^{m} q_i\eta_i\right)\prod_{i=1}^{m}(\lambda - \lambda_i) + (\lambda - \alpha)\sum_{i=1}^{m} q_i\eta_i \prod_{j=1,\, j\neq i}^{m}(\lambda - \lambda_j) \\
&= (\lambda - \alpha)\sum_{i=1}^{m} q_i\eta_i \prod_{j=1,\, j\neq i}^{m}(\lambda - \lambda_j) = 0.
\end{aligned} \tag{5.21}$$

Here we have used the constraint (5.11). Both sides of (5.21) are polynomials of λ. Passage to the limit regarding λ, we see that (5.21) is correct for every λ. Thus all elements of $\sigma(\Xi)$ other than α are derived from the algebraic equation of order $m - 1$:

$$\sum_{i=1}^{m} q_i\eta_i \prod_{j=1,\, j\neq i}^{m}(\lambda - \lambda_j) = 0. \tag{5.22}$$

Set

$$\begin{aligned}
\hat{\Lambda} &= \operatorname{diag}\begin{pmatrix}\lambda_2 & \lambda_3 & \ldots & \lambda_m\end{pmatrix}, \\
\hat{\boldsymbol{q}} &= \begin{pmatrix}q_2 & q_3 & \ldots & q_m\end{pmatrix}^{\mathrm{T}}, \quad \text{and} \\
\hat{\boldsymbol{\eta}} &= \begin{pmatrix}\eta_2 & \eta_3 & \ldots & \eta_m\end{pmatrix}.
\end{aligned}$$

Then, (5.22) is further calculated as

$$\begin{aligned}
0 &= \sum_{i=2}^{m} q_i\eta_i(\lambda-\lambda_1)\prod_{j=2,\, j\neq i}^{m}(\lambda-\lambda_j) + \left(1-\sum_{i=2}^{m} q_i\eta_i\right)\prod_{j=2}^{m}(\lambda-\lambda_j) \\
&= \prod_{i=2}^{m}(\lambda-\lambda_i) + \sum_{i=2}^{m} q_i\eta_i(\lambda_i-\lambda_1)\prod_{j=2,\, j\neq i}^{m}(\lambda-\lambda_j) \\
&= \left(1-\sum_{i=2}^{m}\frac{q_i\eta_i(\lambda_1-\lambda_i)}{\lambda-\lambda_i}\right)\prod_{i=2}^{m}(\lambda-\lambda_i) \\
&= \det\left(1-(\lambda_1-\widehat{\Lambda})\hat{\boldsymbol{q}}\hat{\boldsymbol{\eta}}(\lambda-\widehat{\Lambda})^{-1}\right)\cdot\det(\lambda-\widehat{\Lambda}) \\
&= \det\left(\lambda-\widehat{\Lambda}-(\lambda_1-\widehat{\Lambda})\hat{\boldsymbol{q}}\hat{\boldsymbol{\eta}}\right).
\end{aligned}$$

This shows that

$$\sigma(\Xi) = \{\alpha\}\cup\sigma(\hat{\Xi}), \quad \text{where} \quad \hat{\Xi} = \hat{\Lambda} + (\lambda_1-\hat{\Lambda})\hat{\boldsymbol{q}}\hat{\boldsymbol{\eta}}. \tag{5.23}$$

Since $\eta_i \neq 0,\ 2 \leqslant i \leqslant m$, it is clear that $(\hat{\boldsymbol{\eta}}, \hat{\Lambda})$ is the observable pair. Thus for an arbitrary set $\{\mu_i\}_{i=1}^{m-1}$ of complex numbers, we can find an $(m-1)\times 1$ vector $(\lambda_1-\hat{\Lambda})\hat{\boldsymbol{q}}$, or $\hat{\boldsymbol{q}} = (q_2\ q_3\ \dots\ q_m)^{\mathrm{T}}$ such that $\sigma(\hat{\Xi}) = \{\mu_i\}_{i=1}^{m-1}$ (see [70]). Finally the number q_1 is determined by the relation: $q_1 = \eta_1^{-1}(1-\sum_{i=2}^{m} q_i\eta_i)$. This finishes the proof of Theorem 5.2, and the proof of Theorem 5.1 is thereby completed. □

Remark: One may consider a spectral decomposition associated with the coefficient closed operator in state stabilization to obtain a functional (a Fourier coefficient) decaying faster than the state. As already stated, the term $\langle v, (\alpha-L)q\rangle_H \eta$ is unnecessary only for state stabilization. The setting of the equation then becomes

$$\frac{d}{dt}\begin{pmatrix} u \\ v \end{pmatrix} + \begin{pmatrix} L & -\langle\cdot,\rho\rangle_H\gamma \\ -\langle\cdot,w\rangle_H g & B-\langle\cdot,\rho\rangle_H\gamma \end{pmatrix}\begin{pmatrix} u \\ v \end{pmatrix} = \begin{pmatrix} 0 \\ 0 \end{pmatrix}. \tag{5.1$'$}$$

Then, by assuming that

$$w_i \neq 0, \quad \gamma_i \neq 0, \quad 1 \leqslant i \leqslant \nu,$$

the state stabilization of (5.1′) is achieved with suitable choice of g and ρ in $P_\nu H$. The adjoint operator $\mathscr{A}^*$ of the coefficient operator $\mathscr{A}$ in (5.1′) is

$$\mathscr{A}^* = \begin{pmatrix} L & -\langle\cdot,g\rangle_H w \\ -\langle\cdot,\gamma\rangle_H\rho & B^*-\langle\cdot,\gamma\rangle_H\rho \end{pmatrix}.$$

Let $\{\zeta, (\varphi,\psi)^{\mathrm{T}}\}$ be an eigenpair of $\mathscr{A}^*$. Then, for every solution $(u,v)^{\mathrm{T}}$ to (5.1′), it is clear that

$$\left\langle \begin{pmatrix} u(t) \\ v(t) \end{pmatrix}, \begin{pmatrix} \varphi \\ \psi \end{pmatrix} \right\rangle_H = e^{-\overline{\zeta}t}\left\langle \begin{pmatrix} u_0 \\ v_0 \end{pmatrix}, \begin{pmatrix} \varphi \\ \psi \end{pmatrix} \right\rangle_H, \quad t \geqslant 0.$$

The above left-hand side is $\langle u(t), \varphi \rangle_H + \langle v(t), \psi \rangle_H$. If $\mathscr{A}^*$ has an eigenvector of the form $(\varphi, 0)^{\mathrm{T}}$, we find that the functional $\langle u(t), \varphi \rangle_H$ decays faster than $u(t)$. When does $\mathscr{A}^*$ have an eigenvector of the form $(\varphi, 0)^{\mathrm{T}}$? For this, it is necessary that

$$(\zeta - L)\varphi = 0, \quad \langle \varphi, \gamma \rangle_H = 0.$$

Thus, $\{\zeta, \varphi\}$ must be an eigenpair of L, say $\{\lambda_i, \varphi_{ij}\}$, $\lambda_i \geqslant \lambda_{\nu+1}$, and $\left\langle \varphi_{ij}, \gamma \right\rangle_H = 0$. While a finite number of Fourier coefficients of the actuator γ can be *constructed* (*designed*) for stabilization, however, the others cannot be in general freely assigned. Thus it is almost unplausible to expect an eigenvector of the form $(\varphi, 0)^{\mathrm{T}}$. In fact, in a simple example of one-dimensioanl heat conduction equation, it is easy to illustrate an actuator γ such that all Fourier coefficients differ from 0.

Chapter 7

Stabilization of a class of linear control systems generating C_0-semigroups

7.1 Introduction

We have so far studied stabilization of linear parabolic systems. These systems are characterized by the sectorial operator L such that the resolvent $(\lambda - L)^{-1}$ exists in a sector with angle greater than π. As a result, the semigroup e^{-tL} generated by $-L$ is analytic in $t > 0$ (see Section 4, Chapter 2). We study in this chapter a somewhat more general class of linear systems, and show that the stabilization scheme developed in Chapter 4 effectively works for these systems with some technical changes in the setting. In engineering applications, linear systems other than parabolic systems appear, such that they generate not analytic semigroups but a class of C_0-semigroups, e.g., those appearing in delay-differential equations (see, e.g., [16, 61]). The properties of C_0-semigroups are *less* nicer than those of analytic semigroups. In addition to non-analyticity of semigroups, the infinitesimal generators are not sectorial, that is, the resolvents do not exist in a sector with angle greater than π.

To begin with, let L be a linear closed operator in a Banach space E with the dense domain $\mathscr{D}(L)$, and consider an abstract differential equation in E which is described as

$$\frac{du}{dt} + Lu = 0, \quad t > 0, \quad u(0) = u_0 \in \mathscr{D}(L). \tag{1.1}$$

By the celebrated Hille-Yosida theorem [3, 16, 26, 35, 71], the Cauchy problem for (1.1) is well posed, if and only if there exist constants $M \geqslant 1$ and $\omega \in \mathbb{R}^1$ such that

$$\left\|(\lambda - L)^{-n}\right\|_E \leqslant \frac{M}{(\omega - \mathrm{Re}\,\lambda)^n}, \quad \mathrm{Re}\,\lambda < \omega, \quad n = 1, 2, \ldots. \tag{1.2}$$

In this case, $-L$ generates a C_0-semigroup e^{-tL}, $t \geqslant 0$, which is expressed as

$$\begin{aligned} e^{-tL}u = \lim_{R\to\infty} \frac{-1}{2\pi \mathrm{i}} \int_{\sigma-\mathrm{i}R}^{\sigma+\mathrm{i}R} e^{-t\lambda}(\lambda - L)^{-1} u \, d\lambda, \\ \sigma < \omega, \quad t > 0, \quad u \in \mathscr{D}(L), \end{aligned} \tag{1.3}$$

and satisfies the estimate:

$$\left\|e^{-tL}\right\|_E \leqslant M e^{-\omega t}, \quad t \geqslant 0. \tag{1.4}$$

From the control theoretic viewpoint, condition (1.2) seems too general to implement any control action on (1.1). Thus, we extract a narrow class of linear systems in a Hilbert space H equipped with inner product $\langle \cdot, \cdot \rangle$ and norm $\|\cdot\|$, such that (i) the spectrum $\sigma(L)$ is the union of two disjoint sets σ_1 and σ_2, such that σ_1 is a set of a finite number of poles of L, containing unstable ones, and σ_2 is contained in $\mathbb{C}_+$, and that (ii) the location of σ_2 and some growth rate assumption of the resolvent of L are assumed along each line parallel to the real axis. Owing to assumption (i), the following results are standard: Let P be the projector corresponding to σ_1, i.e., $P = \frac{1}{2\pi \mathrm{i}} \int_C (\lambda - L)^{-1} d\lambda$, where C is a Jordan contour encircling only σ_1 such that σ_2 is located outside C, and set $H_1 = PH$ and $H_2 = (1 - P)H$. The subspace H_1 is invariant relative to L, and the restriction $L_1 = L|_{H_1}$ is a bounded operator such that $\sigma(L_1) = \sigma_1$. The restriction L_2 of L onto the set $\mathscr{D}(L_2) = \mathscr{D}(L) \cap H_2$ acts in H_2, and $\sigma(L_2) = \sigma_2$. Although assumption (ii) looks unclear at this moment, detailed settings and a concrete example satisfying the assumption will be given in Section 2: Owing to assumption (ii), a decay estimate of the semigroup $e^{-tL_2} = e^{-tL}\big|_{H_2}$, $t \geqslant 0$, is ensured. In this sense the situation looks somewhat similar to that of the parabolic case. In fact, it seems difficult to ensure a decay of e^{-tL_2} only by assumption (i).

Based on these assumptions, we show that the feedback control scheme for parabolic control systems in Chapter 4 is effectively generalized to stabilization problems of a more general class of linear systems. It is worthwhile in the generalization process to note that *no* Riesz basis is assumed, and associated with such systems. The control scheme here contains, in the feedback loop, a generalized dynamic compensator of Luenberger type with a linear closed operator B as a coefficient. The operator B is assumed to have a compact resolvent; $-B$ generates a C_0-semigroup, instead of an analytic semigroup, in another Hilbert space $\mathscr{H}$ (see Section 2, Chapter 4 for comparison); and the growth rate assumption of the spectrum $\sigma(B)$ at infinity is a little more restrictive than in the parabolic case, that is, the eigenvalues of B are assumed to grow more *slowly*.

In Section 2, some spectral properties of our C_0-semigroup are studied from the control theoretic viewpoint. Some of them are well known in the abstract framework: It turns out that our semigroup e^{-tL} is classified as an *eventually differentiable semigroup* [16]. A relevant example arising from a one-dimensional mono-tubular heat exchanger problem is also illustrated. Based on these properties, two stabilization schemes are developed in Section 3, where we will present again effectiveness of *generalized* compensators. As emphasized above, we do not assume any Riesz basis associated with L. The compensator in each scheme is finally reduced to a finite dimensional one. These two feedback schemes take different forms. However, they are algebraically *dual* ones with each other. The observability- and the controllability- conditions posed, respectively, on the sensors and the actuators of the controlled plant are the same in each scheme. The difference is that the dimension of the compensator is determined by the actuators, or by the sensors in respective schemes. Thus we can design a stabilizing compensator of *lower* dimension by comparison. The key ideas are the setting of the framework of the spectrum $\sigma(L)$ and the distribution of $\sigma(B)$ at infinity and Carlemen's theorem in Section 1, Chapter 4.

7.2 Basic Properties of the Semigroup

We develop in this section some basic properties of the C_0-semigroup e^{-tL}, $t \geqslant 0$, by posing additional relevant conditions on the distribution of the spectrum $\sigma(L)$ and the growth of the resolvent $(\lambda - L)^{-1}$ for ensuring nice properties of e^{-tL}. As a general result, the Cauchy problem for (1.1) in the Hilbert space H is well posed, so that there exist constants $M > 0$ and $\omega \in \mathbb{R}^1$ such that the estimates (1.2) and (1.3) hold. Our further assumptions are stated as follows:

(i) The spectrum $\sigma(L)$ consists of two disjoint sets σ_1 and σ_2; $\sigma(L) = \sigma_1 \cup \sigma_2$, $\sigma_1 \cap \sigma_2 = \varnothing$, where σ_1 is a set of a finite number of poles, containing unstable ones, of L and σ_2 is contained in $\mathbb{C}_+$. There is a $\beta > 0$ such that

$$\sigma_1 \subset \{\lambda \in \mathbb{C};\ \mathrm{Re}\,\lambda < \beta\}, \quad \sigma_2 \subset \{\lambda \in \mathbb{C};\ \mathrm{Re}\,\lambda > \beta\}. \tag{2.1}$$

Let C be a Jordan contour encircling only σ_1, and P be the projector: $P = \frac{1}{2\pi i}\int_C (\lambda - L)^{-1}\, d\lambda$. Then, H is decomposed into the direct sum of invariant subspaces: $H = H_1 \oplus H_2$, where $H_1 = PH$ and $H_2 = (1-P)H$. Then, set $L_1 = L|_{H_1}$ and $L_2 = L|_{\mathscr{D}(L)\cap H_2}$.

(ii) There is a contour Γ in the complex λ-plane ($\lambda = \sigma + i\tau$) such that the resolvent $(\lambda - L)^{-1}$ exists on and in the left-hand side of Γ, where Γ is symmetric relative to the σ-axis, and is described as

$$\sigma = \frac{1}{\hat{c}} \log(\tau - d) + \beta \ (= \sigma_\tau), \quad \tau \geqslant d+1 \ (\geqslant 0), \quad \hat{c} > 0. \tag{2.2}$$

The contour Γ is continuously connected with a suitable rectifiable contour, such that σ is a function of τ on $[-(d+1), d+1]$. The figure of Γ will be illustrated later.

(iii) There exist constants α $(< \omega)$, $T > 0$, and $C > 0$ such that

$$\begin{aligned} &\left\|(\lambda - L)^{-1}\right\| \leqslant \text{const}\, e^{C\sigma}, \quad \lambda = \sigma + \mathrm{i}\tau, \\ &\alpha \leqslant \sigma \leqslant \sigma_{|\tau|}, \quad |\tau| \geqslant T. \end{aligned} \tag{2.3}$$

Remark: While a more general setting on the class of contours Γ is possible (see, e.g., pages 75 – 79 of [26]), the essense of our arguments is unchanged. Typical examples of L are found, e.g., in a class of delay-differential equations (see, e.g., [16]). However, we illustrate another good example.

Let u be in $\mathscr{D}(L)$. Choose a large $c > 0$ such that $0 < \alpha + c$ $(< \omega + c)$, and set $L_c = L + c$ so that the spectrum $\sigma(L_c)$ lies in the right half-plane. By the above assumption (i), the semigroup e^{-tL} is expressed as the inverse Laplace transform [26, 71]:

$$\begin{aligned} e^{-tL}u &= \frac{-1}{2\pi \mathrm{i}} \int_{\alpha - \mathrm{i}\infty}^{\alpha + \mathrm{i}\infty} e^{-t\lambda} (\lambda - L)^{-1} u \, d\lambda \\ &= \lim_{R \to \infty} \frac{-1}{2\pi \mathrm{i}} \int_{\alpha - \mathrm{i}R}^{\alpha + \mathrm{i}R} e^{-t\lambda} (\lambda - L)^{-1} u \, d\lambda, \\ \text{or} \quad e^{-tL_c}u &= \lim_{R \to \infty} \frac{-1}{2\pi \mathrm{i}} \int_{(\alpha + c) - \mathrm{i}R}^{(\alpha + c) + \mathrm{i}R} e^{-t\lambda} (\lambda - L_c)^{-1} u \, d\lambda, \quad t > 0. \end{aligned} \tag{2.4}$$

Note that, by the decay estimate (1.2),

$$\left\|(-s - L_c)^{-1}\right\| \leqslant \frac{M}{(c + \omega) + s}, \quad s \geqslant 0.$$

Thus there is a sector Σ with small angle containing the non-positive real axis $(-\infty, 0]$ inside such that [26]

$$\left\|(\lambda - L_c)^{-1}\right\| \leqslant \frac{M_1}{(c + \omega) + |\lambda|}, \quad \lambda \in \Sigma. \tag{2.5}$$

Thus fractional powers $L_c^{-\gamma}$, $\gamma > 0$, of L_c is well defined:

$$L_c^{-\gamma} = \frac{-1}{2\pi \mathrm{i}} \int_{\partial \Sigma} \lambda^{-\gamma} (\lambda - L_c)^{-1} d\lambda, \tag{2.6}$$

where the contour $\partial\Sigma$ is oriented according to increasing $\tau = \operatorname{Im}\lambda$. One of our goals of this section is to show the following result:

Proposition 2.1. (i) *The semigroup e^{-tL} is expressed as*

$$e^{-tL}u = \frac{-1}{2\pi \mathrm{i}} \int_{\Gamma} e^{-t\lambda}(\lambda - L)^{-1} u\, d\lambda, \quad u \in H, \quad t > \hat{c} + C. \tag{2.7}$$

(ii) *The restriction L_2 of L onto the set $\mathscr{D}(L_2) = \mathscr{D}(L) \cap H_2$ satisfies a decay estimate*

$$\left\| e^{-tL_2} \right\| \leqslant \text{const}\, e^{-\beta t}, \quad t \geqslant 0. \tag{2.8}$$

Remark: The decay (2.8) can be somewhat improved. In fact, we have, for a $\beta'\,(> \beta)$ which is close to β,

$$\left\| e^{-tL_2} \right\| \leqslant \text{const}\, e^{-\beta' t}, \quad t \geqslant 0. \tag{2.8$_1$}$$

Proof. (i) The expression (2.7) is formally found on page 76 of [26]. Since its derivation is incomplete, however, we give here a complete proof. Let us begin with the inverse Laplace transform (2.4). Choose a γ, $0 < \gamma < 1$. By (2.4) and (2.6), we see—via the standard operational calculus—that

$$\begin{aligned} e^{-tL}u &= \lim_{R\to\infty} \frac{-1}{2\pi \mathrm{i}} \int_{(\alpha+c)-\mathrm{i}R}^{(\alpha+c)+\mathrm{i}R} \lambda^{-\gamma} e^{ct} e^{-t\lambda} (\lambda - L_c)^{-1} L_c^{\gamma} u\, d\lambda \\ &= \lim_{R\to\infty} \frac{-1}{2\pi \mathrm{i}} \int_{\alpha-\mathrm{i}R}^{\alpha+\mathrm{i}R} \frac{e^{-t\mu}}{(\mu + c)^{\gamma}} (\mu - L)^{-1} L_c^{\gamma} u\, d\mu, \quad u \in \mathscr{D}(L). \end{aligned} \tag{2.9}$$

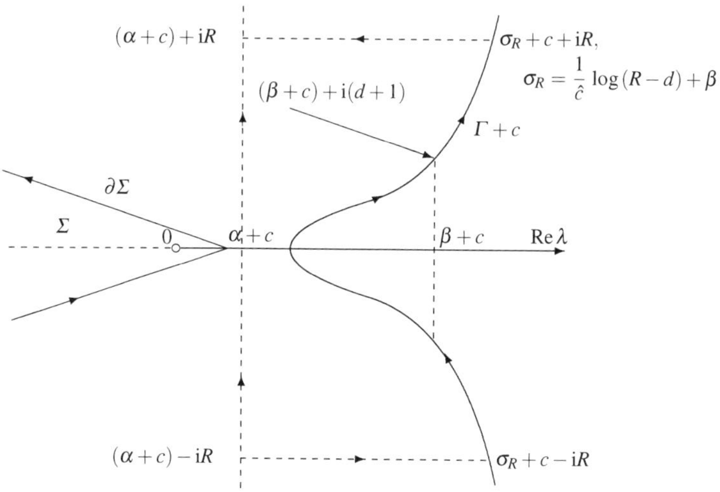

Figure 9

The factor $L_c^\gamma u$ in (2.9) is introduced to change the integral contour from the vertical line, $((\alpha+c)-\mathrm{i}\infty, (\alpha+c)+\mathrm{i}\infty)$ to $\Gamma+c$. Since $(\mu-L)^{-1}$ is analytic on Γ and in the left-hand side of Γ, we have

$$\int_{\alpha-\mathrm{i}R}^{\alpha+\mathrm{i}R} \frac{e^{-t\mu}}{(\mu+c)^\gamma}(\mu-L)^{-1}L_c^\gamma u\,d\mu = \int_{\Gamma_R} + \int_{\alpha-\mathrm{i}R}^{\sigma_R-\mathrm{i}R} + \int_{\sigma_R+\mathrm{i}R}^{\alpha+\mathrm{i}R}, \qquad R>T,$$

where Γ_R denotes a subcontour of Γ such that $|\operatorname{Im}\mu| \leqslant R$ for $\mu\in\Gamma_R$. As for the second term of the right-hand side, we calculate on the segment $[\alpha-\mathrm{i}R, \sigma_R-\mathrm{i}R]$ as

$$\left|\frac{e^{-t\mu}}{(\mu+c)^\gamma}\right| \left\|(\mu-L)^{-1}\right\| \leqslant \text{const}\, R^{-\gamma}e^{-t\sigma}e^{C\sigma} = \text{const}\, R^{-\gamma}e^{-\sigma(t-C)}.$$

Thus, as long as t is greater than C, we see that

$$\begin{aligned}\left\|\int_{\alpha-\mathrm{i}R}^{\sigma_R-\mathrm{i}R} \frac{e^{-t\mu}}{(\mu+c)^\gamma}(\mu-L)^{-1}L_c^\gamma u\,d\mu\right\| &\leqslant \text{const}\, R^{-\gamma}\int_\alpha^{\sigma_R} e^{-\sigma(t-C)}\,d\sigma \\ &< \frac{\text{const}}{R^\gamma}\frac{e^{-\alpha(t-C)}}{t-C} \to 0, \quad R\to\infty.\end{aligned}$$

The third term of the right-hand side is similarly evaluated, and goes to 0 as $R\to\infty$. For $\mu=\sigma_\tau+\mathrm{i}\tau$ on Γ $(|\tau|\geqslant d+1)$,

$$\begin{aligned} d\mu &= \left(\frac{1}{\hat{c}(\tau-d)}+\mathrm{i}\right)d\tau, \quad |d\mu|\leqslant \text{const}\, d\tau, \quad |\tau|\geqslant d+1, \\ \left\|(\mu-L)^{-1}\right\| &\leqslant \text{const}\exp\left(C\sigma_{|\tau|}\right) \\ &\leqslant \text{const}\exp\left(C\left(\frac{1}{\hat{c}}\log\left(|\tau|-d\right)+\beta\right)\right) = \text{const}\left(|\tau|-d\right)^{C/\hat{c}}.\end{aligned}$$

When $|\mu|$ is large enough, the integrand on Γ is evaluated as

$$\begin{aligned}\left\|\frac{e^{-t\mu}}{(\mu+c)^\gamma}(\mu-L)^{-1}L_c^\gamma u\right\| &\leqslant \text{const}\frac{\exp\left(-t\sigma_{|\tau|}\right)}{|\tau|^\gamma}\left(|\tau|-d\right)^{C/\hat{c}} \\ &= \text{const}\frac{\left(|\tau|-d\right)^{-(t-C)/\hat{c}}e^{-t\beta}}{|\tau|^\gamma}.\end{aligned}$$

For t $(\geqslant \hat{c}+C)$ the integral on Γ is absolutely convergent. Thus we have shown that

$$e^{-tL}u = \frac{-1}{2\pi\mathrm{i}}\int_\Gamma \frac{e^{-t\mu}}{(\mu+c)^\gamma}(\mu-L)^{-1}L_c^\gamma u\,d\mu, \quad t\geqslant \hat{c}+C. \tag{2.10}$$

Note that, for such a $t\geqslant \hat{c}+C$,

$$\begin{aligned} &L_c^{-\gamma}\left(\frac{-1}{2\pi\mathrm{i}}\int_{\Gamma+c} e^{-t(\lambda-c)}(\lambda-L_c)^{-1}u\,d\lambda\right) \\ &= \frac{-1}{2\pi\mathrm{i}}\int_{\Gamma+c}\frac{e^{-t(\lambda-c)}}{\lambda^\gamma}(\lambda-L_c)^{-1}u\,d\lambda \in \mathscr{D}(L_c^\gamma),\end{aligned}$$

or

$$L_c^\gamma\left(\frac{-1}{2\pi \mathrm{i}}\int_\Gamma \frac{e^{-t\mu}}{(\mu+c)^\gamma}(\mu-L)^{-1}u\,d\mu\right)=\frac{-1}{2\pi \mathrm{i}}\int_\Gamma e^{-t\mu}(\mu-L)^{-1}u\,d\mu.$$

Thus we finally calculate as

$$\begin{aligned}e^{-tL}u&=\frac{-1}{2\pi \mathrm{i}}\int_\Gamma \frac{e^{-t\mu}}{(\mu+c)^\gamma}(\mu-L)^{-1}L_c^\gamma u\,d\mu\\&=L_c^\gamma\left(\frac{-1}{2\pi \mathrm{i}}\int_\Gamma \frac{e^{-t\mu}}{(\mu+c)^\gamma}(\mu-L)^{-1}u\,d\mu\right)\\&=\frac{-1}{2\pi \mathrm{i}}\int_\Gamma e^{-t\mu}(\mu-L)^{-1}u\,d\mu,\quad t\geqslant \hat{c}+C,\end{aligned}$$

the right-hand side of which clearly defines a bounded operator for $t>\hat{c}+C$. Thus this shows the expression (2.7).

(ii) When u is in $\mathscr{D}(L)\cap H_2$, let us change the contour Γ in (2.7). Let Γ_1 be the contour such that

$$\sigma=\begin{cases}\beta, & |\tau|\leqslant d+1,\\ \dfrac{1}{\hat{c}}\log(|\tau|-d)+\beta, & d+1\leqslant|\tau|.\end{cases}$$

In view of the assumption (2.1), $e^{-t\mu}(\mu-L)^{-1}u=e^{-t\mu}(\mu-L_2)^{-1}u$ is analytic on and in the left-hand side of Γ_1. Thus, we see that

$$\begin{aligned}e^{-tL_2}u=e^{-tL}u&=\frac{-1}{2\pi \mathrm{i}}\int_\Gamma e^{-t\mu}(\mu-L)^{-1}u\,d\mu,\\&=\frac{-1}{2\pi \mathrm{i}}\int_{\Gamma_1} e^{-t\mu}(\mu-L_2)^{-1}u\,d\mu,\quad t\geqslant\hat{c}+C.\end{aligned}\tag{2.11}$$

As long as t is greater than $\hat{c}+C$, the integrand of (2.11) is evaluated as

$$\begin{aligned}\left\|e^{-t\mu}(\mu-L)^{-1}u\right\|&\leqslant \text{const}\exp\left(-t\sigma_{|\tau|}\right)(|\tau|-d)^{C/\hat{c}}\|u\|\\&\qquad\text{for }\mu=\sigma_\tau+\mathrm{i}\tau,\quad |\tau|\geqslant d+1\\&=\text{const}\,(|\tau|-d)^{-t/\hat{c}}e^{-\beta t}(|\tau|-d)^{C/\hat{c}}\|u\|\\&=\text{const}\,e^{-\beta t}\frac{\|u\|}{(|\tau|-d)^{(t-C)/\hat{c}}},\end{aligned}$$

the last term of which is a function of τ in $L^1\left(\mathbb{R}^1_\tau\setminus[-(d+1),d+1]\right)$. It is clear that $\left\|e^{-t\mu}(\mu-L)^{-1}u\right\|$ is bounded from above by $\text{const}\,e^{-\beta t}\|u\|$ on the finite part of Γ_1 with $|\tau|\leqslant d+1$. Thus,

$$\left\|e^{-tL_2}u\right\|\leqslant\text{const}\,e^{-\beta t}\|u\|,\quad \frac{1}{\hat{c}}(t-C)-1\geqslant 1,\quad u\in\mathscr{D}(L_2).\tag{2.12}$$

Since $e^{-tL_2}u$ is, of course, continuous on $[0, 2\hat{c}+C]$, we have

$$\left\|e^{-tL_2}u\right\| \leqslant \text{const}\, e^{-\beta t}\left\|u\right\|, \quad t \geqslant 0, \quad u \in \mathscr{D}(L_2).$$

Since $\mathscr{D}(L_2)$ is dense in H_2, the above estimate is correct for every $u \in H_2$:

$$\left\|e^{-tL_2}u\right\| \leqslant \text{const}\, e^{-\beta t}\left\|u\right\|, \quad t \geqslant 0, \quad u \in H_2. \tag{2.13}$$

Thus we have shown the decay estimate (2.8).

To obtain the decay estimate (2.8_1), we note that the resolvent $(\lambda - L_2)^{-1}$ exists in a neighborhood of the segment; $\lambda = \beta + i\tau$, $|\tau| \leqslant d$. Let d' $(> d)$ be close to d. Setting $\beta' = \frac{1}{\hat{c}}\log(d' - d + 1) + \beta$ $(> \beta)$, we modify Γ_1 a little around this segment: Let Γ_1' be the contour such that

$$\sigma = \begin{cases} \beta', & |\tau| \leqslant d' + 1, \\ \dfrac{1}{\hat{c}}\log\dfrac{|\tau| - d}{1 + (d' - d)} + \beta', & |\tau| \geqslant d' + 1. \end{cases}$$

The resolvent $(\lambda - L_2)^{-1}$ exists on Γ_1' and in the left-hand side of Γ_1'. The expression of e^{-tL_2} in (2.11) is correct with Γ_1 replaced by Γ_1'. Then the estimate (2.8_1) is straightforward by following the above procedure. □

Remark: In view of (2.7), regularity of $e^{-tL}u$, $u \in H$, increases as t grows: As long as t is greater than $\hat{c}(n+1) + C$, the function $(-\lambda)^n e^{-t\lambda}(\lambda - L)^{-1}u$ is integrable on Γ, and thus $e^{-tL}u$ becomes n times differentiable. This fact is discussed in [26], and also known as a standard property of an eventually differentiable semigroup [16].

Besides standard examples of L, such as a class of delay-differential equations, we illustrate here the following example:

Example: Following the study of Sano [61], let us illustrate an example of L arising from a mono-tubular heat exchanger problem. Let a, b, k, and γ be positive constants, and set $H = L^2(0, 1)$. Let L be a differential operator in H defined by

$$\begin{aligned} Lu &= \frac{du}{dx} + au + k\gamma e^{-bx}u(1), \quad u \in \mathscr{D}(L), \\ \mathscr{D}(L) &= \{u \in H^1(0,1);\ u(0) = 0\}. \end{aligned} \tag{2.14}$$

The resolvent $(\lambda - L)^{-1}$ is compact, and the spectrum $\sigma(L)$ consists of λ satisfying the relation

$$1 + k\gamma\frac{e^{-b} - e^{\lambda - a}}{a - \lambda - b} = 0, \tag{2.15}$$

where $\lambda = a - b$ belongs to $\rho(L)$ (see [61]). It is thus apparent that L is a closed operator, since the bounded inverse $(\lambda - L)^{-1}$ exists. For $\lambda \in \rho(L)$, the resolvent is expressed as

$$(\lambda - L)^{-1} f\big|_x = \int_0^x e^{(\lambda-a)(x-y)} \left(k\gamma e^{-by} \left(1 + k\gamma \frac{e^{-b} - e^{\lambda-a}}{a-\lambda-b} \right)^{-1} \times \int_0^1 e^{(\lambda-a)(1-\xi)} f(\xi)\, d\xi - f(y) \right) dy. \tag{2.16}$$

The equation (2.15) for $\lambda = \sigma + i\tau$ is characterized as

$$\sigma(L) = (S_1 \cap S_2) \cup S_3, \tag{2.17}$$

where the sets S_1, S_2, and S_3 are defined by

$$S_1 : \sigma - a = -b + k\gamma e^{-b} + \frac{\tau}{\tan\tau};$$

$$S_2 : \sigma - a = \log\left(\frac{-\tau}{k\gamma \sin\tau} \right),$$

$$\tau \in \bigcup_{n=1}^{\infty} ((2n-1)\pi, 2n\pi) \cup (-2n\pi, (-2n+1)\pi); \quad \text{and}$$

$$S_3 : e^{\sigma} = e^{a} \left(\frac{a-b}{k\gamma} + e^{-b} \right) - \frac{e^{a}}{k\gamma}\sigma,$$

respectively. The distribution of the eigenvalues λ of L at infinity is governed by the set $S_1 \cap S_2$. A part of the behavior of λ for $\tau \in ((2n-1)\pi, 2n\pi)$ is illustrated in Figure 10. Detailed calculations show that there are positive constants c_1 and c_2 such that

$$c_1 e^{\sigma} < |\tau| < c_2 e^{\sigma}, \quad \text{as } \lambda = \sigma + i\tau \to \infty. \tag{2.18}$$

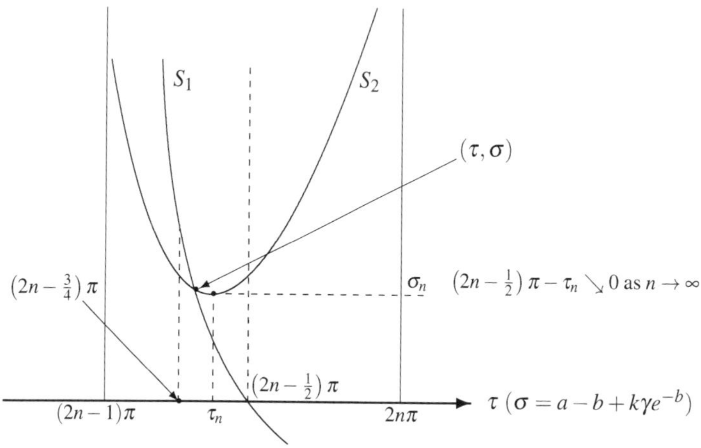

Figure 10

This ensures a contour Γ, a part of which is described by $|\tau| = ce^{\sigma}$, $\sigma \geqslant \beta\ (>0)$. Here, $c > 0$ is chosen large enough, so that $c > \max\,(c_2, k\gamma e^{-a})$. The resolvent exisits on and in the left-hand side of Γ. Since $\sigma(L)$ consists only of eigenvalues, (2.1) is satisfied.

To derive the estimate (2.3), let $\lambda = \sigma + i\tau$ be on and in the left-hand side of Γ such that $|\tau| \geqslant T$, $T > 0$ being chosen large enough. For σ, $0 \leqslant \sigma \leqslant \sigma_\tau = \log(|\tau|/c)$, we calculate in (2.16) as

$$\left| 1 + k\gamma \frac{e^{-b} - e^{\lambda - a}}{a - \lambda - b} \right| \geqslant 1 - k\gamma \frac{e^{\sigma - a} + e^{-b}}{|\lambda| - |a-b|}$$
$$\geqslant 1 - k\gamma \frac{\frac{|\tau|}{c} e^{-a} + e^{-b}}{|\tau| - |a-b|} \geqslant \text{const}\ (>0)$$

as long as $|\tau| \geqslant T$. Thus, for each $x \in (0, 1)$,

$$\begin{aligned} |(\lambda - L)^{-1} f| &\leqslant \text{const} \int_0^x e^{\sigma - a} \left| \int_0^1 e^{(\lambda - a)(1-\xi)} f(\xi)\, d\xi \right| + \int_0^x e^{\sigma - a} |f(y)|\, dy \\ &\leqslant \text{const}\, e^{2\sigma} \|f\| + e^{\sigma} \|f\| \\ &\leqslant \text{const}\, e^{2\sigma} \|f\|, \end{aligned}$$
$$\lambda = \sigma + i\tau, \quad |\tau| \geqslant T, \quad 0 \leqslant \sigma \leqslant \sigma_\tau = \log \frac{|\tau|}{c}.$$

Finally we have

$$\left\|(\lambda - L)^{-1}\right\| \leqslant \text{const}\, e^{2\sigma}, \quad \lambda = \sigma + i\tau, \quad |\tau| \geqslant T, \quad 0 \leqslant \sigma \leqslant \sigma_\tau.$$

The contour Γ at this time is a little different from the contour in (2.2), but has almost same properties. Thus, the assertions in Proposition 2.1 hold in our L in (2.14).

Finally we note -via elementary calculations- that the dimension of each eigenspace is equal to 1, and that there arises *no* generalized eigenspace associated with L.

Brief review of stabilization in static feedback scheme:

We go back to the original setting of the operator L stated in Proposition 2.1. Given a set of actuators $g_k \in H$, $1 \leqslant k \leqslant M$, let $G \in \mathscr{L}(\mathbb{C}^M; H)$ be defined as $G\boldsymbol{f} = \sum_{k=1}^M f_k g_k$ for $\boldsymbol{f} = (f_1\ f_2 \ldots f_M)^{\mathrm{T}} \in \mathbb{C}^M$, where $(\ldots)^{\mathrm{T}}$, as usual, denotes the transpose of vectors or matrices. We briefly review a well known result on stabilization of the system,

$$\begin{aligned} &\frac{du}{dt} + Lu = GWu = \sum_{k=1}^{M} \langle u, \zeta_k \rangle g_k, \quad u(0) = u_0 \in \mathscr{D}(L), \\ &Wu = \langle u, \boldsymbol{\zeta} \rangle = \big(\langle u, \zeta_1 \rangle \quad \langle u, \zeta_2 \rangle \quad \ldots \quad \langle u, \zeta_M \rangle \big)^{\mathrm{T}}. \end{aligned} \tag{2.19}$$

Here, ζ_k denote sensors. Since the right-hand side GW is bounded, the Cauchy problem for (2.19) is well posed, and generates a C_0-semigroup. The framework and the basic ideas of the problem are the same as in Section 2, Chapter 3, where $-L$ was, instead, the infinitesimal generator of an analytic semigroup $e^{-tL}, t > 0$. Based on the preceding properties on our C_0-semigroup e^{-tL}, the situation here is unchanged. Assume that (L_1, PG) is a controllable pair. Then by Corollary 2.2, Chapter 3, given any number b greater than β' $(> \beta)$, we find suitable $\zeta_k \in P^*H$, $1 \leqslant k \leqslant M$ or W such that the decay estimate,

$$\left\| e^{-t(L-GW)} \right\| \leqslant \text{const}\, e^{-\beta' t}, \quad t \geqslant 0 \tag{2.20}$$

is ensured. In our case of C_0-semigroups, small bounded perturbation to the ζ_k affects the decay estimate (2.20) *only a little*. This estimate is a basic one later in Section 3.

Let us describe the controllability condition of the pair (L_1, PG) more concretely. Let $\sigma_1 = \sigma(L_1) = \{\lambda_i\}_{1 \leqslant i \leqslant \nu}$, where $\lambda_i \neq \lambda_j$ for $i \neq j$. Let C_i be a counter-clockwise circle of small radius with center λ_i. Set $P_{\lambda_i} = \frac{1}{2\pi \mathrm{i}} \int_{C_i} (\lambda - L)^{-1}\, d\lambda$, and $m_i = \dim P_{\lambda_i} H$. Then, $P = P_{\lambda_1} + \cdots + P_{\lambda_\nu}$. Let $\{\varphi_{ij};\ 1 \leqslant j \leqslant m_i\}$ be a set of generalized eigenvectors of λ_i such that

$$L\varphi_{ij} = \lambda_i \varphi_{ij} + \sum_{k<j} \alpha^i_{jk} \varphi_{ik}, \quad 1 \leqslant i \leqslant \nu, \quad 1 \leqslant j \leqslant m_i. \tag{2.21$_1$}$$

Then the restriction $L|_{P_{\lambda_i} H}$ of L is equivalent to the $m_i \times m_i$ upper triangular matrix Λ_i (see (1.7), Chapter 3):

$$\Lambda_i|_{(j,k)} = \begin{cases} \alpha^i_{kj}, & j < k, \\ \lambda_i, & j = k, \\ 0, & j > k. \end{cases} \tag{2.22$_1$}$$

For each i, $1 \leqslant i \leqslant \nu$, let $P_{\lambda_i} g_k = \sum_{j=1}^{m_i} g^k_{ij} \varphi_{ij}$, and define the $m_i \times M$ matrix G_i as (see (2.8), Chapter 3)

$$G_i = \left(g^k_{ij};\ \begin{matrix} j & \downarrow & 1, \ldots, m_i \\ k & \rightarrow & 1, \ldots, M \end{matrix} \right). \tag{2.23}$$

The matrices G_i are the so called *controllability matrices* (the corresponding *observability matrices* W_i will be introduced later in Section 3). Then we see that the pair (L_1, PG) is controllable, if and only if the rank conditions,

$$\text{rank} \left(G_i \quad \Lambda_i G_i \quad \ldots \quad \Lambda_i^{m_i - 1} G_i \right) = m_i, \quad 1 \leqslant i \leqslant \nu \tag{2.24}$$

are satisfied (see (2.9), Chapter 3).

Before closing this section, we recall the adjont structure of the operator L. Since $\lambda_i \in \sigma_1$ are the eigenvalues of L with $\dim P_{\lambda_i} H = m_i$, $\overline{\lambda_i}$ are the eigenvalues of L^*, and $P^*_{\lambda_i}$ are the corresponding projectors, such that $\dim P^*_{\lambda_i} H = m_i$ [66]. Let $\{\psi_{ij};\ 1 \leqslant j \leqslant m_i\}$ be a set of generalized eigenvectors for $\overline{\lambda_i}$ such that

$$L^* \psi_{ij} = \overline{\lambda_i} \psi_{ij} + \sum_{k<j} \beta^i_{jk} \psi_{ik}, \quad 1 \leqslant i \leqslant \nu,\ 1 \leqslant j \leqslant m_i. \tag{2.21$_2$}$$

Then the restriction $L^*|_{P^*_{\lambda_i} H}$ of L^* is equivalent to the $m_i \times m_i$ upper triangular matrix $\tilde{\Lambda}_i$:

$$\tilde{\Lambda}_i|_{(j,k)} = \begin{cases} \beta^i_{kj}, & j < k, \\ \overline{\lambda_i}, & j = k, \\ 0, & j > k. \end{cases} \tag{2.22$_2$}$$

Let Π_i, $1 \leqslant i \leqslant \nu$, be the $m_i \times m_i$ non-singular matrices defined as (see (1.9), Chapter 3)

$$\Pi_i = \left(\langle \varphi_{ij}, \psi_{il} \rangle\ ;\ \begin{matrix} j & \rightarrow & 1, \dots, m_i \\ l & \downarrow & 1, \dots, m_i \end{matrix} \right).$$

It is clear that

$$G_i = \Pi_i^{-1} \left(\langle g_k, \psi_{ij} \rangle\ ;\ \begin{matrix} j & \downarrow & 1, \dots, m_i \\ k & \rightarrow & 1, \dots, M \end{matrix} \right).$$

We also note that

$$\tilde{\Lambda}_i = \left(\Pi_i \Lambda_i \Pi_i^{-1} \right)^*. \tag{2.25}$$

Since $P_{\lambda_i} G \in \mathscr{L}(\mathbb{C}^M; P_{\lambda_i} H)$, thus $\left((P_{\lambda_i} G)^* \ (L_1 P_{\lambda_i} G)^* \ \dots \ (L_1^{m_i - 1} P_{\lambda_i} G)^* \right)^{\mathrm{T}}$ belongs to $\mathscr{L}(P_i H; \mathbb{C}^{m_i M})$. In operator notation, the controllability condition (2.24) is interpreted as

$$\ker \left((P_{\lambda_i} G)^* \quad (L_1 P_{\lambda_i} G)^* \quad \dots \quad (L_1^{m_i - 1} P_{\lambda_i} G)^* \right)^{\mathrm{T}} = \{0\}, \quad 1 \leqslant i \leqslant \nu.$$

Similar interpretations can be made in rank conditions of matrices such as those in (3.11) later.

7.3 Stabilization

Based on the spectral properties of the operator L in Section 2, we construct a dynamic feedback scheme for (1.1) to achieve stabilization. Throughout the section, we assume the conditions (2.1) – (2.3) on L, so that Proposition 2.1 is correct. To define a dynamic compensator in the feedback loop, let $\mathscr{H}$ be a separable Hilbert space equipped with inner product $\langle \cdot, \cdot \rangle_{\mathscr{H}}$ and norm $\|\cdot\|_{\mathscr{H}}$. Consider the differential equation with state (u, v) in $H \times \mathscr{H}$, which is described as

$$\begin{cases} \dfrac{du}{dt} + Lu = \displaystyle\sum_{k=1}^{M} \langle v, \rho_k \rangle_{\mathscr{H}} g_k, \quad u(0) = u_0 \in \mathscr{D}(L), \\ \dfrac{dv}{dt} + Bv = \displaystyle\sum_{k=1}^{N} \langle u, w_k \rangle \xi_k + \sum_{k=1}^{M} \langle v, \rho_k \rangle_{\mathscr{H}} \tilde{g}_k, \quad v(0) = v_0 \in \mathscr{D}(B), \end{cases} \tag{3.1}$$

where the first equation denotes the controlled plant with state u, and the second equation the dynamic compensator with state v. In (3.1), the scalar-valued functions $\langle v, \rho_k \rangle_{\mathscr{H}}$ denote outputs of the compensator, and enter the controlled plant as inputs, and g_k and $\tilde{g}_k$ actuators acting in respective spaces. The actuators g_k and the sensors w_k are *given* parameters of the controlled plant, and the space $\mathscr{H}$, the operator B, the sensors ρ_k, and the actuators ξ_k and $\tilde{g}_k$ are parameters *to be designed*. An alternative setting is given as

$$\begin{cases} \dfrac{du}{dt} + Lu = \displaystyle\sum_{k=1}^{M} \langle v, \rho_k \rangle_{\mathscr{H}} g_k, \quad u(0) = u_0 \in \mathscr{D}(L), \\ \dfrac{dv}{dt} + Bv = \displaystyle\sum_{k=1}^{N} \langle u - Yv, w_k \rangle \xi_k, \quad v(0) = v_0 \in \mathscr{D}(B). \end{cases} \tag{3.1$'$}$$

Equation (3.1$'$) is viewed as an algebraic counterpart of (3.1), and corresponds to eqn. (5.1) in Section 5, Chapter 4. Thus, we mainly study (3.1) in the following.

To begin with, we need to characterize the operator B of our compensator: For this, let $\{\eta_{ij}^{\pm};\ i \geqslant 1,\ 1 \leqslant j \leqslant n_i\}$ with $n_i < \infty$ for each i be an orthonormal basis for $\mathscr{H}$. Since $\mathscr{H}$ is one of the designed parameters, there are a variety of choice of the multiplicities n_i. Then every $v \in \mathscr{H}$ is expressed as a Fourier series:

$$v = \sum_{i,j} \left(v_{ij}^{+} \eta_{ij}^{+} + v_{ij}^{-} \eta_{ij}^{-} \right), \quad v_{ij}^{\pm} = \langle v, \eta_{ij}^{\pm} \rangle_{\mathscr{H}}.$$

For each $n \geqslant 1$, let $\mathscr{P}_n$ be the projector defined as

$$\mathscr{P}_n v = \sum_{i(\leqslant n),j} \left(v_{ij}^{+} \eta_{ij}^{+} + v_{ij}^{-} \eta_{ij}^{-} \right) \quad \text{for } v = \sum_{i,j} \left(v_{ij}^{+} \eta_{ij}^{+} + v_{ij}^{-} \eta_{ij}^{-} \right).$$

By (2.1), the vertical line, $\{\lambda;\ \mathrm{Re}\,\lambda = \beta\}$ belongs to $\rho(L)$. Given a sequence of increasing positive numbers $\{\mu_i\}_{i \geqslant 1}$ with $\mu_i \to \infty$ as $i \to \infty$, set $\zeta_i^{\pm} = \beta \pm \mathrm{i}\,\mu_i$, $i \geqslant 1$. Let us define the operator B in (3.1) as

$$\begin{aligned} Bv &= \sum_{i,j} \left(\zeta_i^{+} v_{ij}^{+} \eta_{ij}^{+} + \zeta_i^{-} v_{ij}^{-} \eta_{ij}^{-} \right) = \sum_{i,j} \zeta_i^{\pm} v_{ij}^{\pm} \eta_{ij}^{\pm}, \\ \mathscr{D}(B) &= \left\{ v \in \mathscr{H};\ \sum_{i,j} \left| \zeta_i^{\pm} v_{ij}^{\pm} \right|^2 < \infty \right\}. \end{aligned} \tag{3.2}$$

Then B is a closed operator with dense domain $\mathscr{D}(B)$. It is apparent that

(i) $\sigma(B) = \{\zeta_i^{\pm};\, i \geqslant 1\}$, where $\zeta_i^{\pm} = \beta \pm \mathrm{i}\,\mu_i$;

(ii) $(\zeta_i^{\pm} - B)\eta_{ij}^{\pm} = 0, \quad i \geqslant 1,\ 1 \leqslant j \leqslant n_i$; and

(iii) For $\lambda \in \rho(B)$ and $v = \sum_{i,j} v_{ij}^{\pm}\eta_{ij}^{\pm}$,

$$(\lambda - B)^{-1} v = \sum_{i,j} \frac{v_{ij}^{+}}{\lambda - \zeta_i^{+}}\eta_{ij}^{+} + \sum_{i,j} \frac{v_{ij}^{-}}{\lambda - \zeta_i^{-}}\eta_{ij}^{-} \left(= \sum_{i,j} \frac{v_{ij}^{\pm}}{\lambda - \zeta_i^{\pm}}\eta_{ij}^{\pm} \right).$$

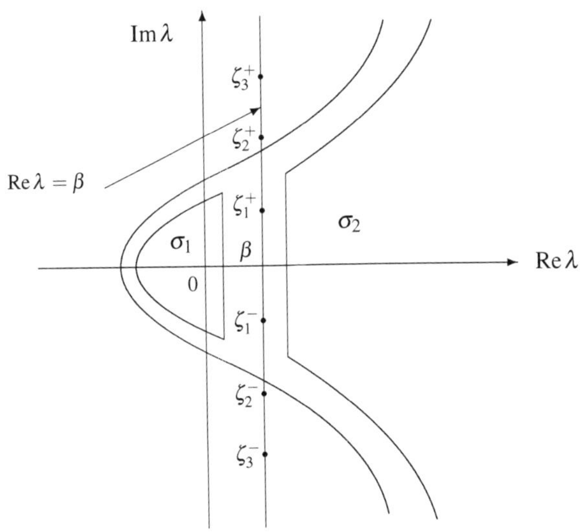

Figure 11

Thus the spectrum $\sigma(B)$ is located on the vertical line, $\{\lambda;\ \mathrm{Re}\,\lambda = \beta\}$ (see Figure 11). We have chosen our B, such that

$$\sigma(L) \cap \sigma(B) = \varnothing. \tag{3.3}$$

The operator $-B$ is the infinitesimal generator of a C_0-semigroup e^{-tB}, $t > 0$, which is described as

$$e^{-tB} v = \sum_{i,j} e^{-\zeta_i^{+} t} v_{ij}^{+}\eta_{ij}^{+} + \sum_{i,j} e^{-\zeta_i^{-} t} v_{ij}^{-}\eta_{ij}^{-} = \sum_{i,j} e^{-\zeta_i^{\pm} t} v_{ij}^{\pm}\eta_{ij}^{\pm}. \tag{3.4}$$

It is clear that

$$\left\| e^{-tB} \right\|_{\mathscr{H}} \leqslant e^{-\beta t}, \quad t \geqslant 0. \tag{3.5}$$

For our stabilization process, we construct the operator B such that

$$\mu_i \leqslant \mathrm{const}\; i^{\gamma}, \quad i \geqslant 1, \quad \text{for } 0 < \exists\gamma < \frac{1}{2}. \tag{3.6}$$

Here we note that the growth rate condition of the sequence $\{\mu_i\}$ is more restrictive than in the parabolic case, where, in fact, γ can be chosen such that $\gamma < 2$ (see Chapter 4).

Let us consider an operator equation on $\mathscr{D}(L)$, Sylvester's equation:

$$XL - BX = C, \quad C = -\sum_{k=1}^{N} \langle \cdot, w_k \rangle \xi_k. \tag{3.7}$$

The actuators ξ_k in (3.1) are such that they are expressed as a Fourier series in terms of the basis $\left\{\eta_{ij}^{\pm}\right\}$,

$$\xi_k = \sum_{i,j} \left(\xi_{ij}^k \eta_{ij}^+ + \overline{\xi_{ij}^k} \eta_{ij}^- \right), \quad 1 \leqslant k \leqslant N, \tag{3.8}$$

where the upper bar means the complex conjugate. The following result is a version of Proposition 3.2, Chapter 4. The difference is just of technical nature. Actually, both operators L and B are *not* sectorial.

Proposition 3.1. *The operator equation* (3.7) *admits a unique operator solution $X \in \mathscr{L}(H; \mathscr{H})$. The solution is expressed as*

$$\begin{aligned} Xu &= \sum_{i,j} \sum_{k=1}^{N} f_k(\zeta_i^+; u) \xi_{ij}^k \eta_{ij}^+ + \sum_{i,j} \sum_{k=1}^{N} f_k(\zeta_i^-; u) \overline{\xi_{ij}^k} \eta_{ij}^-, \quad u \in H, \\ &\textit{where} \quad f_k(\lambda; u) = \left\langle (\lambda - L)^{-1} u, w_k \right\rangle, \quad 1 \leqslant k \leqslant N. \end{aligned} \tag{3.9}$$

Proof. By the assumption (2.3), we first note that

$$\sup_{\substack{\lambda \in \rho(L), \\ \mathrm{Re}\,\lambda = \beta}} \left\| (\lambda - L)^{-1} \right\| < \infty,$$

so that the operator X is well defined, and belongs to $\mathscr{L}(H; \mathscr{H})$. When u is in $\mathscr{D}(L)$, we calculate as

$$\begin{aligned} XLu &= \sum_{i,j} \sum_{k=1}^{N} \left\langle (\zeta_i^+ - L)^{-1} Lu, w_k \right\rangle \xi_{ij}^k \eta_{ij}^+ \\ &\quad + \sum_{i,j} \sum_{k=1}^{N} \left\langle (\zeta_i^- - L)^{-1} Lu, w_k \right\rangle \overline{\xi_{ij}^k} \eta_{ij}^+ \\ &= \sum_{i,j} \sum_{k=1}^{N} \left\langle (-u + \zeta_i^+ (\zeta_i^+ - L)^{-1} u, w_k \right\rangle \xi_{ij}^k \eta_{ij}^+ + \cdots \qquad (\mathrm{Re}\,\zeta_i^{\pm} = \beta) \\ &= -\sum_{i,j} \sum_{k=1}^{N} \langle u, w_k \rangle \xi_{ij}^k \eta_{ij}^+ + \sum_{i,j} \sum_{k=1}^{N} \zeta_i^+ \underbrace{\left\langle (\zeta_i^+ - L)^{-1} u, w_k \right\rangle}_{f_k(\zeta_i^+; u)} \xi_{ij}^k \eta_{ij}^+ + \cdots. \end{aligned}$$

Here, we have used the fact that

$$\sup_{\substack{\lambda\in\rho(L),\\ \mathrm{Re}\,\lambda=\beta}} \left\|\lambda(\lambda-L)^{-1}u\right\| \leqslant \|u\| + \sup_{\substack{\lambda\in\rho(L),\\ \mathrm{Re}\,\lambda=\beta}} \left\|(\lambda-L)^{-1}Lu\right\| < \infty.$$

On the other hand, since $\sup_i \left|\zeta_i^{\pm} f_k(\zeta_i^{\pm};u)\right| < \infty$, and B is closed, we see that

$$BXu = \sum_{i,j}\sum_{k=1}^{N} \zeta_i^{+} f_k(\zeta_i^{+};u)\xi_{ij}^{k}\eta_{ij}^{+} + \sum_{i,j}\sum_{k=1}^{N} \zeta_i^{-} f_k(\zeta_i^{-};u)\overline{\xi_{ij}^{k}}\eta_{ij}^{-}.$$

Thus, (3.7) holds for $u\in\mathscr{D}(L)$.

As for uniqueness, let $XL-BX=0$ on $\mathscr{D}(L)$. Then,

$$X(\lambda-L)=(\lambda-B)X, \qquad (\lambda-B)^{-1}X = X(\lambda-L)^{-1}, \quad \lambda\in\rho(L)\cap\rho(B).$$

For each $\zeta_i^{\pm}$, let $C_i^{\pm}$ be a circle with center $\zeta_i^{\pm}$ and small radius such that the inside of $C_i^{\pm}$ is contained in $\rho(L)$. Then, for any $u\in H$,

$$\begin{aligned} 2\pi\mathrm{i}\,(Xu)_{ij}^{\pm}\eta_{ij}^{\pm} &= \int_{C_i^{\pm}} (\lambda-B)^{-1}Xu\,d\lambda \\ &= \int_{C_i^{\pm}} X(\lambda-L)^{-1}u\,d\lambda = 0, \end{aligned}$$

which shows that $Xu=0$. □

Setting $\tilde{g}_k = Xg_k$, $1\leqslant k\leqslant M$, in (3.1), we have our basic feedback control system in $H\times\mathscr{H}$:

$$\begin{cases} \dfrac{du}{dt} + Lu = \displaystyle\sum_{k=1}^{M} \langle v, \rho_k\rangle_{\mathscr{H}}\, g_k, \quad u(0)=u_0\in\mathscr{D}(L), \\ \dfrac{dv}{dt} + Bv = \displaystyle\sum_{k=1}^{N} \langle u, w_k\rangle\,\xi_k + \sum_{k=1}^{M} \langle v, \rho_k\rangle_{\mathscr{H}}\, Xg_k, \quad v(0)=v_0\in\mathscr{D}(B). \end{cases} \tag{3.1$_1$}$$

Before stating our main result in this section, let us recall that the controllability matrices G_i are defined by (2.23). We need more matrices: Let us furthermore define the matrices Ξ_i, $i\geqslant 1$, and W_i, $1\leqslant i\leqslant \nu$, as

$$\begin{aligned} \Xi_i &= \left(\xi_{ij}^{k};\ \begin{matrix} j & \downarrow & 1,\dots,n_i \\ k & \rightarrow & 1,\dots,N \end{matrix}\right), \quad \text{and} \\ W_i &= \left(w_{ij}^{k};\ \begin{matrix} j & \rightarrow & 1,\dots,m_i \\ k & \downarrow & 1,\dots,N \end{matrix}\right), \quad w_{ij}^{k} = \left\langle \varphi_{ij}, w_k\right\rangle, \end{aligned} \tag{3.10}$$

respectively. The matrices W_i are the so called *observability matrices*. Our main result is stated as follows: The corresponding result regarding eqn. (3.1′) will be stated later at the end of the section:

Theorem 3.2. (i) *Assume that w_k, h_k, and ξ_k satisfiy the rank conditions*

$$\begin{aligned}
&\operatorname{rank}\begin{pmatrix} W_i & W_i\Lambda_i & \dots & W_i\Lambda_i^{m_i-1}\end{pmatrix}^{\mathrm{T}} = m_i, \quad 1 \leqslant i \leqslant \nu, \\
&\operatorname{rank}\begin{pmatrix} G_i & \Lambda_i G_i & \dots & \Lambda_i^{m_i-1} G_i\end{pmatrix} = m_i, \quad 1 \leqslant i \leqslant \nu, \quad \textit{and} \\
&\operatorname{rank} \Xi_i = N, \quad i \geqslant 1,
\end{aligned} \tag{3.11}$$

respectively. Then we find a suitable integer n and $\rho_k \in \mathscr{P}_n\mathscr{H}$, $1 \leqslant k \leqslant M$, such that every solution $(u(t), v(t))$ to (3.1_1) satisfies the decay estimate

$$\|u(t)\| + \|v(t)\|_{\mathscr{H}} \leqslant \text{const}\, e^{-\beta t} \left(\|u_0\| + \|v_0\|_{\mathscr{H}}\right), \quad t \geqslant 0. \tag{3.12}$$

(ii) *Applying the projector $\mathscr{P}_n$ to the equation of v, we derive the equation in $H \times \mathscr{P}_n\mathscr{H}$,*

$$\begin{cases}
\dfrac{du}{dt} + Lu = \displaystyle\sum_{k=1}^{M} \langle v, \rho_k\rangle_{\mathscr{H}}\, g_k, \quad u(0) = u_0 \in \mathscr{D}(L), \\
\dfrac{dv}{dt} + Bv = \displaystyle\sum_{k=1}^{N} \langle u, w_k\rangle\, \mathscr{P}_n\xi_k + \sum_{k=1}^{M} \langle v, \rho_k\rangle_{\mathscr{H}}\, \mathscr{P}_n X g_k, \quad v(0) = v_0 \in \mathscr{P}_n\mathscr{H}.
\end{cases} \tag{3.1$_2$}$$

Equation (3.1_2) is well posed in $H \times \mathscr{P}_n\mathscr{H}$. Every solution (u, v) to (3.1_2) is derived from (3.1_1), and satisfies the decay estimate

$$\|u(t)\| + \|v(t)\|_{\mathscr{P}_n\mathscr{H}} \leqslant \text{const}\, e^{-\beta t} \left(\|u_0\| + \|v_0\|_{\mathscr{P}_n\mathscr{H}}\right), \quad t \geqslant 0. \tag{3.13}$$

Remark 1: Let $W \in \mathscr{L}(H; \mathbb{C}^N)$ be an operator defined as

$$Wu = \begin{pmatrix}\langle u, w_1\rangle & \langle u, w_2\rangle & \dots & \langle u, w_N\rangle\end{pmatrix}^{\mathrm{T}} \quad \text{for } u \in H,$$

and let W_P and W_{P_i} be the restriction of W onto PH and $P_{\lambda_i}H$, respectively. Then the first rank conditions on W_i in (3.11) are rewritten as $\ker\begin{pmatrix} W_{P_{\lambda_i}} & W_{P_{\lambda_i}} L_1 & \dots & W_{P_{\lambda_i}} L_1^{m_i-1}\end{pmatrix}^{\mathrm{T}} = \{0\}$, $1 \leqslant i \leqslant \nu$. This is nothing but the observability condition of the pair (W_P, L_1) [48].

Remark 2: We can apply the theorem to the system in the Example (see (2.14)). As already remarked there, the dimension of each eigenspace is equal to 1, and there arises no generalized eigenspace associated with L. Thus, $m_i = 1$, $i \geqslant 1$, and we choose $M = N = 1$ in (3.1_2), and $n_i = 1$, $i \geqslant 1$ in the setting of B. Then the condition (3.11) is that $w_{i1}^1 \neq 0$, $g_{i1}^1 \neq 0$ for $1 \leqslant i \leqslant \nu$, and that $\xi_{i1}^1 \neq 0$ for $i \geqslant 1$. The problem of finding a suitable vector $\rho_1 \in \mathscr{H}$ is a so called *ill-posed problem* (see Proposition 3.3 below): It is to seek a ρ_1 such that $X^*\rho_1$ approximates a given vector in P^*H arbitrarily.

Proof of Theorem 3.2:

The proof is similar to the proof Theorem 3.1, Chapter 4. The difference, however, consists in Proposition 3.3 stated later.

(i) Since we have set $\tilde{g}_k = X g_k$ in (3.1), it is easily seen that

$$\frac{d}{dt}(Xu - v) + B(Xu - v) = 0.$$

Thus, it means that $Xu(t) - v(t) = e^{-tB}(Xu_0 - v_0)$, $t \geqslant 0$, and $\|Xu(t) - v(t)\| \leqslant e^{-\beta t}\|Xu_0 - v_0\|$, $t \geqslant 0$, by (3.5). The equation for u is rewritten as

$$\frac{du}{dt} + \left(L - \sum_{k=1}^{M} \langle \cdot, X^* \rho_k \rangle h_k\right) u = \sum_{k=1}^{M} \langle v - Xu, \rho_k \rangle_{\mathscr{H}} g_k \tag{3.14}$$

Here we recall the decay estimate (2.20): By assuming the second rank condition in (3.11) (or (2.24)) on g_k, this estimate (2.20) is guaranteed by suitable choice of $\zeta_k \in P^*H$, $1 \leqslant k \leqslant M$. If these ζ_k could be *arbitrarily* approximated by $X^*\rho_k$ in the strong topology of H, we find suitable $\rho_k \in \mathscr{H}$ such that

$$\left\| \exp\left(-t\left(L - \sum_{k=1}^{M} \langle \cdot, X^* \rho_k \rangle g_k\right)\right)\right\| \leqslant \text{const}\, e^{-\beta'' t}, \quad t \geqslant 0, \tag{3.15}$$

where $\beta < \beta'' < \beta'$. Then, in view of this estimate together with the decay estimate of $Xu(t) - v(t)$, the decay of $u(t)$:

$$\|u(t)\| \leqslant \text{const}\, e^{-\beta t}\left(\|u_0\| + \|v_0\|_{\mathscr{H}}\right), \quad t \geqslant 0$$

immediately follows from (3.14). Then the decay of $v(t)$ with the same decay rate also follows, which means (3.12). Since $\mathscr{H}$ is a separable space, the above ρ_k in (3.15) can be chosen in the subspace $\mathscr{P}_n\mathscr{H}$ for some n.

The above approximation of ζ_k by vectors of the form $X^*\rho_k$ is, in fact, ensured by the following proposition. The proof of the proposition will be stated later.

Proposition 3.3. (i) *Under the first and the third conditions of* (3.11) *on* w_k *and* ξ_k, *respectively, we have the inclution relation:*

$$P^*H \subset \overline{X^*\mathscr{H}}. \tag{3.16}$$

(ii) Set $v_1(t) = \mathscr{P}_n v(t)$. Applying the projector $\mathscr{P}_n$ to the equation of v in (3.1_1), we obtain (3.1_2) with v replaced by v_1. Eqn. (3.1_2) is well posed in $H \times \mathscr{P}_n\mathscr{H}$. As long as v_0 belongs to $\mathscr{P}_n\mathscr{H}$, the function $(u(t), v_1(t)) = (u(t), \mathscr{P}_n v(t))$ is a unique solution to (3.1_2). Thus the decay estimate (3.13) is obvious.

Proof of Proposition 3.3:

The result is a non-trivial extension of Proposition 3.3 in Chapter 4 to the case where both L and B are not sectorial. The growth rate γ of $\sigma(B)$ is more restrictive than before (see (3.6)). In fact, we can choose a γ, $0 < \gamma < 2$ in the case where both L and B are sectorial. This restriction is made to avoid a possible effect of $\sigma(L)$ at infinity in consideration of $f_k(\lambda; u)$. The relation (3.16) is equivalent to $\ker X \subset (P^*H)^\perp = \ker P$. Setting $Xu = 0$ in (3.9), we see that

$$\sum_{k=1}^{N} f_k(\zeta_i^+; u)\xi_{ij}^k = \sum_{k=1}^{N} f_k(\zeta_i^-; u)\overline{\xi_{ij}^k} = 0, \quad i \geqslant 1, \quad 1 \leqslant j \leqslant n_i.$$

Since rank $\Xi_i = N$, $i \geqslant 1$, we see that

$$f_k(\zeta_i^\pm; u) = \left\langle (\zeta_i^\pm - L)^{-1}u, w_k \right\rangle = 0, \quad i \geqslant 1, \quad 1 \leqslant k \leqslant N. \tag{3.17}$$

Let β' $(> \beta)$ be close to β such that σ_2 is contained in the right half-plane, $\{\lambda \in \mathbb{C};\ \mathrm{Re}\,\lambda > \beta'\}$ (see (2.1)). The functions $f_k(\lambda; u)$ may be or may not be meromorphic functions. There is a possibility that $f_k(\lambda; u)$ might have poles at respective points of σ_1. Let l_i $(\leqslant m_i)$ be the *ascent* of $\lambda_i - L$ [66] for each $\lambda_i \in \sigma_1$, $1 \leqslant i \leqslant \nu$. To cancel these possible singularities, set

$$\tilde{f}_k(\lambda) = \left(\prod_{i=1}^{\nu} (\lambda - \lambda_i)^{l_i}\right) f_k(\lambda; u), \quad 1 \leqslant k \leqslant N.$$

Then $\tilde{f}_k(\lambda)$ are analytic in the left half-plane, $\mathrm{Re}\,\lambda \leqslant \beta'$. Let us consider the functions

$$g_k(z) = \tilde{f}_k(\beta' + \mathrm{i}z), \quad 1 \leqslant k \leqslant N \tag{3.18}$$

on the upper half-plane, $\mathrm{Im}\, z \geqslant 0$. Set $z_i^\pm = \pm\mu_i + \mathrm{i}\,(\beta' - \beta)$, $i \geqslant 1$. Then g_k are analytic on $\mathrm{Im}\, z \geqslant 0$, and by (3.17)

$$g_k\left(z_i^\pm\right) = 0, \quad 1 \leqslant k \leqslant N, \quad i \geqslant 1. \tag{3.19}$$

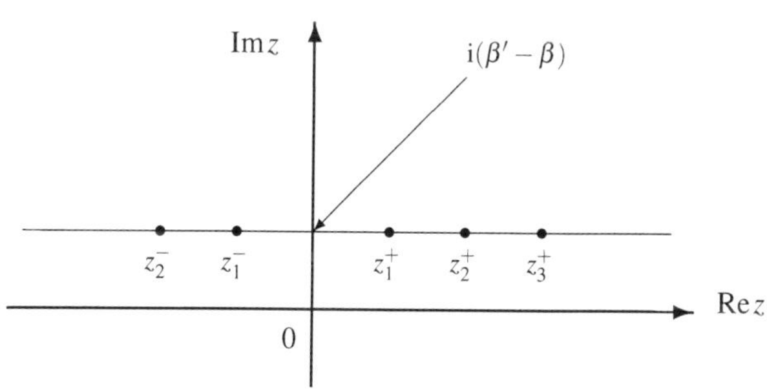

Figure 12

Assuming that $f_k(\lambda; u) \not\equiv 0$, we derive a contradiction. If so, $\tilde{f}_k$ and thus g_k are non-trivial analytic functions. Abbreviate the subscript k for simplicity. Adjusting the parameter β' a little if necessary, we may henceforth assume that $g(0) = \tilde{f}(\beta') \neq 0$. Let us apply Carleman's theorem again (see Theorem 1.1,

Chapter 4) to our g. Let $R > 0$ be chosen large enough. Suppose that $g(z)$ has the zeros $r_k e^{\mathrm{i}\theta_k}$, $1 \leqslant k \leqslant p$, inside the closed contour C_R consisting of the semicircle: $|z| = R$, $0 < \arg z < \pi$, and the segment: $|z| \leqslant R$ on the real axis. Set $\alpha = 1/g(0)$. Then we have the relation:

$$\sum_{k=1}^{p} m_k \left(\frac{1}{r_k} - \frac{r_k}{R^2} \right) \sin\theta_k = \frac{1}{\pi R} \int_0^{\pi} \log\left|\alpha g(Re^{\mathrm{i}\theta})\right| \cdot \sin\theta\, d\theta$$
$$+ \frac{1}{2\pi} \int_0^{R} \log\left|\alpha g(-x)\alpha g(x)\right| \cdot \left(\frac{1}{x^2} - \frac{1}{R^2} \right) dx$$
$$+ \frac{1}{2} \operatorname{Im} \frac{d}{dz} \alpha g(0). \tag{3.20}$$

Let us evaluate each term of (3.20). As for the first term of the right-hand side, we first recall (1.2) and (1.3), and evaluate $(\beta' + \mathrm{i}Re^{\mathrm{i}\theta} - L)^{-1}$, $0 \leqslant \theta \leqslant \pi$, as

$$\left\| (\beta' + \mathrm{i}Re^{\mathrm{i}\theta} - L) \right\|^{-1}$$
$$\leqslant \begin{cases} e^{C(\beta' - R\sin\theta)} \leqslant e^{C\beta'}, & \alpha \leqslant \beta' - R\sin\theta, \\ \dfrac{M}{\omega - (\beta' - R\sin\theta)} \leqslant \dfrac{M}{\omega - \alpha}, & \alpha \geqslant \beta' - R\sin\theta. \end{cases}$$

Then,

$$\left|g\left(Re^{\mathrm{i}\theta}\right)\right| = \left|\tilde{f}(\beta' + \mathrm{i}Re^{\mathrm{i}\theta})\right|$$
$$= \left| \left(\prod_{i=1}^{\nu} (\beta' + \mathrm{i}Re^{\mathrm{i}\theta} - \lambda_i)^{l_i} \right) \right| \left| \left\langle (\beta' + \mathrm{i}Re^{\mathrm{i}\theta} - L)^{-1} u, w \right\rangle \right|$$
$$\leqslant \text{const}\, R^{l_1 + \cdots + l_\nu}, \quad R \to \infty.$$

The first term of the right-hand side of (3.20) is thus bounded from above by

$$\frac{1}{\pi R} \int_0^{\pi} (\text{const} + (l_1 + \cdots + l_\nu)\log R) \sin\theta\, d\theta \to 0 \quad \text{as} \quad R \to \infty.$$

As for the second term of the right-hand side, divide the integral into two as $\int_0^{\delta} + \int_{\delta}^{R}$ for a sufficiently small $\delta > 0$. In view of the estimate $\left|\log\left|\alpha g(-z)\, \alpha g(z)\right|\right| \leqslant \text{const}\, |z|^2$ in a neighborhood of $z = 0$, we see that

$$\frac{1}{2\pi} \int_0^{\delta} \text{const}\, x^2 \left(\frac{1}{x^2} - \frac{1}{R^2} \right) dx \leqslant \text{const}, \quad R \to \infty.$$

Since $\sup_{\operatorname{Re}\lambda = \beta'} \left\| (\lambda - L)^{-1} \right\| < \infty$ by (2.1) and (2.3), we see that

$$g(\pm x) = \tilde{f}(\beta' \pm \mathrm{i}x) = \left(\prod_{i=1}^{\nu} (\beta' \pm \mathrm{i}x - \lambda_i)^{l_i} \right) \left\langle (\beta' \pm \mathrm{i}x - L)^{-1} u, w \right\rangle,$$
$$|g(\pm x)| \leqslant \text{const}\, x^{l_1 + \cdots + l_\nu}, \quad x > 0.$$

Then,

$$\begin{aligned}&\frac{1}{2\pi}\int_\delta^R \log|\alpha g(-x)\alpha g(x)|\cdot\left(\frac{1}{x^2}-\frac{1}{R^2}\right)dx\\&\leqslant \frac{1}{2\pi}\int_\delta^R (\mathrm{const}+2(l_1+\cdots+l_\nu)\log x)\left(\frac{1}{x^2}-\frac{1}{R^2}\right)dx \leqslant \mathrm{const}, \quad R\to\infty.\end{aligned}$$

Thus the second term remains bounded, and so does the right-hand side of (3.20) as $R\to\infty$.

Let us turn to the left-hand side of (3.20). Let $N(x)$, $x\geqslant 0$, be the number of $|z_i^+|<x$, that is, $N(x)=\#\{i\geqslant 0;\ |z_i^+|<x\}$. According to the assumption (3.6), we know that $|z_i^+|=|\mu_i+\mathrm{i}(\beta'-\beta)|\leqslant \mathrm{const}\, i^\gamma$ for $i\geqslant 1$. Then, we find that

$$N(x)=\#\{i\geqslant 0;\ |z_i^+|<x\}\geqslant \mathrm{const}\, x^{1/\gamma}-1.$$

The points $z_i^\pm=\pm\mu_i+\mathrm{i}(\beta'-\beta)$, $i\geqslant 1$, are the zeros of $g(z)$, and $\sin(\arg z_i^\pm)=(\beta'-\beta)/|z_i^\pm|\to 0$ as $i\to\infty$. We evaluate the left-hand side as

$$\begin{aligned}\sum_{k=1}^p m_k\left(\frac{1}{r_k}-\frac{r_k}{R^2}\right)\sin\theta_k &\geqslant \sum_{|z_i^\pm|<R}\left(\frac{1}{|z_i^\pm|}-\frac{|z_i^\pm|}{R^2}\right)\sin(\arg z_i^\pm)\\&=2(\beta'-\beta)\sum_{|z_i^+|<R}\left(\frac{1}{|z_i^+|^2}-\frac{1}{R^2}\right).\end{aligned}$$

Here recall our assumption, $\gamma<1/2$ (see (3.6)). For a sufficiently small $\varepsilon>0$, the above last term is calculated as

$$\begin{aligned}\sum_{|z_i^+|<R}\left(\frac{1}{|z_i^+|^2}-\frac{1}{R^2}\right)&=\int_\varepsilon^R\left(\frac{1}{x^2}-\frac{1}{R^2}\right)dN(x)\\&=\left(\frac{1}{x^2}-\frac{1}{R^2}\right)N(x)\Big|_\varepsilon^R+\int_\varepsilon^R\frac{2}{x^3}N(x)\,dx\\&\geqslant\int_\varepsilon^R\frac{2}{x^3}(\mathrm{const}\,x^{1/\gamma}-1)\,dx\\&\geqslant\frac{\mathrm{const}}{(1/\gamma)-2}R^{1/\gamma-2}-\mathrm{const}\to\infty\quad\text{as } R\to\infty.\end{aligned}$$

This is a contradiction. Thus, we have shown that $g(z)\equiv 0$, or

$$f_k(\lambda;u)=\left\langle(\lambda-L)^{-1}u,w_k\right\rangle\equiv 0,\quad \lambda\in\rho(L),\quad 1\leqslant k\leqslant N. \tag{3.21}$$

Choose a $c>0$ large enough such that $-c\in\rho(L)$, and set $L_c=L+c$. Following Section 3, Chapter 4, let us introduce a series of functions $f_k^l(\lambda;u)$, $l=0,1,\ldots,$ recursively as

$$f_k^0(\lambda;u)=f_k(\lambda;u),\qquad f_k^{l+1}(\lambda;u)=\frac{f_k^l(\lambda;u)}{\lambda+c},\quad l=0,1,\ldots. \tag{3.22}$$

In view of the simple algebraic relation: $(\lambda - L)^{-1} = L_c(\lambda - L)^{-1}L_c{}^{-1} = -L_c{}^{-1} + (\lambda + c)(\lambda - L)^{-1}L_c{}^{-1}$, we easily find that

$$f_k^l(\lambda; u) = \left\langle (\lambda - L)^{-1}L_c^{-l}u, w_k \right\rangle - \sum_{i=1}^{l} \frac{1}{(\lambda + c)^i} \left\langle L_c{}^{-(l+1-i)}u, w_k \right\rangle, \qquad (3.23)$$

and, by (3.17),

$$f_k^l(\zeta_i^{\pm}; u) = 0, \quad i \geqslant 1, \quad 1 \leqslant k \leqslant N, \quad l \geqslant 0. \qquad (3.24)$$

The difference in this case is merely that the point $-c$ is added as a possible singularity. As in the case of $f_k(\lambda; u)$, we find that $f_k^l(\lambda; u) \equiv 0$, or

$$\left\langle (\lambda - L)^{-1}L_c^{-l}u, w_k \right\rangle - \sum_{i=1}^{l} \frac{1}{(\lambda + c)^i} \left\langle L_c{}^{-(l+1-i)}u, w_k \right\rangle \equiv 0. \qquad (3.25)$$

Recall that Laurent's expansion of the resolvent $(\lambda - L)^{-1}$ in a neighborhood of the pole $\lambda_i \in \sigma_1$ is:

$$(\lambda - L)^{-1} = \sum_{j=1}^{l_i} \frac{A_{-j}}{(\lambda - \lambda_i)^j} + \sum_{j=0}^{\infty} (\lambda - \lambda_i)^j A_j,$$

where $l_i \leqslant m_i$, and $A_{-1} = P_{\lambda_i}$. Thus, we see that

$$\begin{aligned} f_k(\lambda; u) &= \sum_{j=1}^{l_i} \frac{\langle A_{-j}u, w_k \rangle}{(\lambda - \lambda_i)^j} + \sum_{j=0}^{\infty} (\lambda - \lambda_i)^j \langle A_j u, w_k \rangle = 0, \\ f_k^l(\lambda; u) &= \sum_{j=1}^{l_i} \frac{\langle A_{-j}L_c^{-l}u, w_k \rangle}{(\lambda - \lambda_i)^j} + \sum_{j=0}^{\infty} (\lambda - \lambda_i)^j \left\langle A_j L_c^{-l}u, w_k \right\rangle \\ &\quad - \sum_{i=1}^{l} \frac{1}{(\lambda + c)^i} \left\langle L_c^{-(l+1-i)}u, w_k \right\rangle = 0, \quad l \geqslant 1. \end{aligned} \qquad (3.26)$$

Setting $P_{\lambda_i}u = \sum_{j=1}^{m_i} u_{ij}\varphi_{ij} = A_{-1}u$, and calculating the residue of $f_k(\lambda; u)$ at λ_i, we see that

$$\begin{aligned} \left\langle P_{\lambda_i}u, w_k \right\rangle &= \sum_{j=1}^{m_i} \langle \varphi_{ij}, w_k \rangle u_{ij} = \begin{pmatrix} w_{i1}^k & w_{i2}^k & \dots & w_{im_i}^k \end{pmatrix} \boldsymbol{u}_i \\ &= 0, \quad 1 \leqslant i \leqslant \nu, \quad 1 \leqslant k \leqslant N, \end{aligned} \qquad (3.27)$$

where $\boldsymbol{u}_i = \begin{pmatrix} u_{i1} & u_{i2} \dots u_{im_i} \end{pmatrix}^{\mathrm{T}}$. As for $f_k^l(\lambda; u)$, $l \geqslant 1$, we note that $A_{-1}L_c^{-l}u = P_{\lambda_i}L_c^{-l}u = L_c^{-l}P_{\lambda_i}u$, and that the restriction $L_c^{-l}|_{P_{\lambda_i}H}$ of L_c^{-l} is equivalent to the matrix $(\Lambda_i + c)^{-l}$. Thus, $A_{-1}L_c^{-l}u = \sum_{j=1}^{m_i}((\Lambda_i + c)^{-l}\boldsymbol{u}_i)_j\varphi_{ij}$.

Calculating the residue of $f_k^l(\lambda;u)$ at λ_i, we see that

$$\begin{aligned} 0 &= \left\langle P_{\lambda_i} L_c^{-l} u, w_k \right\rangle = \left\langle \sum_{j=1}^{m_i} \left((\Lambda_i + c)^{-l} \boldsymbol{u}_i \right)_j \varphi_{ij}, w_k \right\rangle \\ &= \sum_{j=1}^{m_i} \left((\Lambda_i + c)^{-l} \boldsymbol{u}_i \right)_j \langle \varphi_{ij}, w_k \rangle \\ &= \begin{pmatrix} w_{i1}^k & w_{i2}^k & \dots & w_{im_i}^k \end{pmatrix} (\Lambda_i + c)^{-l} \boldsymbol{u}_i, \\ &\qquad \text{for } 1 \leqslant i \leqslant \nu, \quad 1 \leqslant k \leqslant N, \quad l \geqslant 1. \end{aligned} \tag{3.28}$$

This, combined with (3.27), yields that

$$\begin{pmatrix} W_i & W_i(\Lambda_i + c)^{-1} & \dots & W_i(\Lambda_i + c)^{-(m_i - 1)} \end{pmatrix}^{\mathrm{T}} \boldsymbol{u}_i = \boldsymbol{0}, \quad 1 \leqslant i \leqslant \nu.$$

It is apparent that the rank of the above coefficient matrix is equal to the rank of $\begin{pmatrix} W_i & W_i\Lambda_i & \dots & W_i\Lambda_i^{m_i-1} \end{pmatrix}^{\mathrm{T}} = m_i$, $1 \leqslant i \leqslant \nu$, by (3.11). Thus, we find that

$$\boldsymbol{u}_i = \boldsymbol{0} \quad \text{for } 1 \leqslant i \leqslant \nu,$$

that is, $Pu = (P_{\lambda_1} + \cdots + P_{\lambda_\nu})u = 0$. This is nothing but the relation (3.16). The proof of Proposition 3.3, and thus the proof of Theorem 3.2 is thereby complete. □

Finally, let us briefly sketch an alternative feedback scheme for eqn. (3.1′) without proof. The operator $Y \in \mathscr{L}(\mathscr{H}; H)$ is a unique solution to the operator equation,

$$LY - YB = D \quad \text{on } \mathscr{D}(B), \quad D = \sum_{k=1}^{M} \langle \cdot, \rho_k \rangle_{\mathscr{H}} g_k. \tag{3.29}$$

Let us express ρ_k as $\rho_k = \sum_{i,j} \left(\rho_{ij}^k \eta_{ij}^+ + \overline{\rho_{ij}^k} \eta_{ij}^- \right)$, $1 \leqslant k \leqslant M$, and define $n_i \times M$ matrices R_i as

$$R_i = \begin{pmatrix} \rho_{ij}^k; & \begin{matrix} j & \downarrow & 1, \dots, n_i \\ k & \rightarrow & 1, \dots, M \end{matrix} \end{pmatrix}.$$

A unique solution Y to (3.29) exists, and is then described as

$$\begin{aligned} Yv = &-\sum_{i,j} \sum_{k=1}^{M} \overline{\rho_{ij}^k} \langle v, \eta_{ij}^+ \rangle_{\mathscr{H}} (\zeta_i^+ - L)^{-1} g_k \\ &- \sum_{i,j} \sum_{k=1}^{M} \rho_{ij}^k \langle v, \eta_{ij}^- \rangle_{\mathscr{H}} (\zeta_i^- - L)^{-1} g_k. \end{aligned} \tag{3.30}$$

Note that $Y = Z^*$, where $Z \in \mathscr{L}(H; \mathscr{H})$ denotes a unique solution to the operator equation, $ZL^* - B^*Z = D^*$ on $\mathscr{D}(L^*)$, and $D^* = \sum_{k=1}^{M} \langle \cdot, g_k \rangle \rho_k$ (compare it with (3.7)). Thus, eqn. (3.1′) is found well posed in $H \times \mathscr{H}$. By noting that $u - Yv$ is subject to a well posed differential equation in H,

$$\frac{d}{dt}(u - Yv) + L(u - Yv) + \sum_{k=1}^{N} \langle u - Yv, w_k \rangle Y\xi_k = 0,$$

a result corresponding to Theorem 3.2 is stated as follows:

Theorem 3.2′. (i) *In addition to the rank conditions on* w_k *and* g_k *in* (3.11), *assume that* rank $R_i = M$, $i \geqslant 1$. *Then we find a suitable integer* n *and* $\xi_k \in \mathscr{P}_n\mathscr{H}$, $1 \leqslant k \leqslant N$, *such that every solution* $(u(t), v(t))$ *to* (3.1′) *satisfies the decay estimate*

$$\|u(t)\| + \|v(t)\|_{\mathscr{H}} \leqslant \text{const}\, e^{-\beta t} \left(\|u_0\| + \|v_0\|_{\mathscr{H}} \right), \quad t \geqslant 0. \tag{3.31}$$

(ii) *As long as* v_0 *stays in* $\mathscr{P}_n\mathscr{H}$, $v(t)$ *also stays in* $\mathscr{P}_n\mathscr{H}$. *Thus, eqn.* (3.1′) *is regarded as a well posed differential equation in* $H \times \mathscr{P}_n\mathscr{H}$.

Chapter 8

A Computational Algorhism for an Infinite-Dimensional Sylvester's Equation

8.1 Introduction

In stabilization studies of linear parabolic boundary control systems, e.g., (1.3) or (2.7), Chapter 4, and more general systems, e.g., (3.1), Chapter 7, Sylvester's equation (1.1) below and its unique operator solution X plays a central role. By the property (1.2) below, we can approximate a given $u \in P_\nu L^2(\Omega)$ by a suitable sequence of $X^* \rho_n$, $\rho_n \in H$. This leads to the existence of a stabilization scheme. Let $X \in \mathscr{L}(L^2(\Omega); H)$ be a unique operator solution to Sylvester's equation:

$$XL - BX = C \quad \text{on } \mathscr{D}(L), \quad \text{where} \quad C = -\sum_{k=1}^{N} \langle \cdot, w_k \rangle_\Gamma \, \xi_k. \tag{1.1}$$

We have seen in the previous chapters a geometric property of X (see, e.g., Proposition 3.3, Chapter 4):

$$\ker X \subset \left\{ u \in L^2(\Omega);\ P_\nu u = 0 \right\}, \tag{1.2}$$

or $P_\nu^* L^2(\Omega) \subset \overline{X^* H}$, where P_ν denotes the projector coresponding to the first ν eigenvalues of L such that $\mathrm{Re}\, \lambda_{\nu+1} > 0$. Thus, given a $u \in P_\nu^* L^2(\Omega)$, relation (1.2) ensures a suitable sequence $\{\rho_n\}$ of approximation in H such that $X^*\rho_n \to u$ as $n \to \infty$ in the $L^2(\Omega)$-topology. These ρ_n belong to a finite-dimensional subspace of H, the dimension of which determines the dimension of the compensator. In this sense, the approximation is important from an engineering viewpoint, too. It is uncertain, however, how fast or effectively the vector u could be approximated. The approximation problem is seen from another viewpoint: We are seeking an approximated solution to the equation $X^*\rho = u$ for a given u. Even in the best case where $\overline{X^*H} = L^2(\Omega)$ $(\nu = \infty)$ in (1.2) and the inverse X^{*-1} exists, the Riesz-Schauder theory implies that X^{*-1} necessarily *unbounded*. Thus the problem is an *ill-posed* problem, and decisively differes from finite-dimensional problems discussed in Chapter 1. There is a broad literature on other types of ill-posed problems. We refer the readers, e.g., to [36] and the references therein.

We study in this chapter a computational algorhism of finding a suitable sequence $\{\rho_n\}$, by limiting ourselves to the case where L is a self-adjoint operator with compact resolvent, and where C is replaced by $C = -\sum_{k=1}^{N} \langle \cdot, w_k \rangle \xi_k$ with $w_k \in L^2(\Omega)$. Thus, we consider in this chapter Sylvester's equation

$$XL - BX = C \quad \text{on } \mathscr{D}(L), \quad \text{where} \quad C = -\sum_{k=1}^{N} \langle \cdot, w_k \rangle \xi_k. \tag{1.1'}$$

The result in this chapter is based on a somewhat improved version of [40]. However, the algorhism in Section 2 below is not satisfactory enough for boundary control systems at present, and developments of more effective algorhisms are hoped in the future. But, it provides us a sufficient condition of an actual computational tool.

The setting of (1.1′) is the same as in Chapter 4 except that L is a self-adjoint operator with compact resolvent. First of all, let $\{\lambda_i, \varphi_{ij}\}$ be a set of eigenpairs of L, ensured by the Hilbert-Schmidt theory, such that

(i) $\sigma(L) = \{\lambda_i\}_{i \geqslant 1}$; $\lambda_1 < \lambda_2 < \cdots < \lambda_i < \cdots < \to \infty$, $(\lambda_\nu \leqslant 0 < \lambda_{\nu+1})$;

(ii) $(\lambda_i - L)\varphi_{ij} = 0$, $i \geqslant 1$, $1 \leqslant j \leqslant m_i$ $(< \infty)$; and

(iii) the set $\{\varphi_{ij}\}$ forms an orthonormal basis for $L^2(\Omega)$.

Let H be a separable Hilbert space equipped with an orthonormal basis $\left\{\eta_{ij}^{\pm};\, i \geqslant 1,\, 1 \leqslant j \leqslant n_i\right\}$, $n_i < \infty$ for each i. In (1.1′), let $\xi_k = \sum_{i,j} \xi_{ij}^k \eta_{ij}^+ + \sum_{i,j} \overline{\xi_{ij}^k} \eta_{ij}^-$, $1 \leqslant k \leqslant N$. Let a, $0 < a < 1$, be a constant; $\{\mu_i\}_{i \geqslant 1}$ a sequence of increasing positive numbers: $0 < \mu_1 < \mu_2 < \cdots \to \infty$; and define an operator B as (see (2.2), Chapter 4)

$$Bv = \sum_{i,j} \left(\mu_i \omega^+ v_{ij}^+ \eta_{ij}^+ + \mu_i \omega^- v_{ij}^- \eta_{ij}^- \right) \tag{1.3}$$

for vectors $v = \sum_{i,j} \left(v_{ij}^{+} \eta_{ij}^{+} + v_{ij}^{-} \eta_{ij}^{-} \right) \in \mathscr{D}(B)$, where

$$\mathscr{D}(B) = \left\{ v \in H;\ \sum_{i,j} |\mu_i v_{ij}^{\pm}|^2 < \infty \right\}, \quad \text{and } \omega^{\pm} = a \pm \mathrm{i}\sqrt{1-a^2}.$$

It is easily seen that B is a closed operator with a dense domain $\mathscr{D}(B)$. In addition,

(i) $\sigma(B) = \{\mu_i \omega^{\pm};\ i \geqslant 1\}$; and

(ii) $(\mu_i \omega^{\pm} - B)\eta_{ij}^{\pm} = 0, \quad i \geqslant 1,\ 1 \leqslant j \leqslant n_i.$

Since $\sigma(L) \cap \sigma(B) = \varnothing$, a unique existence of an operator solution X to $(1.1')$ is ensured. The solution X is expressed as

$$Xu = \sum_{i,j} \sum_{k=1}^{N} f_k(\mu_i \omega^{+}; u) \xi_{ij}^{k} \eta_{ij}^{+} + \sum_{i,j} \sum_{k=1}^{N} f_k(\mu_i \omega^{-}; u) \overline{\xi_{ij}^{k}} \eta_{ij}^{-}, \quad u \in L^2(\Omega),$$
$$\text{where} \quad f_k(\lambda; u) = \left\langle (\lambda - L)^{-1} u, w_k \right\rangle, \quad 1 \leqslant k \leqslant N. \tag{1.4}$$

We see that, for $\rho = \sum_{i,j} \left(\rho_{ij}^{+} \eta_{ij}^{+} + \rho_{ij}^{-} \eta_{ij}^{-} \right) \in H$

$$X^* \rho = \sum_{i,j} \sum_{k=1}^{N} \rho_{ij}^{+} \overline{\xi_{ij}^{k}} (\mu_i \omega^{-} - L)^{-1} w_k + \sum_{i,j} \sum_{k=1}^{N} \rho_{ij}^{-} \xi_{ij}^{k} (\mu_i \omega^{+} - L)^{-1} w_k. \tag{1.5}$$

Then, we see that

$$\left\langle X^* \rho, \varphi_{pq} \right\rangle = \sum_{i,j} \sum_{k=1}^{N} \rho_{ij}^{+} \overline{\xi_{ij}^{k}} \frac{w_{pq}^{k}}{\mu_i \omega^{-} - \lambda_p} + \sum_{i,j} \sum_{k=1}^{N} \rho_{ij}^{-} \xi_{ij}^{k} \frac{w_{pq}^{k}}{\mu_i \omega^{+} - \lambda_p},$$
$$\text{where} \quad w_{pq}^{k} = \left\langle w_k, \varphi_{pq} \right\rangle, \quad p \geqslant 1, \quad 1 \leqslant q \leqslant m_p.$$

Especially, let $\rho = \sum_{i,j} \left(\rho_{ij}^{+} \eta_{ij}^{+} + \overline{\rho_{ij}^{+}} \eta_{ij}^{-} \right) \in H$. By setting

$$\boldsymbol{\rho} = \left(\rho_{ij}^{+};\ (i,j) \downarrow \right) = \left(\boldsymbol{\rho}^{i};\ i \downarrow \right), \quad \boldsymbol{\rho}^{i} = \left(\rho_{i1}^{+} \dots \rho_{in_i}^{+} \right)^{\mathrm{T}}, \tag{1.6}$$

the above relation is rewritten as

$$\left\langle X^* \rho, \varphi_{pq} \right\rangle = \sum_{i} \sum_{k=1}^{N} (\Xi_i^* \boldsymbol{\rho}^{i})\big|_k \frac{w_{pq}^{k}}{\mu_i \omega^{-} - \lambda_p} + \sum_{i} \sum_{k=1}^{N} \overline{(\Xi_i^* \boldsymbol{\rho}^{i})}\big|_k \frac{w_{pq}^{k}}{\mu_i \omega^{+} - \lambda_p},$$

or

$$\begin{aligned} &\left(\left\langle X^* \rho, \varphi_{pq} \right\rangle;\ q \downarrow 1, \dots, m_p \right) \\ &= \sum_{i} \frac{W_p^{\mathrm{T}} \Xi_i^* \boldsymbol{\rho}^{i}}{\mu_i \omega^{-} - \lambda_p} + \sum_{i} \frac{W_p^{\mathrm{T}} \overline{\Xi_i^* \boldsymbol{\rho}^{i}}}{\mu_i \omega^{+} - \lambda_p} \\ &= W_p^{\mathrm{T}} \left(\sum_{i} \frac{\Xi_i^* \boldsymbol{\rho}^{i}}{\mu_i \omega^{-} - \lambda_p} + \sum_{i} \frac{\overline{\Xi_i^* \boldsymbol{\rho}^{i}}}{\mu_i \omega^{+} - \lambda_p} \right), \end{aligned} \tag{1.7}$$

for $p \geqslant 1$, where Ξ_i are the $n_i \times N$ matrices defined by (3.4), Chapter 4, and

$$W_p = \begin{pmatrix} w^k_{pq}; & \begin{matrix} k & \downarrow & 1,\dots,N \\ q & \rightarrow & 1,\dots,m_p \end{matrix} \end{pmatrix}, \quad p \geqslant 1.$$

Let $\{\alpha_i\}_{i\geqslant 1}$ be a monotonically increasing sequence of positive numbers which will be determined later, and set

$$\begin{aligned} F &= \left(\frac{1}{(\mu_i\omega^- - \lambda_p)\alpha_p} I_N; \begin{matrix} i & \rightarrow & 1,\dots \\ p & \downarrow & 1,\dots \end{matrix} \right); \\ \Xi^* &= \operatorname{diag}\left(\Xi_1^* \quad \Xi_2^* \quad \dots \quad \Xi_i^* \quad \dots\right); \qquad \text{and} \\ W_\alpha &= \operatorname{diag}\left(\alpha_1 W_1^{\mathrm{T}} \quad \alpha_2 W_2^{\mathrm{T}} \quad \dots \quad \alpha_p W_p^{\mathrm{T}} \quad \dots\right). \end{aligned} \tag{1.8}$$

Suppose that w_k satisfy the additional conditions

$$\sum_{i,j} \alpha_i^2 |w^k_{ij}|^2 < \infty, \quad 1 \leqslant k \leqslant N. \tag{1.9}$$

Note that W_α belongs to $\mathscr{L}(\ell^p;\, \ell^2)$, $2 \leqslant p \leqslant \infty$. In fact, for $\boldsymbol{x} \in \ell^p$, $2 \leqslant p < \infty$

$$\begin{aligned} W_\alpha \boldsymbol{x} &= \begin{pmatrix} \alpha_1 W_1^{\mathrm{T}} & 0 & \dots & 0 & \dots \\ 0 & \alpha_2 W_2^{\mathrm{T}} & \dots & 0 & \dots \\ \vdots & \vdots & \ddots & \vdots & \vdots \\ 0 & 0 & \dots & \alpha_i W_i^{\mathrm{T}} & \dots \\ \vdots & \vdots & \vdots & \vdots & \vdots \end{pmatrix} \begin{pmatrix} \boldsymbol{x}_1 \\ \boldsymbol{x}_2 \\ \vdots \\ \boldsymbol{x}_i \\ \vdots \end{pmatrix} \\ &= \begin{pmatrix} \alpha_1 W_1^{\mathrm{T}} \boldsymbol{x}_1 \\ \alpha_2 W_2^{\mathrm{T}} \boldsymbol{x}_2 \\ \vdots \\ \alpha_i W_i^{\mathrm{T}} \boldsymbol{x}_i \\ \vdots \end{pmatrix}, \qquad \boldsymbol{x}_i = \begin{pmatrix} x_{i1} \\ x_{i2} \\ \vdots \\ x_{iN} \end{pmatrix}. \end{aligned}$$

Hölder's inequality implies that $(1/p + 1/q = 1)$,

$$\begin{aligned} \left| \sum_{k=1}^N \alpha_i w^k_{ij} x_{ik} \right|^2 &\leqslant \left(\sum_{k=1}^N 1 \cdot |\alpha_i w^k_{ij}|^q \right)^{2/q} \left(\sum_{k=1}^N |x_{ik}|^p \right)^{2/p} \\ &\leqslant \left(\left(\sum_{k=1}^N 1 \right)^{1-q/2} \left(\sum_{k=1}^N |\alpha_i w^k_{ij}|^2 \right)^{q/2} \right)^{2/q} \|\boldsymbol{x}\|_p^2 \\ &= N^{2/q-1} \left(\sum_{k=1}^N |\alpha_i w^k_{ij}|^2 \right) \|\boldsymbol{x}\|_p^2, \end{aligned}$$

from which we obtain

$$\|W_\alpha \boldsymbol{x}\|_2 \leqslant N^{1/q-1/2} \left(\sum_{i,j} \sum_{k=1}^{N} \alpha_i^2 |w_{ij}^k|^2 \right) \|\boldsymbol{x}\|_p, \quad \boldsymbol{x} \in \ell^p.$$

A similar estimate is also obtained when $p = \infty$.

Relation (1.7) is rewritten as

$$\begin{aligned} \left(\langle X^* \rho, \varphi_{pq} \rangle ; q \downarrow 1, \ldots, m_p \right) &= \alpha_p W_p^{\mathrm{T}} \left(F \Xi^* \boldsymbol{\rho} + \overline{F \Xi^* \boldsymbol{\rho}} \right) \Big|_p, \quad \text{or} \\ \left(\langle X^* v, \varphi_{pq} \rangle ; (p,q) \downarrow \right) &= W_\alpha \left(F \Xi^* \boldsymbol{\rho} + \overline{F \Xi^* \boldsymbol{\rho}} \right). \end{aligned} \tag{1.10}$$

Thus, $X^* \rho$ is identified with $W_\alpha \left(F \Xi^* \boldsymbol{\rho} + \overline{F \Xi^* \boldsymbol{\rho}} \right) \in \ell^2$. Any real-valued function $y \in P_v L^2(\Omega)$ is identified with

$$\tilde{y} = \Big(\underbrace{y_{11}\ y_{12}\ \cdots\ y_{1m_1}}_{m_1}\ y_{21}\ \cdots\ y_{vm_v}\ 0\ 0\ \ldots \Big)^{\mathrm{T}}, \quad y_{ij} = \langle y, \varphi_{ij} \rangle .$$

Suppose further that w_k and ξ_k, $1 \leqslant k \leqslant N$, satisfy the conditions:

$$\text{rank } W_i = m_i, \quad 1 \leqslant i \leqslant K, \quad \text{and} \quad \text{rank } \Xi_i = N, \quad i \geqslant 1, \tag{1.11}$$

respectively: The former is apparently the observability condition for w_k. Then, we find a z such that

$$z = \Big(\underbrace{z_{11}\ \cdots\ z_{1N}\ \cdots\ z_{v1}\ \cdots\ z_{vN}}_{(=z_0^{\mathrm{T}})}\ 0\ 0\ \ldots \Big)^{\mathrm{T}} = (z_0^{\mathrm{T}}\ 0\ 0\ \ldots)^{\mathrm{T}}; \quad W_\alpha z + W_\alpha \bar{z} = \tilde{y}.$$

Our first result in this chapter is the following:

Proposition 1.1. *Suppose that w_k and ξ_k satisfy the conditions (1.11), and let $2 \leqslant p \leqslant \infty$. If we find a sequence $\{\sigma_n\}_{n \geqslant 1} \subset \ell^p$ such that $F\sigma_n \to z$ in ℓ^p, then,*

$$W_\alpha \left(F\sigma_n + \overline{F\sigma_n} \right) \to W_\alpha (z + \bar{z}) = \tilde{y} \text{ in } \ell^2. \tag{1.12}$$

The latter condition of (1.11) on ξ_k enables us to regard $\sigma = \Xi^* \boldsymbol{\rho}$ as a new variable. Thus, (1.11) means that there is a sequence $\{\rho_n\}$ such that $X^* \rho_n \to y \in P_v L^2(\Omega)$ in the topology of $L^2(\Omega)$.

8.2 An Algorhism

Based on Proposition 1.1, we seek a suitable sequence $\{\sigma_n\}_{n \geq 1} \subset \ell^p$ such that $F\sigma_n \to z$ in ℓ^p. Let $\tilde{P}_n$ be the projector in ℓ^p such that $\tilde{P}_n \sigma =$

$(\sigma_1\ \sigma_2 \dots \sigma_n\ 0\ 0 \dots)^{\mathrm{T}}$ for $\sigma = (\sigma_1\ \sigma_2 \dots \sigma_n \dots)^{\mathrm{T}} \in \ell^p$, and $\tilde{Q}_n = 1 - \tilde{P}_n$. The vector $\tilde{P}_n\sigma$ may be viewed as a vector in $\mathbb{C}^n$ without any confusion. The operator $\tilde{P}_{nN} F \tilde{P}_{nN}$ is identified with an $nN \times nN$ non-singular matrix F_n:

$$F_n = \tilde{P}_{nN} F \tilde{P}_{nN} = \left(\frac{1}{(\mu_i \omega^- - \lambda_p)\alpha_p} I_N; \quad \begin{matrix} i & \rightarrow & 1, \dots, n \\ p & \downarrow & 1, \dots, n \end{matrix} \right) \tag{2.1}$$

Since $\tilde{P}_{KN} z = z$, we uniquely solve an equation

$$F_n \sigma_n = \tilde{P}_{nN} F \tilde{P}_{nN} \sigma_n = \tilde{P}_{nN} z = \Big(z_0^{\mathrm{T}}\ \underbrace{0 \dots 0}_{N} \dots \underbrace{0 \dots 0}_{N}\Big)^{\mathrm{T}} \in \mathbb{C}^{nN} \tag{2.2}$$

for each $n > \nu$. Let $\tilde{\sigma}_n \in \ell^p$ be such that $\tilde{P}_{nN} \tilde{\sigma}_n = \tilde{\sigma}_n = \sigma_n$. Note that

$$\begin{aligned} F\tilde{\sigma}_n &= \tilde{P}_{nN} F \tilde{\sigma}_n + \tilde{Q}_{nN} F \tilde{\sigma}_n = \tilde{P}_{nN} z + \tilde{Q}_{nN} F \tilde{P}_{nN} \tilde{\sigma}_n \\ &= z + \tilde{Q}_{nN} F \tilde{P}_{nN} \tilde{\sigma}_n. \end{aligned}$$

Thus, as long as $\tilde{Q}_{nN} F \tilde{P}_{nN} \tilde{\sigma}_n \to 0$ in ℓ^p, the convergence of $F\tilde{\sigma}_n$ to z in ℓ^p is ensured. Note that a vector ρ_n determined by the relation $\tilde{\sigma}_n = \Xi^* \boldsymbol{\rho}_n$ (see (1.6) for $\boldsymbol{\rho}$) stays in a finite-dimensional subspace spanned by $\eta_{ij}^{\pm}$, $1 \leqslant i \leqslant n$, $1 \leqslant j \leqslant n_i$.

It is assumed throughout the section that

$$|\lambda_n| \geqslant c_1 n^{\alpha}, \quad n \geqslant 1 \tag{2.3}$$

for some constant $\alpha \in (0, 1]$. When L is induced, for example, by a uniformly elliptic self-adjoint operator in a bounded domain, the asymptotic distribution of the eigenvalues is well known, as Weyl's formula [1], by counting multiplicities. Thus, (2.1) is satisfied, where α is determined by the order of the elliptic operator and the spatial dimension. Since we are given a fairly arbitrary choice of the sequence $\{\mu_n\}_{n\geqslant 1}$, we set

$$\mu_n = c_2 n^{\alpha}, \quad n \geqslant 1, \tag{2.4}$$

where the constant c_2 is specified later. Set

$$c_3 = \max_{1\leqslant i \leqslant \nu} |\lambda_i|. \tag{2.5}$$

The constant c_3 is an important factor in our approximation algorhism. Another setting of μ_n is also possible: $\mu_n = c_2 n^{\beta}$, $n \geqslant 1$, where $\beta \neq \alpha$. We show later that the setting: $\beta = \alpha$ gives the best algorhism in our approach. Let $\kappa(x, y)$ be an auxiliary function defined as

$$\kappa(x, y) = \left(\frac{x}{y}\right)^{1/(1-\alpha)}, \quad x, y > 0, \quad 0 < \alpha < 1.$$

Theorem 2.1. *Let $\{\beta_n\}_{n\geqslant 1}$ be a positive increasing sequence such that β_n tends to infinity. For each $n > \nu$, let σ_n be a unique solution to the equation* (2.2).

(i) *The case where $\frac{1}{2} < \alpha < 1$: Set $c_2 = \sqrt{c_1 c_3}$ in* (2.4). *Suppose that w_k, $1 \leqslant k \leqslant N$, satisfy the convergence condition* (1.8) *with α given by*

$$\log \alpha_n = \left(\log \frac{1}{\alpha} + 2(1-\alpha)\log\left(1+\sqrt{\kappa(c_3, c_1)}\right) + (1+2\alpha)\log 2 + 1 - \alpha \right) n \qquad (2.6)$$
$$+ \frac{c_3}{c_1(1-\alpha)}\left(n^{1-\alpha} - \nu^{1-\alpha}\right) + \log \beta_n.$$

(ii) *The case where $\alpha = 1$: Let $c_2 \in [c_1, c_3]$. Suppose*

$$\log \alpha_n = \left(\log \frac{8c_3}{c_1} \right) n + \frac{c_3}{c_1} \log \frac{n}{\nu} + \log(\log n) + \log \beta_n. \qquad (2.7)$$

Then, the sequence of solutions $\{\sigma_n\}_{n>\nu}$ ensures the convergence of $F\tilde{\sigma}_n$ to z in ℓ^∞ [1]. Thus, the corresponding sequence $\{\rho_n\}_{n>\nu} \subset H$ satisfies the estimate

$$\|X^* \rho_n - y\| \leqslant \frac{\text{const}}{\beta_n}, \quad n \to \infty. \qquad (2.8)$$

Proof. Let K_n be an $n \times n$ matrix whose (p, i)th component is given by $(\mu_i \omega^- - \lambda_p)^{-1} \alpha_p^{-1}$. It is clear that $\det K_n$, the determinant of K_n contains factors $\lambda_i - \lambda_j$ and $(\mu_i - \mu_j)\omega^-$, $i \neq j$. Thus, $\det K_n$ is calculated in a straightforward manner as

$$\det K_n = (-1)^{n(n+1)/2} \frac{\prod_{1\leqslant i<j\leqslant n}(\lambda_i - \lambda_j)(\nu_i - \nu_j)}{\prod_{1\leqslant i,j\leqslant n}(\lambda_i - \nu_j)} \cdot \frac{1}{\prod_{1\leqslant i\leqslant n}\alpha_i},$$

$$\text{where} \quad \nu_i = \mu_i \omega^-.$$

[1] The convergence in the topology of ℓ^p, $p < \infty$ is also examined in the proof below. It is shown, however, that the case where $p = \infty$ gives the best result.

The cofactor Δ_{ij} of K_n has a structure similar to det K_n, and is calculated as

$$\Delta_{ij} = (-1)^{i+j}(-1)^{(n-1)n/2} \frac{\prod_{\substack{1\leqslant p<q\leqslant n,\\ p,q\neq i}}(\lambda_p-\lambda_q)\prod_{\substack{1\leqslant p<q\leqslant n,\\ p,q\neq j}}(\nu_p-\nu_q)}{\prod_{\substack{1\leqslant p,q\leqslant n,\\ p\neq i,q\neq j}}(\lambda_p-\nu_q)} \times \frac{1}{\prod_{\substack{1\leqslant p\leqslant n,\\ p\neq i}}\alpha_p}$$

$$= (-1)^{i+j}(-1)^{(n-1)n/2}\frac{\det K_n}{(-1)^{n(n+1)/2}} \times \frac{\alpha_i\prod_{1\leqslant p\leqslant n}(\lambda_i-\nu_p)(\lambda_p-\nu_j)}{(\lambda_i-\nu_j)(-1)^{i-1}\prod_{\substack{1\leqslant k\leqslant n,\\ k\neq i}}(\lambda_i-\lambda_k)\,(-1)^{j-1}\prod_{\substack{1\leqslant k\leqslant n,\\ k\neq j}}(\nu_j-\nu_k)}$$

$$= \frac{(-1)^n\alpha_i\prod_{1\leqslant k\leqslant n}(\lambda_i-\nu_k)(\lambda_k-\nu_j)}{(\lambda_i-\nu_j)\prod_{\substack{1\leqslant k\leqslant n,\\ k\neq i}}(\lambda_i-\lambda_k)\prod_{\substack{1\leqslant k\leqslant n,\\ k\neq j}}(\nu_j-\nu_k)}\det K_n.$$

The $nN\times nN$ matrix $F_n = \tilde{P}_{nN}F\tilde{P}_{nN}$ has the same structure as that of K_n. The inverse K_n^{-1} is then calculated as

$$F_n^{-1} = \left(\frac{\Delta_{ij}}{\det K_n}I_N;\ \begin{matrix} i & \rightarrow & 1,\dots,n \\ j & \downarrow & 1,\dots,n\end{matrix}\right)$$

$$= (-1)^n\left(\frac{\alpha_i\prod_{1\leqslant k\leqslant n}(\lambda_i-\nu_k)(\lambda_k-\nu_j)}{(\lambda_i-\nu_j)\prod_{\substack{1\leqslant k\leqslant n,\\ k\neq i}}(\lambda_i-\lambda_k)\prod_{\substack{1\leqslant k\leqslant n,\\ k\neq j}}(\nu_j-\nu_k)}I_N;\ \begin{matrix} i & \rightarrow & 1,\dots,n \\ j & \downarrow & 1,\dots,n\end{matrix}\right).$$

In order to obtain the solution $\sigma_n = F_n^{-1}\tilde{P}_{nN}z$ in practice, only the (j,i)th block, $j\geqslant 1,\ 1\leqslant i\leqslant \nu$, is necessary. Let us express F as

$$F = \begin{pmatrix} F_n & B_n \\ A_n & D_n\end{pmatrix}, \quad A_n = \tilde{Q}_{nN}F\tilde{P}_{nN},$$

where the remaining operators are self-explanatory. Let us calculate $A_n\sigma_n = A_nF_n^{-1}\tilde{P}_{nN}z$. The (k,i)th block $\rho_{ki}^{(n)}I_N$ of $A_nF_n^{-1}$ is given by

$$\rho_{ki}^{(n)}I_N = (-1)^n\frac{\alpha_i}{\alpha_k}\sum_{j=1}^{n}\frac{1}{\nu_j-\lambda_k}\frac{\prod_{\substack{1\leqslant p\leqslant n,\\ p\neq j}}(\lambda_i-\nu_p)\prod_{1\leqslant p\leqslant n}(\lambda_p-\nu_j)}{\prod_{\substack{1\leqslant p\leqslant n,\\ p\neq i}}(\lambda_i-\lambda_p)\prod_{\substack{1\leqslant p\leqslant n,\\ p\neq j}}(\nu_j-\nu_p)}I_N, \qquad (2.9)$$

$$k\geqslant n+1, \quad 1\leqslant i\leqslant \nu.$$

In view of the form of the vector z, however, we only have to show that

$$\|A_n\sigma_n\|_p \leqslant \begin{cases} \text{const}\left(\sum_{k=n+1}^{\infty}\sum_{i=1}^{\nu}|\rho_{ki}^{(n)}|^p\right)^{1/p} \leqslant \frac{\text{const}}{\beta_n}, & 2 \leqslant p < \infty, \quad \text{and} \\ \text{const} \sup_{\substack{k \geqslant n+1, \\ 1 \leqslant i \leqslant \nu}} |\rho_{ki}^{(n)}| \leqslant \frac{\text{const}}{\beta_n}, & p = \infty \end{cases} \tag{2.10}$$

(i) Let us show (2.10) in the case where $\frac{1}{2} < \alpha < 1$. The other case: $\alpha = 1$ is similarly examined. We hope to pose the possible smallest α_n on our w_k. We will see below that the choice of $c_2 = \sqrt{c_1 c_3}$ and $p = \infty$ gives the best result in our approach ($c_2 \in [c_1, c_3]$ and $p = \infty$ in the case where $\alpha = 1$). Now let us evaluate each factor of $\rho_{ki}^{(n)}$ term by term. First of all,

$$(*) = \left|\frac{\prod_{1 \leqslant p \leqslant n}(\lambda_p - \nu_j)}{\prod_{1 \leqslant p \leqslant n, \, p \neq i}(\lambda_i - \lambda_p)}\right| \leqslant \frac{\prod_{1 \leqslant p \leqslant n}(|\lambda_p| + \mu_j)}{\prod_{1 \leqslant p \leqslant n, \, p \neq i}|\lambda_i - \lambda_p|}. \tag{2.11}$$

The numerator on the right side of (2.11) is bounded from above by

$$\begin{aligned} \prod_{1 \leqslant p \leqslant n}|\lambda_p|\left(1 + \frac{\mu_j}{|\lambda_p|}\right) &\leqslant \prod_{1 \leqslant p \leqslant n}|\lambda_p|\left(1 + \frac{c_2 j^\alpha}{c_1 p^\alpha}\right) \\ &\leqslant \prod_{1 \leqslant p \leqslant n}|\lambda_p| \prod_{1 \leqslant p \leqslant n}\frac{c_1 p^\alpha + c_2 j^\alpha}{c_1 p^\alpha}. \end{aligned}$$

By noting that $c_1 p^\alpha + c_2 j^\alpha \leqslant c_4 (p+j)^\alpha$, $c_4 = c_1 (1 + \kappa(c_2, c_1))^{1-\alpha}$ for $p, j \geqslant 1$, it is further bounded from above by

$$\prod_{1 \leqslant p \leqslant n}|\lambda_p| \prod_{1 \leqslant p \leqslant n}\frac{c_4 (p+j)^\alpha}{c_1 p^\alpha} \leqslant \prod_{1 \leqslant p \leqslant n}|\lambda_p| \left(\frac{c_4}{c_1}\right)^n \left(\frac{(2n)!}{(n!)^2}\right)^\alpha, \quad 1 \leqslant j \leqslant n,$$

and thus

$$(*) = \left|\frac{\prod_{1 \leqslant p \leqslant n}(\lambda_p - \nu_j)}{\prod_{1 \leqslant p \leqslant n, \, p \neq i}(\lambda_i - \lambda_p)}\right| \leqslant \left(\frac{c_4}{c_1}\right)^n \left(\frac{(2n)!}{(n!)^2}\right)^\alpha |\lambda_i| \prod_{\substack{1 \leqslant p \leqslant n, \\ p \neq i}}\left|\frac{\lambda_p}{\lambda_i - \lambda_p}\right|.$$

Taking the logarithm of the last term, we see that

$$\begin{aligned}\log \prod_{\substack{1\leqslant p\leqslant n,\\ p\neq i}} \left|\frac{\lambda_p}{\lambda_i-\lambda_p}\right| &= \sum_{\substack{1\leqslant p\leqslant n,\\ p\neq i}} \log\left|\frac{\lambda_p}{\lambda_i-\lambda_p}\right| \leqslant \sum_{\substack{1\leqslant p\leqslant n,\\ p\neq i}} \log\left(1+\left|\frac{\lambda_i}{\lambda_p-\lambda_i}\right|\right)\\ &\leqslant \sum_{\substack{1\leqslant p\leqslant n,\\ p\neq i}} \left|\frac{\lambda_i}{\lambda_p-\lambda_i}\right| \leqslant c_3 \sum_{\substack{1\leqslant p\leqslant n,\\ p\neq i}} \frac{1}{|\lambda_p-\lambda_i|}\\ &\leqslant c_3\left(c_5+\sum_{p=\nu+1}^{n}\frac{1}{\lambda_p-\lambda_\nu}\right) = c_3(c_5+\zeta_n),\end{aligned}$$

where

$$c_5 = \max_{1\leqslant i\leqslant \nu} \sum_{\substack{1\leqslant p\leqslant \nu,\\ p\neq i}} \frac{1}{|\lambda_p-\lambda_i|}, \quad \text{and} \quad \zeta_n = \sum_{p=\nu+1}^{n} \frac{1}{\lambda_p-\lambda_\nu}.$$

We have shown that

$$\begin{aligned}(*) &= \left|\frac{\prod_{1\leqslant p\leqslant n}(\lambda_p-\nu_j)}{\prod_{1\leqslant p\leqslant n,\, p\neq i}(\lambda_i-\lambda_p)}\right|\\ &\leqslant \text{const}\,(1+\kappa(c_2,c_1))^{(1-\alpha)n}\left(\frac{(2n)!}{(n!)^2}\right)^{\alpha} e^{c_3\zeta_n},\\ &\quad 1\leqslant i\leqslant \nu, \quad 1\leqslant j\leqslant n.\end{aligned} \tag{2.12}$$

Let us turn to the next factor. The other product in (2.9) is evaluated as

$$(**) = \left|\frac{\prod_{1\leqslant p\leqslant n,\, p\neq j}(\lambda_i-\nu_p)}{\prod_{1\leqslant p\leqslant n,\, p\neq j}(\nu_j-\nu_p)}\right| \leqslant \prod_{\substack{1\leqslant p\leqslant n,\\ p\neq j}} \frac{|\lambda_i|+\mu_p}{|\mu_j-\mu_p|}. \tag{2.13}$$

As for the denominator of (2.13), note that

$$|\mu_j-\mu_p| = c_2|j^\alpha - p^\alpha| \geqslant \frac{\alpha c_2}{\max(j,p)^{1-\alpha}}|j-p|,$$

which implies that

$$\prod_{\substack{1\leqslant p\leqslant n,\\ p\neq j}} |\mu_j-\mu_p| \geqslant (\alpha c_2)^{n-1}\left(\frac{j!}{j^{j-1}n!}\right)^{1-\alpha} \prod_{\substack{1\leqslant p\leqslant n,\\ p\neq j}} |j-p|.$$

Let us recall the classic Stirling's formula to evaluate the last term above as well as the terms below. The formula gives an accurate estimate for $n!$, and is stated as follows:

$$\sqrt{2\pi}\, n^{n+1/2} e^{-n} < n! < \sqrt{2\pi}\, n^{n+1/2} e^{-n}\left(1+\frac{1}{4n}\right), \quad n\geqslant 1.$$

Since

$$\prod_{\substack{1\leqslant p\leqslant n,\\ p\neq j}} |j-p|\Bigg|_{j=1,n} = (n-1)! > (n-2)! = \prod_{\substack{1\leqslant p\leqslant n,\\ p\neq j}} |j-p|\Bigg|_{j=2},$$

we note that

$$\prod_{\substack{1\leqslant p\leqslant n,\\ p\neq j}} |j-p| \geqslant \min_{2\leqslant j\leqslant n-1}\left(\prod_{\substack{1\leqslant p\leqslant n,\\ p\neq j}} |j-p|\right), \quad 1\leqslant j\leqslant n.$$

Thus, we seek a lower bound of $\prod_{1\leqslant p\leqslant n,\, p\neq j}|j-p|$ for $2\leqslant j\leqslant n-1$. By Stirling's formula, we see that, for $2\leqslant j\leqslant n-1$ and thus for $1\leqslant j\leqslant n$,

$$\prod_{\substack{1\leqslant p\leqslant n,\\ p\neq j}} |j-p| = (j-1)!\,(n-j)!$$

$$\geqslant 2\pi e^{1-n}(j-1)^{j-1+1/2}(n-j)^{n-j+1/2} \geqslant 2\pi e^{1-n}\left(\frac{n-1}{2}\right)^{n}.$$

The numerator of (2.13) is bounded from above by

$$\prod_{\substack{1\leqslant p\leqslant n,\\ p\neq j}} (|\lambda_i|+\mu_p) \leqslant \prod_{\substack{1\leqslant p\leqslant n,\\ p\neq j}} (c_3+c_2p^{\alpha}) \leqslant \prod_{\substack{1\leqslant p\leqslant n,\\ p\neq j}} c_3(1+\kappa(c_2,c_3))^{1-\alpha}(1+p)^{\alpha}$$

$$\leqslant \left(c_3(1+\kappa(c_2,c_3))^{1-\alpha}\right)^{n-1}\left(\frac{(n+1)!}{j+1}\right)^{\alpha}, \quad 1\leqslant j\leqslant n.$$

Substituting these estimates into (2.13) and applying Stirling's formula again, we come to an estimate

$$\begin{aligned}\left|\prod_{\substack{1\leqslant p\leqslant n,\\ p\neq j}} \frac{\lambda_i-\nu_p}{\nu_j-\nu_p}\right| &\leqslant \frac{\left(c_3(1+\kappa(c_2,c_3))^{1-\alpha}\right)^{n-1}\left(\frac{(n+1)!}{j+1}\right)^{\alpha}}{2\pi e(n-1)^n e^{-(1+\log 2)n}(\alpha c_2)^{n-1}\left(\frac{j!}{j^{j-1}n!}\right)^{1-\alpha}} \\ &\leqslant \text{const}\left(\frac{1}{\alpha}\left(\frac{1+\kappa(c_2,c_3)}{\kappa(c_2,c_3)}\right)^{1-\alpha}\right)^{n} \\ &\quad\times (n-1)^{-n}j^{-(3-\alpha)/2}e^{(\log 2)n+j(1-\alpha)}n^{n+\alpha+1/2},\end{aligned} \tag{2.14}$$

for $1\leqslant j\leqslant n$. According to estimates (2.12) and (2.14), we are able to evaluate

$\rho_{ki}^{(n)}$ in (2.9) as

$$
\begin{aligned}
&\left|\rho_{ki}^{(n)}\right| \\
&\leqslant \frac{\text{const}}{\alpha_k} \sum_{j=1}^{n} \frac{1}{|\nu_j - \lambda_k|} \left(\frac{1}{\alpha} \left(\frac{1+\kappa(c_2,c_3)}{\kappa(c_2,c_3)} \right)^{1-\alpha} \right)^{n} (1+\kappa(c_2,c_1))^{(1-\alpha)n} \\
&\qquad \times (n-1)^{-n} j^{-(3-\alpha)/2} e^{(\log 2)n + j(1-\alpha)} n^{n+\alpha+1/2} \left(\frac{(2n)!}{(n!)^2} \right)^{\alpha} e^{c_3 \zeta_n} \\
&\leqslant \frac{\text{const}}{\alpha_k} \frac{1}{\alpha^n} \left(\frac{(1+\kappa(c_2,c_1))(1+\kappa(c_2,c_3))}{\kappa(c_2,c_3)} \right)^{(1-\alpha)n} \\
&\qquad \times (n-1)^{-n} e^{(\log 2)n + c_3 \zeta_n} n^{n+\alpha+1/2} \frac{(2n)^{(2n+1/2)\alpha} e^{-2n\alpha}}{n^{(2n+1)\alpha} e^{-2n\alpha}} \\
&\qquad \times \sum_{j=1}^{n} \frac{1}{|\nu_j - \lambda_k|} j^{-(3-\alpha)/2} e^{(1-\alpha)j}.
\end{aligned}
$$

Note that $|\nu_j - \lambda_k| = |\mu_j \omega^- - \lambda_k| \geqslant \lambda_k \sin\theta$, where $\arg \omega^- = -\theta$. When n is large enough, i.e., $n > \frac{3-\alpha}{2(1-\alpha)}$, the above last term is estimated as follows:

$$
\begin{aligned}
\sum_{j=1}^{n} \frac{1}{|\nu_j - \lambda_k|} j^{-(3-\alpha)/2} e^{(1-\alpha)j} &\leqslant \text{const} \sum_{j=1}^{n} \frac{1}{\lambda_k} n^{-(3-\alpha)/2} e^{(1-\alpha)n} \\
&= \text{const}\, \frac{n^{-(1-\alpha)/2} e^{(1-\alpha)n}}{\lambda_k}, \qquad k \geqslant n+1.
\end{aligned}
$$

Consequently, we have an estimate:

$$
\begin{aligned}
\left|\rho_{ki}^{(n)}\right| &\leqslant \frac{\text{const}}{\alpha^n} \left(\frac{(1+\kappa(c_2,c_1))(1+\kappa(c_2,c_3))}{\kappa(c_2,c_3)} \right)^{(1-\alpha)n} \\
&\qquad \times \frac{(2n)^{(2n+1/2)\alpha}}{n^{(2\alpha-1)n-\alpha/2}} \frac{1}{(n-1)^n} e^{(1-\alpha+\log 2)n + c_3 \zeta_n} \frac{1}{\alpha_k \lambda_k} \\
&= \text{const}\, \frac{C_n}{\alpha_k \lambda_k},
\end{aligned}
$$

where

$$
\begin{aligned}
C_n = \frac{1}{\alpha^n} &\left(\frac{(1+\kappa(c_2,c_1))(1+\kappa(c_2,c_3))}{\kappa(c_2,c_3)} \right)^{(1-\alpha)n} \\
&\times \frac{(2n)^{(2n+1/2)\alpha}}{n^{(2\alpha-1)n-\alpha/2}} \frac{1}{(n-1)^n} e^{(1-\alpha+\log 2)n + c_3 \zeta_n}.
\end{aligned}
$$

We are ready to evaluate $\|A_n\sigma_n\|_p$ in (2.10). When $p < \infty$, we calculate as

$$\sum_{k=n+1}^{\infty}\sum_{i=1}^{\nu}|\rho_{ki}^{(n)}|^p \leqslant \text{const}\, C_n^p \sum_{k=n+1}^{\infty}\frac{1}{\alpha_k^p\lambda_k^p} \leqslant \text{const}\,\frac{C_n^p}{\alpha_{n+1}^p}\sum_{k=n+1}^{\infty}\frac{1}{c_1^p k^{\alpha p}}$$
$$\leqslant \text{const}\,\frac{C_n^p}{\alpha_{n+1}^p}\,\frac{1}{n^{\alpha p-1}},$$

$$\frac{1}{p}\log\left(\sum_{k=n+1}^{\infty}\sum_{i=1}^{\nu}|\rho_{ki}^{(n)}|^p\right)$$
$$\leqslant n\log\frac{1}{\alpha}+(1-\alpha)n\log\frac{(1+\kappa(c_2,c_1))(1+\kappa(c_2,c_3))}{\kappa(c_2,c_3)}+\left(2n+\frac{1}{2}\right)\alpha\log(2n)$$
$$-\left((2\alpha-1)n-\frac{\alpha}{2}\right)\log n - n\log(n-1)+(1-\alpha+\log 2)n + c_3\zeta_n$$
$$-\log\alpha_{n+1}-\frac{\alpha p-1}{p}\log n+\text{const}$$
$$\leqslant\left(\log\frac{1}{\alpha}+(1-\alpha)\log\frac{(1+\kappa(c_2,c_1))(1+\kappa(c_2,c_3))}{\kappa(c_2,c_3)}+(1+2\alpha)\log 2\right.$$
$$\left.+1-\alpha\right)n+\frac{1}{p}\log n + c_3\zeta_n-\log\alpha_{n+1}+\text{const}.$$

The constant c_2 is a parameter to be designed. The function of c_2:

$$\frac{(1+\kappa(c_2,c_1))(1+\kappa(c_2,c_3))}{\kappa(c_2,c_3)} = 1+\frac{c_3^{1/(1-\alpha)}}{c_2^{1/(1-\alpha)}}+\frac{c_2^{1/(1-\alpha)}}{c_1^{1/(1-\alpha)}}+\left(\frac{c_3}{c_1}\right)^{1/(1-\alpha)}$$

attains its minimum $\left(1+\sqrt{\kappa(c_3,c_1)}\right)^2$ when $c_2 = \sqrt{c_1c_3}$. The choice of the constant c_2 as $\sqrt{c_1c_3}$ in this theorem is thus justified. c_2 is chosen as $\sqrt{c_1c_3}$ in this theorem. Thus,

$$\frac{1}{p}\log\left(\sum_{k=n+1}^{\infty}\sum_{i=1}^{\nu}|\rho_{ki}^{(n)}|^p\right)\leqslant\left(\log\frac{1}{\alpha}+2(1-\alpha)\log(1+\sqrt{\kappa(c_3,c_1)})\right.$$
$$\left.+(1+2\alpha)\log 2+1-\alpha\right)n \qquad (2.15)$$
$$+\frac{1}{p}\log n+c_3\zeta_n-\log\alpha_{n+1}+\text{const}.$$

It remains to obtain an estimate of ζ_n. Recalling that $\lambda_\nu \leqslant 0$, we estimate as

$$\zeta_n = \sum_{p=\nu+1}^{n}\frac{1}{\lambda_p-\lambda_\nu}\leqslant\sum_{p=\nu+1}^{n}\frac{1}{c_1p^\alpha-\lambda_\nu}$$
$$<\int_\nu^n\frac{dx}{c_1x^\alpha}=\frac{n^{1-\alpha}-\nu^{1-\alpha}}{c_1(1-\alpha)}.$$

Substituting this estimate into (2.15), we obtain

$$\frac{1}{p}\log\left(\sum_{k=n+1}^{\infty}\sum_{i=1}^{\nu}|\rho_{ki}^{(n)}|^p\right) \leqslant \left(\log\frac{1}{\alpha}+\cdots+1-\alpha\right)n+\frac{1}{p}\log n$$
$$+\frac{c_3}{c_1(1-\alpha)}(n^{1-\alpha}-\nu^{1-\alpha})$$
$$-\log\alpha_{n+1}+\text{const}.$$

For a positive increasing sequence $\beta_n \nearrow \infty$, let α_n be defined as

$$\begin{aligned}&\log\alpha_n\\ &=\left(\log\frac{1}{\alpha}+2(1-\alpha)\log\left(1+\sqrt{\kappa(c_3,c_1)}\right)+(1+2\alpha)\log 2+1-\alpha\right)n\\ &\quad+\frac{c_3}{c_1(1-\alpha)}(n^{1-\alpha}-K^{1-\alpha})+\frac{1}{p}\log n+\log\beta_n.\end{aligned}$$

Then $\{\alpha_n\}$ is a monotonically increasing sequence, which ensures the estimate (2.10).

Let us proceed to the other case where $p=\infty$. As before, set $c_2=\sqrt{c_1c_3}$. We seek a sequence $\{\alpha_n\}$ such that

$$\|A_n\sigma_n\|_\infty \leqslant \text{const}\sup_{\substack{k\geqslant n+1,\\ 1\leqslant i\leqslant\nu}}\left|\rho_{ki}^{(n)}\right| \leqslant \text{const}\,\frac{C_n}{\alpha_{n+1}\lambda_{n+1}} \to 0, \qquad n\to\infty.$$

The preceding argument immediately implies that

$$\begin{aligned}&\log\frac{C_n}{\alpha_{n+1}\lambda_{n+1}}\\ &\leqslant n\log\frac{1}{\alpha}+2(1-\alpha)n\log\left(1+\sqrt{\kappa(c_3,c_1)}\right)+\left(2n+\frac{1}{2}\right)\alpha\log(2n)\\ &\quad-\left((2\alpha-1)n-\frac{\alpha}{2}\right)\log n-n\log(n-1)+(1-\alpha+\log 2)n+c_3\zeta_n\\ &\quad-\log\alpha_{n+1}-\alpha\log(n+1)-\log c_1\\ &\leqslant\left(\log\frac{1}{\alpha}+2(1-\alpha)\log\left(1+\sqrt{\kappa(c_3,c_1)}\right)+(1+2\alpha)\log 2+1-\alpha\right)n\\ &\quad+\frac{c_3}{c_1(1-\alpha)}(n^{1-\alpha}-K^{1-\alpha})-\log\alpha_{n+1}+\text{const}.\end{aligned}$$

A sequence $\{\alpha_n\}$ given by

$$\log \alpha_n = \left(\log \frac{1}{\alpha} + 2(1-\alpha) \log \left(1 + \sqrt{\kappa(c_3, c_1)}\right) + (1+2\alpha)\log 2 + 1 - \alpha \right) n + \frac{c_3}{c_1(1-\alpha)} \left(n^{1-\alpha} - \nu^{1-\alpha}\right) + \log \beta_n \tag{2.16}$$

is then an increasing sequence which ensures the estimate (2.10). Comparing these two cases, we see that the choice of $p = \infty$ gives the optimal result, i.e., the smallest α_n in our approach.

(ii) The proof in the case where $\alpha = 1$ is carried out in a similar manner with a little modification. By choosing c_2 on the interval $[c_1, c_3]$, we see that

$$\left|\rho_{ki}^{(n)}\right| \leqslant \text{const} \frac{C_n}{\alpha_k \lambda_k}, \quad k \geqslant n+1, \quad 1 \leqslant i \leqslant \nu,$$

where the constant C_n is replaced at this time by

$$C_n = \left(\frac{c_3}{c_1}\right)^n (n-1)^{-n} e^{(\log 2)n + c_3 \zeta_n} \frac{(2n)^{2n+1/2}(n+1)^{n+3/2}}{n^{2n+1}} (1 + \log n).$$

Then, by noting an estimate: $\zeta_n < \frac{1}{c_1} \log \frac{n}{\nu}$,

$$\begin{aligned} &\log \frac{C_n}{\alpha_{n+1}\lambda_{n+1}} \\ &\leqslant n \log \frac{c_3}{c_1} - n \log(n-1) + (\log 2)n + c_3 \zeta_n + \left(2n + \frac{1}{2}\right) \log(2n) \\ &\quad + \left(n + \frac{3}{2}\right) \log(n+1) - (2n+1)\log n + \log(1 + \log n) \\ &\quad - \log \alpha_{n+1} - \log \lambda_{n+1} \\ &\leqslant \left(\log \frac{8c_3}{c_1}\right) n + \log(\log n) + \frac{c_3}{c_1} \log \frac{n}{\nu} - \log \alpha_{n+1} + \text{const}. \end{aligned} \tag{2.17}$$

Thus, the sequence $\{\alpha_n\}$ defined by (2.7) ensures the estimate (2.10). □

Remark on the setting of the sequence μ_n:

We mentioned just before Theorem 2.1 that there is another possibility of setting of the sequence μ_n as $\mu_n = c_2 n^\beta$, $\beta \neq \alpha$. We comment briefly on this. Let us consider again the case where $\alpha < 1$, and let $\beta < \alpha$ in μ_n. The estimate (2.12) is unchanged. However, the estimate (2.14) is changed into

$$\left| \prod_{\substack{1\leqslant p\leqslant n,\\ p\neq j}} \frac{\lambda_i - \nu_p}{\nu_j - \nu_p} \right| \leqslant \text{const} \left(\frac{1}{\beta} \left(\frac{1+\hat{\kappa}(c_2,c_3)}{\hat{\kappa}(c_2,c_3)} \right)^{1-\beta} \right)^n \times (n-1)^{-n} j^{-(3-\beta)/2} e^{(\log 2)n + j(1-\beta)} n^{n+\beta+1/2}, \tag{2.18}$$

for $1 \leqslant j \leqslant n$, where $\hat{\kappa}(c_2,c_3) = \left(\frac{c_2}{c_3}\right)^{1/(1-\beta)}$. Then $\rho_{ki}^{(n)}$ is evaluated, when $n > \frac{3-\beta}{2(1-\beta)}$, as

$$\begin{aligned}
&\left|\rho_{ki}^{(n)}\right| \\
&\leqslant \frac{\text{const}}{\alpha_k} \frac{1}{\beta^n} \left\{ \left(\frac{1+\hat{\kappa}(c_2,c_3)}{\hat{\kappa}(c_2,c_3)} \right)^{1-\beta} (1+\kappa(c_2,c_1))^{1-\alpha} \right\}^n (n-1)^{-n} e^{(\log 2)n} \\
&\qquad \times n^{n+\beta+1/2} e^{c_3\zeta_n} \frac{(2n)^{(2n+1/2)\alpha} e^{-2n\alpha}}{n^{(2n+1)\alpha} e^{-2n\alpha}} \sum_{j=1}^{n} \frac{1}{|\nu_j - \lambda_k|} j^{-(3-\beta)/2} e^{(1-\beta)j} \\
&\leqslant \frac{\text{const}}{\alpha_k} \frac{1}{\beta^n} \left\{ \left(\frac{1+\hat{\kappa}(c_2,c_3)}{\hat{\kappa}(c_2,c_3)} \right)^{1-\beta} (1+\kappa(c_2,c_1))^{1-\alpha} \right\}^n (n-1)^{-n} e^{(\log 2)n} \\
&\qquad \times n^{n+\beta+1/2} e^{c_3\zeta_n} \frac{(2n)^{(2n+1/2)\alpha} e^{-2n\alpha}}{n^{(2n+1)\alpha} e^{-2n\alpha}} \frac{n^{-(1-\beta)/2} e^{(1-\beta)n}}{\lambda_k}.
\end{aligned}$$

By setting $p = 1/(1-\alpha)$ and $q = 1/(1-\beta)$, the function of the parameter c_2:

$$\left(\frac{1+\hat{\kappa}(c_2,c_3)}{\hat{\kappa}(c_2,c_3)} \right)^{1-\beta} (1+\kappa(c_2,c_1))^{1-\alpha}, \quad \beta < \alpha$$

attains its minimum, when $c_2 = (c_1^p c_3^q)^{1/(p+q)}$,

$$\begin{aligned}
&\min \left(\frac{1+\hat{\kappa}(c_2,c_3)}{\hat{\kappa}(c_2,c_3)} \right)^{1-\beta} (1+\kappa(c_2,c_1))^{1-\alpha} \\
&= \left(1 + \left(\frac{c_3}{c_1} \right)^{\frac{1}{2-(\alpha+\beta)}} \right)^{2-(\alpha+\beta)} = (\ddagger).
\end{aligned}$$

In this setting of c_2, we have an estimate

$$\left|\rho_{ki}^{(n)}\right| \leqslant \text{const} \frac{C_n}{\alpha_{n+1}\lambda_{n+1}}, \quad k \geqslant n+1, \quad 1 \leqslant i \leqslant \nu,$$

where

$$C_n = \frac{1}{\beta^n} (\ddagger)^n \frac{(2n)^{(2n+1/2)\alpha}}{n^{(2\alpha-1)n+\alpha-3\beta/2}} \frac{1}{(n-1)^n} e^{(1-\beta+\log 2)n + c_3\zeta_n}.$$

Thus,

$$\log\frac{C_n}{\alpha_{n+1}\lambda_{n+1}}$$
$$\leqslant\left(\log\frac{1}{\beta}+(2-(\alpha+\beta))\log\left(1+\left(\frac{c_3}{c_1}\right)^{\frac{1}{2-(\alpha+\beta)}}\right)+(1+2\alpha)\log 2+1-\beta\right)n$$
$$+\frac{c_3}{c_1(1-\alpha)}(n^{1-\alpha}-\nu^{1-\alpha})-\frac{3(\alpha-\beta)}{2}\log n-\log\alpha_{n+1}+\text{const}.$$

The corresponding setting of α_n then becomes

$$\log\alpha_n=\left(\log\frac{1}{\beta}+(2-(\alpha+\beta))\log\left(1+\left(\frac{c_3}{c_1}\right)^{\frac{1}{2-(\alpha+\beta)}}\right)\right.$$
$$\left.+(1+2\alpha)\log 2+1-\beta\right)n \tag{2.19}$$
$$+\frac{c_3}{c_1(1-\alpha)}(n^{1-\alpha}-\nu^{1-\alpha})-\frac{3(\alpha-\beta)}{2}\log n+\log\beta_n.$$

Let us compare it with α_n in (2.6). In (2.18), we have a better factor $-\frac{3(\alpha-\beta)}{2}\log n$. As for the coefficient of n, however, we note that

$$2(1-\alpha)\log\left(1+\left(\frac{c_3}{c_1}\right)^{\frac{1}{1-\alpha}}\right)<(2-(\alpha+\beta))\log\left(1+\left(\frac{c_3}{c_1}\right)^{\frac{1}{2-(\alpha+\beta)}}\right),$$

which poses a severer condition on our w_k.

When $\mu_n=c_2n^{\beta},\ \ n\geqslant 1$ with $\beta>\alpha$ instead, we proceed to similar evaluations: In (2.11), we calculate as

$$(*)=\left|\frac{\prod_{1\leqslant p\leqslant n}(\lambda_p-\nu_j)}{\prod_{1\leqslant p\leqslant n,\, p\neq i}(\lambda_i-\lambda_p)}\right|\leqslant\text{const}\,(1+\hat{\kappa}(c_2,c_1))^{(1-\beta)n}\frac{((2n)!)^{\beta}}{(n!)^{\alpha+\beta}}e^{c_3\zeta_n}.$$

In (2.13), the estimate of $(**)$ is the same as in (2.18). By setting $c_2=\sqrt{c_1c_3}$, a resultant estimate of $\rho_{ki}^{(n)}$ then becomes

$$\left|\rho_{ki}^{(n)}\right|\leqslant\text{const}\,\frac{C_n}{\alpha_{n+1}\lambda_{n+1}},\quad k\geqslant n+1,\quad 1\leqslant i\leqslant\nu,$$

where

$$C_n=\frac{1}{\beta^n}\left(1+\sqrt{\hat{\kappa}(c_3,c_1)}\right)^{2(1-\beta)n}(n-1)^{-n}e^{(\log 2+1+\alpha-2\beta)n}e^{c_3\zeta_n}$$
$$\times n^{n(\beta-\alpha+1)+3\beta/2-\alpha/2}\,2^{(2n+1/2)\beta}.$$

Thus,

$$
\begin{aligned}
&\log \frac{C_n}{\alpha_{n+1}\lambda_{n+1}} \\
&\leqslant (\beta-\alpha) n \log n + \left(\log \frac{1}{\beta} + 2(1-\beta)\log\left(1+\sqrt{\hat{\kappa}(c_3,c_1)}\right) \right. \\
&\qquad\qquad \left. + (1+2\beta)\log 2 + 1 + \alpha - 2\beta \right) n \\
&\quad + \frac{c_3}{c_1(1-\alpha)}\left(n^{1-\alpha} - \nu^{1-\alpha}\right) + \frac{3(\beta-\alpha)}{2}\log n - \log \alpha_{n+1} + \text{const}.
\end{aligned}
$$

Based on the above estimate, $\log \alpha_n$ must contains factor $(\beta-\alpha)n\log n$, which also poses a severer assumption on our w_k.

References

[1] S. Agmon, "Lectures on Elliptic Boundary Value Problems," Van Nostrand, Princeton, 1965.

[2] S. Agmon and L. Nirenberg, 1963. Properties of solutions of ordinary differential equations in Banach space, *Commn. Pure Appl. Math.* **16**, 121 – 239.

[3] A. V. Balakrishnan, "Applied Functional Analysis," Springer-Verlag, New York, 1981.

[4] V. Barbu, 2003. Feedback stabilization of Navier-Stokes equations. *ESIAM Contr., Optim., and Calc. Var.* **9**, 197 – 206.

[5] V. Barbu, 2010. Exponential stabilization of the linearized Navier-Stokes equation by pointwise feedback noise controllers, *Automatica* **46**, 2022 – 2027.

[6] R. Bhatia and P. Rosenthal, 1997. How and why to solve the operator equation $AX - XB = Y$, *Bull. London Math. Soc.* **29**, 1 – 21.

[7] S. P. Bhattacharyya and E. de Souza, 1982. Pole assignment via Sylvester's equation, *Syst. Contr. Lett.* **1**, 261 – 263.

[8] C. I. Byrnes, I. G. Laukó, D. S. Gilliam, and V. I. Shubov, 2000. Output regulation for linear distributed parameter systems, *IEEE Automatic Control*, **AC-45** 2236 – 2252.

[9] C. I. Byrnes, D. S. Gilliam, V. I. Shubov, and G. Weiss, 2002. Regular linear systems governed by a boundary controlled heat equation, *J. Dynamical and Control Systems*, **8**, 341 – 370.

[10] E. K. Chu, 1986. A pole-assignment algorithm for linear state feedback, *Syst. Contr. Lett.* **7**, 289 – 299.

[11] R. Courant and D. Hilbert, "Methods of Mathematical Physics, I," Wiley Interscience, New York, 1953.

[12] R. F. Curtain, 1984. Finite dimensional compensators for parabolic distributed systems with unbounded control and observation, *SIAM J. Control Optim.* **22**, 255 – 276.

[13] R. F. Curtain and H. J. Zwart, 1995. "An Introduction to Infinite-Dimensional Linear Systems Theory," Springer-Verlag, 1995.

[14] K. Datta, 1988. The matrix equation $XA - BX = R$ and its applications, *Lin. Alg. Applic.* **109**, 91 – 105.

[15] N. Dunford and J. T. Schwartz, "Linear Operators," Parts 2 and 3, Wiley Interscience, New York, 1963.

[16] K. -J. Engel and R. Nagel, "One-Parameter Semigroups for Linear Evolution Equations," Springer-Verlag, New York, 2000.

[17] H. O. Fattorini, 1967. On complete controllability of linear systems, *J. Differential Equations* **3**, 391 – 402.

[18] H. O. Fattorini, 1968. Boundary control systems, *SIAM J. Control* **6**, 349 – 385.

[19] D. Fujiwara, 1967. Concrete characterization of the domain of fractional powers of some elliptic differential operators of the second order, *Proc. Japan Acad. Ser. A Math Sci.* **43**, 82 – 86.

[20] D. Gilbarg and N. S. Trudinger, "Elliptic Partial Differential Equations of Second Order," 2nd ed., Springer-Verlag, New York, 1983.

[21] P. Grisvard, 1967. Caractérisation de quelques espaces d'interpolation, *Arch. Rational Mech. Anal.* **25**, 40 – 63.

[22] D. Henry, "Geometric Theory of Semilinear Parabolic Equations," Lectute Notes in Mathematics #840, Springer-Verlag, Berlin, 1981.

[23] S. Ito, 1957. Fundamental solutions of parabolic differential equations and boundary value problems, *Japan J. Math.* **27**, 55 – 102.

[24] S. Itô, "Diffusion Equations," Amer. Math. Soc., Providence, 1992.

[25] T. Kato, 1961. A generalization of the Heinz inequality, *Proc. Japan Acad. Ser. A Math. Sci.* **37**, 305 – 308.

[26] S. G. Krein, "Linear Differential Equations in Banach Space," Amer. Math. Soc. Transl. Math. Monographs, Providence, 1971.

[27] M. Krstic and A. Smyshlyaev, "Boundary Control of PDEs: A Course on Backstepping Designs," SIAM, Philadelphia, 2008.

[28] I. Lasiecka and R. Triggiani, "Control Theory for Partial Differential Equations I: Abstract Parabolic Systems," Cambridge Univ. Press, Cambridge, 2000.

[29] M. Léautaud, 2010. Spectral inequalities for non-selfadjoint elliptic operators and application to the null- controllability of parabolic systems, *J. Functional Analysis* **258**, 2739 – 2778.

[30] N. Levan, 1980. Controllability, $*$-controllability and stabilizability, *J. Differential Equations* **38**, 61 – 79.

[31] N. Levinson, "Gap and Density Theorems," Amer. Math. Soc. Colloquium Publications, New York, 1940.

[32] J. L. Lions and E. Magenes, "Non-Homogeneous Boundary Value Problems and Applications," vol. I, Springer-Verlag, New York, 1972.

[33] D. G. Luenberger, 1966. Observers for multivariable systems, *IEEE Trans. Automat. Contr.* **AC-11**, 190 – 197.

[34] A. S. Markus, "Introduction to the Spectral Theory of Polynomial Operator Pencils," Transl. Math. Monogr. Amer. Math. Soc., Providence, 1988.

[35] S. Mizohata, "The Theory of Partial Differential Equations," Cambridge Univ. Press, Cambridge, 1973.

[36] V. A. Morozov, "Methods for Solving Incorrectly Posed Problems," Springer-Verlag, Berlin, 1984.

[37] T. Nambu, 1985. On stabilization of partial differential equations of parabolic type: boundary observation and feedback, *Funkcial. Ekvac.* **28**, 267 – 298.

[38] T. Nambu, Output stabilisation for a class of linear parabolic differential equations, *Proc. Roy. Soc. Edinburgh, Sec. A* **110A**, 125 – 133 1988.

[39] T. Nambu, 1989. An extension of stabilizing compensators for boundary control systems of parabolic type, *J. Dynamics and Differential Equations* **1**, 327 – 346.

[40] T. Nambu, 1994. Approximation algorithm for an infinite-dimensional operator equation $XL - BX = C$, *Math. Control, Signals, and Systems*, **7**, 76 – 93.

[41] T. Nambu, 1997. Characterization of the domain of fractional powers of a class of elliptic differential operators with feedback boundary conditions, *J. Differential Equations* **136**, 294 – 324.

[42] T. Nambu, 2001. An algebraic method of stabilization for a class of boundary control systems of parabolic type, *J. Dynamics and Differential Equations* **13**, 59 – 85.

[43] T. Nambu, 2003. Stability enhancement of output for a class of linear parabolic systems, *Proc. Roy. Soc. Edinburgh, Sec. A* **133A**, 157 – 175.

[44] T. Nambu, 2004. An $L^2(\Omega)$-based algebraic approach to boundary stabilization for linear parabolic systems, *Quarterly of Applied Mathematics* **62**, 711 – 748.

[45] T. Nambu, 2005. A new algebraic approach to stabilization for boundary control systems of parabolic type, *J. Differential Equations* **218**, 136 – 158.

[46] T. Nambu, 2007. Stability analysis of linear parabolic systems and removement of singularities in substructure: Static feedback, *J. Differential Equations* **238**, 257 – 288.

[47] T. Nambu, 2010. Stabilization and a class of functionals for linear parabolic control systems, *Proc. Roy. Soc. Edinburgh, Sec. A* **140A**, 153 – 174.

[48] T. Nambu, 2011. A note on the minimum number of the actuators for stabilization in linear parabolic boundary control systems, *SICE JCMSI* **4**, 349 – 352.

[49] T. Nambu, 2012. Equivalence of two stabilization schemes for a class of linear parabolic boundary control systems, *Bull. Polish Acad. Sci. Math.* **60**, 187 – 199.

[50] T. Nambu, 2014. Alternative algebraic approach to stabilization for linear parabolic boundary control systems, *Math. Control, Signals, and Systems* **26**, 119 – 144.

[51] T. Nambu, 2014. Remarks on the stabilization problem for linear finite-dimensional systems, *Bull. Polish Acad. Sci. - Math.* **62**, 87 – 99.

[52] T. Nambu, 2014. Algebraic multiplicities arising from static feedback control systems of parabolic type, *Numer. Func. Anal. Optim.* **35**, 1359 – 1381.

[53] T. Nambu, 2014. Stabilization for boundary control systems of parabolic type, *System, Control, and Information* **58**, 358 – 364 (in Japanese).

[54] R. E. A. C. Paley and N. Wiener, "Fourier Transforms in the Complex Domain," Amer. Math. Soc. Colloquium Publications, Providence, 1934.

[55] Jean-Pierre Raymond and L. Thevenet, 2010. Boundary feedback stabilization of the two dimensional Navier-Stokes equations with finite dimensional controllers, *Discrete Contin. Dyn. Syst.* **27**, 1159 – 1187.

[56] J. W. Polderman and J. C. Willems, "Introduction to Mathematical Systems Theory," Springer, New York, 1998.

[57] Y. Sakawa, 1974. Controllability for partial differential equations of parabolic type, *SIAM J. Control* **12**, 389 – 400.

[58] Y. Sakawa and T. Matsushita, 1975. Feedback stabilization of a class of distributed systems and construction of a state estimator, *IEEE Trans. Automat. Contr.* **AC-20**, 748 – 753.

[59] Y. Sakawa, 1983. Feedback stabilization of linear diffusion systems, *SIAM J. Control Optim.* **21**, 667 – 676.

[60] D. Salamon, 1987. Infinite dimensional linear systems with unbounded control and observation: A functional analytic approach, *Trans. Amer. Math. Soc.*, **300**, 383 – 431.

[61] H. Sano, 2003. Exponential stability of a mono-tubular heat exchanger equation with output feedback, *Systems & Control Letters* **50**, 363 – 369.

[62] J. M. Schumacher, 1983. Finite-dimensional regulators for a class of infinite-dimensional systems, *Systems & Control Letters*, **3**, 7 – 12.

[63] A. Smyshlyaev and M. Krstic, 2004. Closed form boundary state feedbacks for a class of 1D partial integro-differential equations, *IEEE Trans. Automat. Contr.* **AC-49**, 2185 – 2202.

[64] O. Szaśz, 1916. Über die approximation stetiger funktionen durch lineare aggregate von potenzen, *Math. Ann.* **77**, 482 – 496.

[65] B. Sz-Nagy, and C. Foias, "Harmonic Analysis of Operators on Hilbert Spaces," North Holland, New York, NY, 1970.

[66] A. E. Taylor, "Introduction to Functional Analysis," John Wiley & Sons, New York, 1958.

[67] E. C. Titchmarsh, "The Theory of Functions," The Clarendon Press, Oxford, 1939.

[68] H. L. Trentelman, A. A. Stoorvogel, and M. L. J. Hautus, "Control Theory for Linear Systems," Springer, London, 2001.

[69] G. Weiss and R. F. Curtain, 1997. Dynamic stabilization of regular linear systems, *IEEE Trans. Automat. Control*, **AC-42**, 4 – 21.

[70] W. M. Wonham, 1967. On pole assignment in multi-input controllable linear systems, *IEEE Trans. Automat. Control*, **12**, 660 – 665.

[71] K. Yosida, "Functional Analysis," 6th ed., Springer-Verlag, Berlin, 1980.

[72] R. M. Young, "An Introduction to Nonharmonic Fourier Series," Academic Press, New York, 1980.

Index